Municipal Wastewater Reclamation and Reuse

Nitin Vitthalrao Thombre
Editor

Municipal Wastewater Reclamation and Reuse

A Machine-Generated Literature Overview

Springer

Editor
Nitin Vitthalrao Thombre
VA Tech Wabag Ltd.
Pune, Maharashtra, India

ISBN 978-981-96-9111-1 ISBN 978-981-96-9112-8 (eBook)
https://doi.org/10.1007/978-981-96-9112-8

This Springer imprint is published by the registered company Springer Nature Singapore Pte Ltd.
The registered company address is: 152 Beach Road, #21-01/04 Gateway East, Singapore 189721, Singapore

Preface

Springer Nature wrote an email to me with an interesting idea to collaborate authors and AI to write a book on the subject of my interest. The idea of working with an AI tool was confusing at first but was rewarding after I get to know the process. We cannot deny the power AI holds when it comes to help humans by reducing the time and efforts in an efficient way and I am very delighted to see this book as the end product of this collaboration. I sincerely congratulate Springer Nature publication house to put so much efforts into publishing such Human-Machine collaboration books that will help researchers around the globe by reaching out to millions of readers through these books.

As a researcher and an employee at VA Tech Wabag, I am involved with water and wastewater treatment studies and the improvements in the established treatment technologies. First thing popped up in my head was to write a book on the water reuse. Rapid urbanization in the recent decades have increased the consumption of potable water in the municipalities by multiple folds. This is putting a lot of pressure on the freshwater resources as the demand is increasing with each passing day. Hence, municipal wastewater reclamation and reuse have emerged as crucial strategies in the pursuit of sustainable water management worldwide. This book aims to provide a comprehensive overview of this rapidly evolving field, uniting the expertise of professionals, academics, and researchers in water resources and environmental engineering. Water scarcity poses a significant challenge in both urban and rural areas, exacerbated by population growth, urbanization, industrialization, and climate change. The efficient and responsible use of water resources is of paramount importance in this context. Municipal wastewater, once considered a liability, is now recognized as a valuable resource that can mitigate water scarcity and enhance overall water availability.

The significance of municipal wastewater reuse cannot be overstated. Through the treatment and reclamation of wastewater, we can harness its potential for non-potable applications such as irrigation, industrial processes, and environmental restoration. This approach not only reduces the strain on freshwater sources but also minimizes pollution discharge, safeguarding ecosystems and human health. Nevertheless, implementing wastewater reclamation and reuse systems presents a range of challenges. Technological advancements, regulatory frameworks, financial considerations, and societal acceptance must all be carefully addressed. Developing effective treatment processes that meet stringent water quality standards, ensuring the safety and public perception of reclaimed water, and integrating reuse systems into existing urban infrastructure are just a few of the multifaceted challenges faced by practitioners and policymakers.

The potential benefits of municipal wastewater reuse extend beyond water availability, encompassing the promotion of a circular economy, reduction of carbon footprint, and enhancement of climate change resilience. This book aims to foster collaboration, inspire further research, and encourage innovative approaches that contribute to the advancement of wastewater reclamation and reuse as an integral part of sustainable water management worldwide.

This Human-Machine collaboration book endeavors to illuminate these challenges and provide insights into the innovative solutions that have emerged from research and practical experiences. It covers a wide range of topics, including wastewater treatment technologies, water quality assessment and monitoring, risk management, some case studies worth mentioning from around the globe, and socio-economic considerations related to wastewater reuse. Through the contributions of esteemed experts in the field, this book seeks to facilitate a deeper understanding of municipal wastewater reclamation and reuse as a viable and sustainable solution to the global water crisis. Our hope is that this comprehensive compilation of knowledge will serve as a valuable resource for professionals, researchers, policymakers, and students engaging with this critical subject.

I express my sincere gratitude towards my co-editors and mentors, Prof. Aniruddha B. Pandit and Prof. Anand V. Patwardhan, who have contributed their expertise and insights to this endeavor. Their dedication and commitment have been instrumental in creating a comprehensive and forward-thinking resource that I believe will make a valuable contribution to the field of municipal wastewater reclamation and reuse.

Dr. Nitin Vitthalrao Thombre
Principal Engineer (R&D, Process and Technology)
VA Tech Wabag Ltd.
Pune, India

About This book

Auto-summaries can be generated by either an abstractive or extractive auto-summarization:

- An extraction-based summarizer identifies the most important sentences of a text and uses the original sentences to create the summary.
- An abstraction-based summarizer creates new text based on deep learning. New phrases are created to summarize the content.

The auto-summaries you will find in this book have been generated via an extractive summarization approach.

Each chapter was carefully edited by Nitin Vitthalrao Thombre. The editors selected the papers which were then auto-summarized. The editors have not edited the auto-summaries due to the extraction-based approach, and have not changed the original sentences. You will find the editors' reviews and guidance on the auto-summaries in their chapter introductions.

In machine-generated books, editors are defined as those who curate the content for the book by selecting the papers to be auto-summarized and by organizing the output into a meaningful order. Next to the thoughtful curation of the papers, editors should guide the readers through the auto-summaries and make transparent why they selected the papers.

The ultimate goal is to provide a current literature review of Springer Nature publications on a given topic in order to support readers in overcoming information overload and to help them dive into a topic faster; to identify interdisciplinary overlaps; and to present papers which might not have been on the readers' radar.

Please note, that the selected papers are not used to train a LLM while the auto-summaries are created.

Contents

About the Editor

Dr. Nitin Vitthalrao Thombre is a dedicated researcher and an accomplished author specializing in water and wastewater treatment, with a keen focus on advanced technologies. He currently serves as Principal Engineer in the Research & Development and Process and Proposal divisions at VA Tech Wabag, where he is instrumental in pioneering innovative solutions in wastewater treatment and resource recovery.

Dr. Nitin's academic journey began with a Bachelor's degree in Chemical Engineering from JNEC, Aurangabad. He went on to earn a Master's degree in Petroleum Engineering from the Indian School of Mines (now IIT-ISM), Dhanbad. While his early academic path was rooted in the energy sector, it was the complex challenges of water treatment that ultimately captured his passion and defined his career trajectory.

Driven by a deep-seated curiosity and a desire to create meaningful change, Dr. Nitin pursued a Ph.D. in Chemical Engineering at the Institute of Chemical Technology (ICT), Mumbai. His doctoral research focused on wastewater reclamation, where he made significant contributions that continue to influence practice and policy in the sector. One of the outcomes of his work was the development of novel ceramic membranes for treating petroleum wastewater, for which he filed a patent at the Indian Patent Office, Chennai—a collaborative effort between ICT Mumbai and VA Tech Wabag.

Dr. Nitin has authored numerous research papers, presented his work at various national and international conferences, and contributed four chapters to the *Handbook of Water and Used Water Purification* published by Springer Nature, Switzerland. In recognition of his contributions, he was honoured with the M. P. Chary Memorial Award for Best Young Chemical Engineer by the Indian Institute of Chemical Engineers (IIChE) in 2018 and NRC Best paper award by Indian Chemical Engineer by Indian Institute of Chemical Engineers (IICHE) in 2024.

With a unique ability to bridge scientific research and real-world application, Dr. Nitin brings both depth and practical insight to his writing. His forthcoming book aims to provide a comprehensive overview of wastewater treatment and reuse technologies. Drawing from his academic foundation and industry experience, this work is poised to serve as a vital resource for students, researchers, and professionals committed to building a sustainable water future.

Chapter 1
Overview

Nitin Vitthalrao Thombre

1.1 Introduction by Editor

Municipal wastewater, also known as sewage or sanitary sewage, refers to the water-borne waste generated by households, commercial establishments, and institutions in urban areas. It comprises domestic wastewater from toilets, sinks, showers, and laundry, as well as industrial wastewater from manufacturing processes, and storm water runoff. Proper management of municipal wastewater is crucial to protect water resources, public health, and the sustainability of urban areas. Municipal wastewater is a complex mixture that includes various components:

- Organic Matter: Wastewater contains organic substances derived from human and animal waste, food remnants, and other biological sources. These organic compounds contribute to the biochemical oxygen demand (BOD) and chemical oxygen demand (COD) of the wastewater, affecting its overall quality and oxygen levels.
- Nutrients: Nutrients like nitrogen and phosphorus, originating from human waste, detergents, and food residues are present in the wastewater. Excessive nutrient levels can lead to eutrophication, causing harmful algal blooms and oxygen depletion in water bodies.
- Pathogens: Pathogenic microorganisms such as bacteria, viruses, and parasites are often present in municipal wastewater. These pathogens enter the wastewater through human and animal excreta, posing risks to public health if the wastewater is not treated properly.
- Suspended Solids: The sewage may contain suspended particles, including soil, debris, and organic matter. These solids contribute to turbidity and can affect the efficiency of treatment processes if not effectively removed.

N. V. Thombre (✉)
VA Tech Wabag Ltd., Pune, Maharashtra, India
e-mail: nitin01thombre@gmail.com

N. V. Thombre (ed.), *Municipal Wastewater Reclamation and Reuse*,
https://doi.org/10.1007/978-981-96-9112-8_1

- Toxic Substances: Industrial and commercial activities in urban areas introduce various toxic chemicals and pollutants into municipal wastewater. These may include heavy metals, organic compounds, pharmaceuticals, and personal care products, necessitating specialized treatment for their removal or degradation.

This municipal wastewater, if go untreated to the environment can cause many problems. Nature has tendency to self-clean this water by natural means. But this applies only for less quantities, to treat in larger quantities, we require a dedicated municipal sewage treatment plant to quickly remove the organic and inorganic contaminants and dispose the water into the natural bodies. Due to rapid industrialization, a large chunk of population has shifted to urban areas causing a sudden increase in load on municipal wastewater treatment plants.

Improper treatment and disposal of municipal wastewater can contaminate water sources used for drinking, recreation, and irrigation. This contamination can lead to the spread of waterborne diseases such as gastrointestinal illnesses, cholera, and hepatitis. Discharging untreated wastewater into water bodies can have severe ecological consequences. High nutrient levels can trigger algal blooms, leading to oxygen depletion and the death of aquatic organisms. Toxic substances in wastewater can harm fish, wildlife, and sensitive ecosystem. The discharge of untreated wastewater into rivers, lakes, or oceans contributes to water pollution, degrading overall water quality and diminishing the aesthetic value of water bodies. This pollution can have cascading effects on biodiversity, ecosystem services, and the long-term sustainability of aquatic habitats.

Reclamation and reuse of this wastewater is the process of treating and repurposing wastewater generated by urban areas for various beneficial purposes. It is a sustainable approach that helps to address water scarcity issues, reduce the strain on freshwater resources, and minimize the environmental impact of wastewater discharge.

- Water Scarcity: With increasing population and urbanization, water scarcity has become a significant concern. Reusing the reclaimed water helps alleviate the pressure on freshwater supplies and ensures a sustainable water source for municipalities.
- Environmental Conservation: Reusing wastewater minimizes the pollution of water bodies, protects ecosystems, and preserves biodiversity by reducing the discharge of untreated wastewater into rivers, lakes, and oceans.
- Energy Conservation: Treating wastewater for reuse requires less energy compared to treating freshwater sources, leading to energy conservation and reduced greenhouse gas emissions.
- Economic Benefits: Reclaimed wastewater can be a cost-effective alternative to freshwater sources for various non-potable applications, reducing the overall water bill for municipalities.

Various technologies have been used and developed over a few decades to treat the sewage coming from households and agricultural lands in the vicinity of the urban areas. These technologies are common everywhere but can be changed slightly if the effluent is changed over a period of time in some cases. The process can be divided as follows:

1. Primary Treatment: This step involves the removal of large debris, grit, and solids through physical processes such as screening and grit chambers. Utilizes sedimentation to separate suspended solids and organic matter from wastewater, producing sludge as a byproduct.
2. Secondary Treatment: Involves biological processes where microorganisms break down organic pollutants through aeration, digestion of the biological content and creating a clarified effluent. The solid part can be sent for composting which hold a great value in farming and agriculture.
3. Tertiary Treatment: Applies advanced processes like filtration, disinfection, and chemical treatments to further purify the effluent and remove nutrients and pathogens.
4. Advanced Treatment Technologies: Techniques such as reverse osmosis, and ultraviolet disinfection can be employed to achieve higher levels of water quality for specific reuse applications.

1.2 Machine Generated Summaries

Disclaimer: The summaries in this chapter were generated from Springer Nature publications using extractive AI auto-summarization: An extraction-based summarizer aims to identify the most important sentences of a text using an algorithm and uses those original sentences to create the auto-summary (unlike generative AI). As the constituted sentences are machine selected, they may not fully reflect the body of the work, so we strongly advise that the original content is read and cited. The auto generated summaries were curated by the editor to meet Springer Nature publication standards. To cite this content, please refer to the original papers.

Machine generated keywords: wwtp, discharge, micropollutant, influent, aquatic, carry, stage, sample, river, municipal wwtp, management, detect, estimate, risk, maintenance

Contaminants of Emerging Concerns (CECs) in a Municipal Wastewater Treatment Plant in Indonesia [59]

This is a machine-generated summary of:

Astuti, Maryani Paramita; Notodarmojo, Suprihanto; Priadi, Cindy Rianti; Padhye, Lokesh P.: Contaminants of emerging concerns (CECs) in a municipal wastewater treatment plant in Indonesia [59]

Published in: Environmental Science and Pollution Research (2022)

Link to original: https://doi.org/https://doi.org/10.1007/s11356-022-23567-8

If you want to cite the papers, please refer to the original.

For technical reasons we could not place the page where the original quote is coming from.

Abstract-Summary

"This study provides the first set of quantitative data on the occurrence and fate of a wide range of contaminants of emerging concerns (CECs) in Indonesia's largest wastewater treatment plant (WWTP)."

"Fourteen out of twenty-two monitored CECs were detected in the wastewater influent, and seven were present in the effluent, with a total concentration of 29.8 ± 0.4 μg/L and 0.5 ± 0.0 μg/L, respectively."

"Caffeine (CAF) at 12.2 ± 0.1 μg/L, acetaminophen (ACT) at 9.1 ± 0.1 μg/L, N,N-diethyl-m-toluamide (DEET) at 5.0 ± 0.1 μg/L, ibuprofen (IBU) at 2.3 ± 0.0 μg/L, and triclosan (TCS) at 470 ± 64 ng/L were discovered as the five most prevalent CECs, followed by bisphenol A (BPA), trimethoprim (TMP), Tris(2-chloroethyl) phosphate (TCEP), sulfamethazine (SMZ), carbamazepine (CBZ), fluoxetine (FLX), benzotriazole (BTA), sulfamethoxazole (SMX), and metformin (METF)."

"Biodegradable CECs (SMX, SMZ, ACT, IBU, TCS, BPA, CAF, DEET, and TMP) were efficiently removed (83–100%) by the WSP."

"Recalcitrant CECs achieved poor removal efficiencies (e.g., FLX at 24%), and for others, treatment processes even resulted in elevated concentrations in the effluent (CBZ by 85%, TCEP by 149%, and BTA by 92%)."

"The CECs' influent concentrations were determined to pose a moderate aquatic cumulative risk, while no such risk was associated with their effluent concentrations."

Introduction

"This work presents the first study on the occurrence and fate of CECs in a municipal WWTP in Indonesia, the world's fourth most populous country, employing waste stabilization pond (WSP) system."

"Apart from investigating CECs' occurrence concentrations and removal efficiencies in a WWTP, the study has critically reviewed and discussed (1) the possible sources of the detected CECs and their consumption nature in the city and the country, (2) the comparison of occurrence and removal of the CECs from this study with their reported concentrations in surface waters in Indonesia as well as with those reported in other countries and other WWTPs employing WSP, and (3) the removal mechanism of the detected CECs by the WWTP."

"This study would fill a significant knowledge gap on CECs in Indonesia and contribute to global knowledge development, as WSP is one of the most prevalent treatments practiced worldwide, especially in developing countries."

Material and Methods

"For spiked samples, 500 mL of filtrate samples was spiked with 2 μg/L of target CECs prior to SPE."

"A separate method was developed for E1 analysis employing the same column using 0.4 mM of ammonium fluoride buffer in Milli-Q water (A) and a mixture of acetonitrile and methanol (1:1) (B) at a flow rate of 0.2 mL/min with the total program of 7 min (85% B) as reported in our earlier work (Astuti and others 1)."

"The extraction recoveries were determined by the matrix spike method, which were calculated by subtracting the analyzed concentrations of the unspiked samples from the analyzed concentrations of the spiked samples and dividing that difference by the known concentration of spiked CEC (Kumar and others 2)."

"Low SPE recoveries of these CECs ($\leq$ 20%) in wastewater using the same SPE cartridge have been reported previously (Kumar and others 2)."

Results and Discussions

"The predominant occurrence of ACT (8.6 μg/L) and CAF (9.1 μg/L) was also evident in seven WWTPs' influents in Bangkok, Thailand, at similar concentrations detected in this study (Li and others 3)."

"TCS was detected in the influent at 470 ng/L, which was around the median of its concentrations in Southeast Asian wastewater reported values."

"TCEP concentrations reported in this study (101 ng/L in influent and 251 ng/L effluent) were within the range of its reported values in Asia."

"The CECs which were removed efficiently, such as SMZ, ACT, IBU, TCS, BPA, and CAF, have K_{bio} greater than 0.1 L/g SS/day (biodegradable)."

"Negative removal or increased concentrations of CECs after treatment stages in WWTPs have been reported by a number of studies which have been attributed to several reasons."

Conclusions

"Fourteen out of twenty-two CECs were detected in the wastewater influent, with a total concentration of 29.8 ± 0.4 μg/L. Eight had remained in the effluent at a total concentration of 0.5 ± 0.0 μg/L. The occurrence of the CECs was found to be well correlated with their consumption and their known detection in surface waters in Indonesia."

"Recalcitrant CECs with low adsorption capacities achieved poor removal in WSP."

"This study demonstrated the significance of having even conventional WWTP in removing or reducing the occurrence of CECs to minimize their contamination of the aquatic environment, particularly for countries with very limited coverage of centralized wastewater treatment like Indonesia."

"Although the WWTP achieved more than 98% removal of the detected CECs, certain CECs were still present in the effluent."

Wastewater Treatment Systems for City-Based Municipal Drains for Achieving Sustainability [60]

This is a machine-generated summary of:

Dasgupta, Pinaki; Kumar, Vivek; Malik, Anushree; Kumar, Madan: Wastewater Treatment Systems for City-Based Municipal Drains for Achieving Sustainability [60]

Published in: Circular Economy and Sustainability (2022)

Link to original: https://doi.org/https://doi.org/10.1007/s43615-022-00163-z

If you want to cite the papers, please refer to the original.

For technical reasons we could not place the page where the original quote is coming from.

Abstract-Summary

"Drains in several cities carry both rainwater and untreated greywater with black water from settlements nearby."

"In emerging economies, cities often become hubs of illegal and unauthorised colonies which thrive in the vicinity of stormwater drains."

"The flow characteristics (variable, seasonal, minimum or continuous flow) and structural constraints (the bed and site complexity) in the design of these drains restrict the options for implementing large wastewater treatment plants."

"In these scenarios, human health is a critical factor, as frequent exposure to sewage without any protective equipment during maintenance could lead to health hazards and high-stress levels."

"The utilisation of decentralised and distributed wastewater treatment systems offers an in situ choice for achieving the desired result in quality and nutrient removal in the influent."

"The selection, design and implementation with maintenance are essential for improved efficiency and productivity of the system."

"This paper aims to establish various concepts and schemes that municipal corporations could adopt sustainably for efficient treatment within the limited spatial and temporal boundaries offered by city drains."

Introduction

"In 2020, archaeologists unearthed a drain estimated at least 500 years old or more, 30 m in length, at the iconic Red Fort in Delhi [4]."

"Steadily, the stormwater drains in Delhi started functioning predominantly as sewage drains."

"The sewerage system serves 55% of the population of Delhi, while drains also carry the sewage flow from entering stormwater drains and laying sewers in unauthorised colonies."

"The subject of this paper assumes importance in establishing a circular approach focussed on municipal drains."

"It is worth briefly covering the nexus between circular economy, sustainability and wastewater."

"The management of natural resources such as water in a circular way has been considered essential for furthering circular economy in the urban context [5] and utilising 3Rs (Reduce, Recycle and Reuse) [6]."

"Wastewater as part of circular economy and water reuse as an aid to attaining sustainability have been discussed through PESTEL analysis, including barriers to water use and possible solution pathways [7]."

Interlinking with SDGs

"Sustainable Development Goal (SDG) 6 relates to water and sanitation with 8 targets and associated 11 indicators."

"The target number 6.3 intends that (sic) [8] "By 2030, improve water quality by reducing pollution, eliminating dumping and minimising release of hazardous chemicals and materials, halving the proportion of untreated wastewater and substantially increasing recycling and safe reuse globally.""

"There are two indicators for Target 6.3, namely, the proportion of wastewater safely treated and the proportion of water bodies with good health."

"It is essential to understand the present context of the drains for developing an effective strategy for monitoring and treating wastewater flowing in the drains for minimising the impact of pollution load in basins and rivers."

Present Status of Drains in Delhi

"Lack of proper infrastructure and policy measures forces households to discharge untreated grey and black water into the nearest stormwater drains."

"The drains in Delhi fall into Barapullah, Najafgarh and Trans Yamuna basins [9]."

"The complexities of the topography and geography in Delhi have been a significant challenge in mapping and constructing standard size drains."

"Due to political oversight and the absence of suitable connectors for ensuring wastewater flow to the sewage treatment plants, the sole option for the colonies and dwellers near drains is to dump their grey and black water in storm drains."

"These result in untreated grey and black water outlets in the stormwater drain."

"Drains in a cosmopolitan city like Delhi are monitored by multi-disciplinary and diverse agencies governed by different political powers within a more complex role and responsibility web."

Materials and Methods

"The overstudy involved two phases for deriving and suggesting feasible in situ treatment schemes for wastewater treatment in municipal drains."

"A physical survey of major and minor drains, along with their final outfall and catchment areas, was completed."

"Standard android mobile with Google Photo and Google Maps for tracking locations of drains was utilised."

"The video from the drone and image was referenced and analysed to assess physical conditions and surroundings concerning a drain."

"The videos assisted the research team in understanding the landscape and extent of development near the drains."

"The coordinates obtained from imaging and tracking were processed on Google Earth for locating the drains under study."

"The results were analysed for understanding control points, every 500 m with additional parameters such as water level, gradient, slopes and culverts for a specific drain section."

Wastewater Treatment for Municipal Drains

"Let us consider the categorisation of wastewater treatment systems for municipal drains."

"Wastewater treatment plants could be categorised based on a spatial basis, such as: Centralised Decentralised The centralised treatment facilities are typically advantageous where a suitable area is available to implement large-scale treatment plants."

"A second categorisation could be suggested, based on source treatment such as: Off-site In situ Off-site treatment facilities generally constitute transportation of the wastewater from the source of generation to a separate treatment scheme."

"In situ treatment involves installing a suitable remediation system engineered to be constructed inside a drain or a water body (existing or proposed), considering a similar series of steps leading to the desired water quality on the other side of the module."

Challenges

"The earliest recorded health data pertains to deaths due to toxic gases from sewage systems in the 1880s [10], and a Jaundice outbreak in 1956 [11], attributed to water contamination by sewer drains."

"There are several challenges in implementing a standard scheme solution for treatment in stormwater and municipal wastewater drains."

"Typical are the varying flow characteristics due to mixing grey and black water, source point, constraints in space, shock loading, maintenance aspects, stability and cost implications."

"Flow characteristics The flow in the drains is erratic depending on the population served and micrometeorology."

"There are pockets where the flow remains steady due to resident population or serving multiple smaller drains."

"The more planned areas have drains with defined boundary walls, culverts and a stable base enabling easy maintenance."

"Maintenance The maintenance of these systems requires ample time and human resources."

Technological Interventions

"There are several controlling conditions while customising treatment schemes for wastewater for municipal drains."

"The selection of the treatment technology could follow the proposed functionwhere T is the proposed treatment scheme for a given drain; q is the quality of wastewater to be treated obtained from its physicochemical and biological analysis; v is the volumetric flow rate, cu.m/h; A is the area, m^2; a is the aeration or oxygen available; s is the seasonality of the flow; and $a = 1$ when conditions are aerobic, $a = 0$ in anaerobic situations and $a = ½$ in semi-open conditions, The design of the system should preferably consider all these factors quantitatively wherever possible."

"External influences warrant coverage, such as the political, economic, aesthetics and management of the process deployed."

Outlet or Discharge Standards

"Several agencies suggest and prescribe suitable standards for the discharge of treated wastewater from sewage treatment plants (STP) in India."

"The Indian Standard (IS) 2296:1992 [12] provides five different water quality standards from A to E. The former represents water fit for drinking and the latter suitable for irrigation and industrial purpose."

"The National Environmental Standards of 2000 under Environmental Protection (EP) Rules, 1989 of India, specify the discharge standards applicable to inland surface waters [13]."

"It is assumed that the outlet parameters would conform to acceptable standards for permitting the treated municipal wastewater into either a river body or fit for irrigation or construction and industrial purposes."

Suggested Treatment Schemes

"A primary survey and subsequent study of drains under two municipalities (South Delhi and North Delhi) by the CRDT, IIT Delhi, classified drains depending on their width."

"Depending on their final discharge point, the drains are often classified into primary, secondary, tertiary or main and branched."

"A study by IIT Delhi in 2020 [14] as a response to the call of the Hon'ble National Green Tribunal (NGT) on the objective of ensuring an acceptable quality of water into the Yamuna led to compact decentralised designs for treatment of wastewater in city drains."

"Tertiary treatment for primary drains is essential for ensuring that proper treatment occurs before discharge to a surface water body."

Quality Control

"These schemes could be utilised for two purposes: replenishing a river's flow or reutilising it for industrial and horticulture activities."

"An ideal design for a small urban area would be to circulate the treated water for flushing or toilet purpose to nearby settlements, subject to strict monitoring for ensuring zero microorganisms in the treated water."

"The performance of the in situ treatment plant would be subject to stringent monitoring of physicochemical and biological parameters as per relevant discharge standards and adequate supervision for ensuring streamlined delivery of the desired effluent characteristics."

Advantages of In Situ Approach

"The in situ approach offers several features such as ease of installation for shallow depths, customised designs, appropriate for organic and mineral loads, low-energy requirements and ability to recover nutrients depending on the intensity and efficiency of the system."

"Easy to install for shallow depths The treatment options could conveniently be engineered for drains with shallow depths and low water flow rates."

"Customised design The systems can easily be customised as per the overall water quality predominant in the drains."

"Desired quality for safe discharge to rivers These designs can ensure that the discharge parameters meet the prescribed or desirable parameters set by regional or national standards for safe discharge to natural water bodies."

"Natural systems Several natural systems could offer a cheap, viable and require minimal energy and chemical dosing."

Overall Management Strategy

"The complexities of managing in situ wastewater treatment for urban drains require a robust management strategy to ensure sustainability."

"This step would ensure proper planning at the design stage."

"Schemes could only be effective in multiple scenarios when all variables are considered in the design, thereby considerably reducing the number of design iterations required to manage drains."

Opportunities

"There are opportunities through schemes and systems, some requiring minimal energy without compromising efficiency, especially in shallow water depth."

"Steps coupled with emerging technologies such as bio-ropes (or bio-chords), which are low on chemical loading, where anaerobic and aerobic organisms can thrive, could effectively reduce pollutant loads."

"Availability of low-energy systems Design of systems that can be efficient in shallow water depth and low flow Systems for co-existence of anaerobic and aerobic organisms Use of biofilms—natural and artificial Exploring new technology such as Bio-ropes (or Bio-chords), algal-based solutions and others for cold weather conditions In situ integration possibility Low on chemical loading"

Suggested Implementation Framework

"The steps in the flow chart could include several other components and decision trees based on drain characteristics and other deciding parameters local to the selected drain."

"The framework should also consist of socio-political, economic and environmental considerations duly incorporated either at the feasibility state or the detailed project report."

Circular Economy Approach

"This philosophy would ensure maximum efficiency and recover beneficial nutrients and provide other services, including livelihood."

"The philosophy effectively would comprise of all phases from ensuring adequate checks and controls before wastewater discharge, proper treatment, recovering nutrients and utilising the solid and liquid fractions for suitable end-use applications."

"The circular model proposed for achieving sustainability in urban or city drains consists of checking the inflow to the stormwater drain, controlling the pollutant load, treating pollutants, recovering nutrients and reusing the treated water or replenishing our water bodies with the treated water."

Future Research Scope

"The deliberations covered in the paper lead to several opportunities for research and in-depth investigation."

"Some of these are as follows: Performance evaluation of low cost and compact treatment systems for shallow depth The efficiency of biofilms in different wastewater types along with seasonal variations Optimising and standardising process sequence for municipal drains Innovation for management of wastewater in anaerobic conditions (covered drains or partially covered drains) The utilisation of decentralised bio-based systems, including algal systems [15] for treatment of drain wastewater Maintenance of drains and sludge management Use of AI and machine language for enhancement of wastewater quality monitoring and treatment thereof"

Conclusions

"Toxicity of rivers due to untreated wastewater from urban drains in and around the NCR has been established [16–18]."

"It becomes imperative that planners and urban local bodies consider a sound framework for in situ management of wastewater."

"The paper's primary objective was to establish that nature-based or zero-energy in situ wastewater treatment systems are technically feasible for urban municipal drains that are or might have become sewers due to several geopolitical or other reasons."

"The paper dwells on the fact that with the advancement of science and technology, in situ treatment of wastewater is technically and economically viable in a decentralised manner."

Sludge Minimization in Municipal Wastewater Treatment by Polyhydroxyalkanoate (PHA) Production [61]

This is a machine-generated summary of:

Valentino, Francesco; Morgan-Sagastume, Fernando; Fraraccio, Serena; Corsi, Giovanna; Zanaroli, Giulio; Werker, Alan; Majone, Mauro: Sludge minimization in municipal wastewater treatment by polyhydroxyalkanoate (PHA) production [61]

Published in: Environmental Science and Pollution Research (2014)

Link to original: https://doi.org/https://doi.org/10.1007/s11356-014-3268-y

If you want to cite the papers, please refer to the original.

For technical reasons we could not place the page where the original quote is coming from.

Abstract-Summary

"An innovative approach has been recently proposed in order to link polyhydroxyalkanoates (PHA) production with sludge minimization in municipal wastewater treatment, where (1) a sequencing batch reactor (SBR) is used for the simultaneous municipal wastewater treatment and the selection/enrichment of biomass with storage ability and (2) the acidogenic fermentation of the primary sludge is used to produce a stream rich in volatile fatty acids (VFAs) as the carbon source for the following PHA accumulation stage."

"The reliability of the proposed process has been evaluated at lab scale by using substrate synthetic mixtures for both stages, simulating a low-strength municipal wastewater and the effluent from primary sludge fermentation, respectively."

"In every SBR run, despite the low VFA content (10% chemical oxygen demand, COD basis) of the substrate synthetic mixture, a stable feast–famine regime was established, ensuring the necessary selection/enrichment of the sludge and soluble COD removal to 89%."

"Even though the PHA production performance still requires optimization, the proposed process has a good potential especially if coupled to minimization of both primary sludge (by its use as the VFA source for the PHA accumulation, via previous fermentation) and excess secondary sludge (by its use as the biomass source for the PHA accumulation)."

Introduction

"The effectiveness of MMC PHA production processes is strongly dependent on selection principles of operating conditions in an open bioreactor that impose a strong pressure in favour of microorganisms that are most able to store PHA."

"MMC PHA processes have been commonly investigated in a three-stage arrangement: (1) the acidogenic fermentation stage, (2) the culture selection stage (SBR) and (3) the PHA accumulation stage."

"As for sludge minimization, the proposed process could be advantageous in two ways: the primary sludge is strongly reduced by converting organic solids into VFA through acidogenic fermentation and the excess secondary sludge with enhanced PHA-storing capacity from the SBR is fully conveyed to the PHA accumulation step."

"For each SBR run, the PHA accumulation capacity of the selected MMC was then evaluated by using batch accumulation tests, which were carried out with a synthetic mixture of acetic and propionic acid, used as a reference for mimicking the VFA-rich effluent from acidogenic fermentation of excess primary sludge."

Materials and Methods

"The SBR performance during the steady-state was monitored by the measurement of biomass concentration, as volatile suspended solid (VSS, sample taken at the end of the feast phase) and PHA (sample taken at the end of the feast phase and at the end of the cycle)."

"The amount of stored PHA (ΔPHA) during an SBR cycle was calculated as the difference between PHA concentration in the mixed liquor at the end of feast phase (i.e. the RBCOD depletion time) and at the end of the previous famine phase."

"As consequence: The polymer content in the biomass was calculated (in terms of COD) at RBCOD depletion as the ratio of PHA concentration and VSS concentration (i.e. the sum of active biomass and produced polymer): The COD removal efficiency was calculated as ratio between the soluble COD of the effluent and the soluble COD fed in each cycle."

Results and Discussions

“In all SBR run, the enriched biomass exhibited a decrease of the substrate uptake rate when exposed to the substrate shift: the$-qS_{VFA}{}^{batch}$ being lower than the corresponding$-qS_{RBCOD}$ values, with closer values only in Run V (198 mg COD_{VFA} g $COD_{Xa}{}^{-1}$ h^{-1} in the batch test vs 272 mg RBCOD g $COD_{Xa}{}^{-1}$ h^{-1} in the SBR).”

“The PHA production rate in batch (qP^{batch}) was also lower than in SBR (qP^{feast}), with the exception of Runs V and VI, being the respective values 63 vs 67 mg COD_{PHA} g $COD_{Xa}{}^{-1}$ h^{-1} in Run V and 48 vs 54 mg COD_{PHA} g $COD_{Xa}{}^{-1}$ h^{-1} in Run VI.”

“Due to the positive effect of the applied sOLR on the selection of storing biomass in SBR, an increasing storage response during accumulation was observed when the selected biomass was taken from SBR runs at increasing sOLR.”

Conclusions

“Dynamic feeding using a VFA-poor, low-strength substrate synthetic mixture allowed to establish a stable feast–famine regime (6.7–15% of the aerobic phase of the cycle), based on the feed RBCOD, which in turn ensured the selection/enrichment of MMCs with PHA storage properties.”

“Based on the averaged data from the different SBR runs, the storage response was directly linked to the SRT and sOLRs, influenced by the biomass settling properties and effluent VSS.”

“This was further evidenced by the storage activity exhibited in the following batch accumulation tests, especially during the first 6 h. This “uncontrolled” variation of SRT and sOLR appears as the principal factor affecting the storage properties of MMC in spite of no significant effect was exerted on the main known PHA-storing microbes enriched.”

“Higher PHA content of 34% g PHA g VSS^{-1} has been achieved after 20-h fed-batch accumulation using acetate (provided with feed on demand strategy) and biomass selected with VFA-poor real municipal wastewater (Morgan-Sagastume and others 19).”

Triclosan in Treated Wastewater from a City Wastewater Treatment Plant and Its Environmental Risk Assessment [62]

This is a machine-generated summary of:

Mohan, S.; Balakrishnan, P.: Triclosan in Treated Wastewater from a City Wastewater Treatment Plant and its Environmental Risk Assessment [62]

Published in: Water, Air, & Soil Pollution (2019)

Link to original: https://doi.org/https://doi.org/10.1007/s11270-019-4098-9

If you want to cite the papers, please refer to the original.

For technical reasons we could not place the page where the original quote is coming from.

Abstract-Summary

"Triclosan (TCS) is a potential endocrine-disrupting compound (EDC), which produces an adverse impact on aquatic life and human beings."

"Wastewater discharge is considered as the primary source of triclosan in water bodies."

"The study is aimed to investigate the occurrence and environmental risk of triclosan released by municipal wastewater treatment plants (WWTP)."

"The TCS concentrations in samples were found in the range of 443 to 1757 ng/L. The Risk Quotient (RQ) method was performed to evaluate the environmental (ecotoxicological and human health) risk associated with the exposure of TCS-containing wastewater."

"The results of the study revealed that primary producer (algae) was highly vulnerable to exposure of TCS in the aquatic environment."

"The estimated daily intake of TCS was much lower than the reference dosage, and this indicates that TCS did not produce any considerable risk to human health."

Introduction

"The carcinogenic effect of TCS on human beings has not been studied so far."

"They clearly state that even the presence of low concentrations of triclosan is capable of producing adverse effects on human and aquatic life."

"It is essential to analyze the risk associated with the discharge of TCS-containing wastewater to aquatic life and human health."

"A few risk assessment studies have been conducted for triclosan in surface water and wastewater (Brazil, Santos and others 20; US, Reiss and others 21; Lyndall and others 22; Europe, Thomaidi and others 23)."

"The present study aims to assess the environmental risk potential and possible effects caused by the discharge of triclosan-containing treated as well as untreated wastewater into the environment, especially on surface water bodies (lakes/rivers)."

"RQ method was applied to evaluate the impacts of triclosan in three trophic levels of aquatic species (algae, invertebrate, and fish) and human health risk."

Materials and Methods

"To preserve organic chemical degradation and for complete recovery of TCS, the filtered sample pH was reduced and maintained to 2 ± 0.5 adding concentrated sulfuric acid (H_2SO_4, 1:1 v/v) and stored in amber glass bottles at 4 ± 2 °C in the dark until the analysis was completed."

"Hazard assessment of chemical is estimated by the exposure pathway and toxicity (EC_{50}/LC_{50}) or reference dosage concentration for an ecosystem and human health."

"PEC_{water}Predicted environmental concentration in surface water (mg/L) MEC_{eff}Measured environmental concentration in the effluent (mg/L) Kp_{susp}Solids-water partitioning coefficient of suspended matter (L/Kg) SUSP $_{water}$Concentration of suspended matter in the river (mg/L) DFDilution factor (1, 2, 10, 100, 1000) Five different dilution factors (1, 2, 10, 100, and 1000) were used in the calculation to consider the water scarcity and seasonal variation in surface water flow."

Results and Discussion

"The mean PEC_{water} concentration of triclosan in water bodies is found 425 ng/L and 165 ng/L in scenarios I and II respectively."

"The Risk Quotient (RQ) method is successfully applied for ecotoxicological risk by evaluating the exposure (PEC_{water}) and hazardous (PNEC) concentration of TCS to the most sensitive aquatic species for scenarios I and II."

"Under the low-flow condition into the surface water bodies, TCS poses adverse effects in the aquatic environment."

"In scenario I and II, mean fish TCS concentration is 2.83 μg/g and 1.38 μg/g for DF = 1."

"Detected mean triclosan concentration in bluegill fish is 0.017–0.031 μg/g in Texas, USA (Mottaleb and others 24)."

"The observed RQ is less than 1 for both scenarios, which clearly states that human health risk via triclosan-accumulating fish is not alarming."

Conclusions

"The present study has examined the occurrence of triclosan in two wastewater treatment plants in Chennai, India, using LC/ESI/MS."

"It indicates that the secondary treatment level is not adequate to remove triclosan in wastewater."

"Risk assessment was carried out using risk quotient methodology to evaluate the exposure of triclosan released from treatment plants in three trophic level aquatic species (Algae, Crustacean, and Fish) and dietary exposure to human."

"The crustacean and fish have only the risk of low dilution conditions (DF = 1 and DF = 2)."

"The results highlighted that additional treatment process is required to remove triclosan from wastewater completely."

Estimating Levels of Micropollutants in Municipal Wastewater [63]

This is a machine-generated summary of:

Rowsell, Victoria F.; Tangney, Peter; Hunt, Claire; Voulvoulis, Nikolaos: Estimating Levels of Micropollutants in Municipal Wastewater [63]

Published in: Water, Air, & Soil Pollution (2009)

Link to original: https://doi.org/https://doi.org/10.1007/s11270-009-0112-y

If you want to cite the papers, please refer to the original.

For technical reasons we could not place the page where the original quote is coming from.

Abstract-Summary

"Although sewage treatment works (STWs) aim to reduce levels of pollutants in municipal wastewater discharges, they have become a significant point source of dangerous substances to the aquatic environment."

"This paper has therefore focussed on developing calculations to estimate micropollutant levels in STW influents."

"Results demonstrated that, where monitoring data were available for metals and organic pollutants, STW influent could be adequately estimated and validated, with accuracy between 77 and 100%."

"Based on these calculations and using data for over 600 STWs in England, our analysis showed that compounds such as di(2-ethylhexyl) phthalate, lead and linear alkylbenzenesulfonate could reach influent levels that could be challenging for conventional wastewater treatment removal."

Introduction

"A potential opportunity to reduce levels in the environment could be to control substances at source, through a better understanding of the linkages between different sources and pathways and levels of micropollutants at individual STWs (Bound and Voulvoulis 25)."

"Within the UK, only a handful of studies of this nature have been carried out; Bowen and others (26) and Rule and others (27–29) monitored levels of micropollutants from specific source pathways (domestic, commercial and rainfall runoff) entering STWs."

"The idea of modelling levels at individual STWs has largely not been attempted due to restricted access to STW data and the limited understanding of micropollutant sources and pathways with regard to STWs influent concentrations."

"This paper aimed to review all available data in order to provide calculations for estimating micropollutant influent levels for individual STWs."

"Our work therefore aimed to assess the relative importance of these pathways and specific sources for different micropollutants for individual STWs."

Methodology

"To identify the significance of each pathway for site-specific STWs further information on flow, domestic populations, trade populations and runoff data were utilised for each STW."

"By collating this information, the total influent concentration per compound at individual STW could be calculated as 'the sum of each pathway flow multiplied by its corresponding concentration value (as derived from our systematic review of the literature), divided by the average daily flow'."

"Micropollutant levels from domestic and trade pathways were achieved by multiplying site-specific STW 'population equivalents'."

"Population equivalents are based on the daily biological oxygen demand (BOD) load, i.e. influent BOD multiplied by average flow (grams per day) divided by 60, i.e. 60 g of BOD per head of population (EA 30)."

"Calculating the runoff was achieved by subtracting the dry weather flow (DWF) from the average daily flow (ADF) for each STW, i.e. the remainder represents rainfall additions to the system."

Calculation Accuracy Assessment

"Only a small number of compounds had been monitored in STWs' influents in the selected UK region."

"Only four compounds could be assessed for accuracy: cadmium, lead, nickel and total polyaromatic hydrocarbons (PAHs)."

"For these compounds, influent monitoring data were gathered over a 5-year period, and the mean value (x) was selected to represent the average level of micropollutant at each STW influent."

"The assessment demonstrated that cadmium, lead, nickel and 'total PAH' influent estimations achieved between 77 and 100% of values within the accepted accuracy range of 3 when evaluated against monitored values."

"The influent calculations displayed a propensity to overestimate compound concentrations for lead, nickel and total PAHs (values > 1), whereas cadmium showed a relatively even distribution between over- and underestimated values."

Results

"The range in micropollutant level (grams per day) between STWs is dependent on the size of the flow of the individual STWs, which in turn is determined by the size of the domestic and trade populations."

"Dichloromethane (DCM) was also found in domestic pathways on average at 4 g/day; this compound is used for a number of domestic applications such as paints, PVC, polish and domestic cleaners (Thornton and others 31)."

"The largest increase in micropollutant levels from DWF to ADF rainfall in the region was linked to lead (43 to 69 g/day) owing to its association with vehicle emissions and roofing materials (Comber and Gunn 32)."

"LAS, a common surfactant in domestic and trade sources, would only be expected to be found in runoff from car washing products, and so, inputs are expected to be minimal (Fauser and others 33)."

"This compound was more strongly associated with domestic and trade inputs."

Overall Discussion

"This is obviously not the case as within each domestic population there will be variations in micropollutant levels due to the use of different products, the age of household appliances and pipes and the varied behaviour of individuals' ablution regimes, which all give rise to a host of compound inputs and an almost infinite level of variability."

"The assumption made on trade inputs is possibly the most significant assumption within the calculations, but without serious research into micropollutant levels associated with individual trade practices and data on the number and prevalence of each practice associated with each individual STW, it was the most expedient method with the limited data."

"The first of these is source management, and although this may not be appropriate for all micropollutants, assessing compound inputs from source pathways will inform whether these compounds are specific to certain pathways, present at high levels and/or ubiquitous."

Conclusions

"The purpose of this study was to estimate micropollutant influent levels at specific STWs in order to inform wastewater treatment options and identify opportunities for source management."

"This study demonstrates a tool to estimate STW influent levels for micropollutants, and where calculations could be evaluated, an acceptable level of accuracy (77–100%) was established."

"Its use can be complimentary, as a way of prioritising monitoring needs for detailed source assessment of specific compounds and at specific STWs whilst assisting with treatment selection and optimisation, as a way of improving STW effluent quality."

Post-ozonation in a Municipal Wastewater Treatment Plant Improves Water Quality in the Receiving Stream [64]

This is a machine-generated summary of:

Ashauer, Roman: Post-ozonation in a municipal wastewater treatment plant improves water quality in the receiving stream [64]

Published in: Environmental Sciences Europe (2016)

Link to original: https://doi.org/https://doi.org/10.1186/s12302-015-0068-z

If you want to cite the papers, please refer to the original.

For technical reasons we could not place the page where the original quote is coming from.

Abstract-Summary

"The SPEAR (species at risk) indicator was used to analyse macroinvertebrate abundance data that were collected in the receiving stream before, during and after ozonation to investigate whether ozonation improved the water quality."

"The SPEAR values indicate a better water quality downstream the WWTP during ozonation."

"With ozonation the relative abundance of vulnerable macroinvertebrates in the stream receiving the treated wastewater increases from 18% (CI 15–21%) to 30% (CI 28–32%)."

"The change in the macroinvertebrate community downstream the WWTP indicates that toxicity by pollutants decreased by more than one order of magnitude during ozonation."

"Ozonation followed by sand filtration has favourable impacts on the composition of the macroinvertebrate community and can improve the water quality in the receiving stream."

Background

"Ozonation followed by sand filtration was shown to remove most of the micropollutants [34]."

"A complementary study using an in vitro mode-of-action-based bioassay battery also demonstrated that ozonation reduced the toxicity of the mixture of micropollutants in the effluent in this experiment [36]."

"Another complementary study using fish early life stage toxicity tests (FELST) [37] found that the ozonation step led to reduced growth and development in the FELST, although post-treatment with sand filtration eliminated such toxic effects."

"These three studies showed reduced micropollutant loads and toxicity in the wastewater after ozonation together with sand filtration compared to conventionally treated wastewater."

"My objective is to use the macroinvertebrate data that were collected as part of the full-scale ozonation experiment [34, 36, 37] and investigate whether ozonation followed by sand filtration improved the water quality as indicated by the abundance of vulnerable macroinvertebrates."

Results and Discussion

"Ozonation increases the abundance of vulnerable macroinvertebrates in the stream receiving the treated wastewater from 18% (CI 15–21%) to 30% (CI 28–32%)."

"Other researchers have found that WWTP effluents change the assemblages of macroinvertebrates in receiving stream mesocosms [38], although they attributed the effect to increased nutrients and reduced dissolved oxygen."

"Ozonation followed by sand filtration increases the relative abundance of vulnerable species present downstream of the WWTP and leads to an improved water quality classification."

"An improvement of the ecological status in the receiving stream due to the additional ozonation step followed by sand filtration as indicated by SPEAR is consistent with the reduction of overall micropollutant load found by chemical analysis [34] and monitoring with bioassays [36]."

"The change in SPEAR in the macroinvertebrate community downstream the WWTP indicates that toxicity by pollutants decreased by more than one order of magnitude during ozonation."

Methods

"Before, during and after ozonation followed by sand filtration was installed at the wastewater treatment plant, the macroinvertebrates in the receiving stream were sampled, identified to species or family level (according to [39]) and abundances recorded [40]."

"The approach taken here, using the SPEAR concept, assumes that the species traits that make SPEAR indicative of chemical stress by pesticides are also defining the vulnerability of macroinvertebrates to micropollutants present in WWTP effluent."

"As the macroinvertebrate samples consist of only few replicates I followed recent developments in statistical reasoning and calculated the confidence interval (CI) of the difference that ozonation makes [41, 42]."

"I answered one question with the analysis of the SPEAR data: How much of a difference did ozonation make for the proportion of vulnerable species in the macroinvertebrate community in the receiving stream?"

"The results are very similar to those given above: ozonation increases the abundance of vulnerable macroinvertebrates downstream the WWTP from 17% (CI 12–22%) to 30% (CI 28–32%)."

Characterization and Assessment of a Large-Scale Domestic Advanced Wastewater Treatment Plant in Turkey [65]

This is a machine-generated summary of:

Özkan, Oktay; Oğuz, Merve; Özdemir, Özgür: Characterization and assessment of a large-scale domestic advanced wastewater treatment plant in Turkey [65]

Published in: Environmental Monitoring and Assessment (2011)

Link to original: https://doi.org/https://doi.org/10.1007/s10661-011-2338-6

If you want to cite the papers, please refer to the original.

For technical reasons we could not place the page where the original quote is coming from.

Abstract-Summary

"The paper presents new studies about the removal efficiencies and characterization of the Municipal Wastewater Treatment Plant (MWWTP) in Kayseri, Turkey, which is serving for 800,000 population equivalents with the capacity of 110,000 m^3/day, between 2006 and 2009."

"Influent and effluent samples were collected daily and suspended solids (SS), chemical oxygen demand (COD), biochemical oxygen demand in 5 days (BOD_5), total nitrogen (TN), and total phosphorus (TP) were measured, aiming to study the evolution of the removal efficiencies for each year."

"The following yearly mean removal efficiencies were performed in the treatment plant: 94%, 97%, 97%, 95% for SS; 94%, 96%, 95%, 95% for COD; 98%, 98%, 98%, 98% for BOD_5; 84%, 87%, 89%, 82% for TN; and 71%, 86%, 80%, 86% for TP, in 2006, 2007, 2008, and 2009, respectively."

"The plant performances concerning these parameters were excellent and Kayseri MWWTP was operating efficiently according to the conventional control parameters."

Introduction

"In order to remove nutrients like nitrogen and phosphorus compounds in the secondary treatment stage, nitrification–denitrification processes are also expected to occur."

"The municipal wastewater treatment plant (WWTP) of Kayseri (Turkey) has an important mission for the removal of the pollutants that are discharged as effluent in the River of Kizilirmak."

"The purpose of this study was to assess the performance of Kayseri wastewater treatment plant as one of the big treatment plants in Turkey which discharges into Kizilirmak Basin and to present original results concerning the removal efficiencies related to suspended solids (SS), chemical oxygen demand (COD), biochemical oxygen demand (BOD_5), nitrogen and phosphorus, between 2006 and 2009."

"The significance of this study is that the effluent characteristics of Kayseri WWTP are very important because of discharging the effluents to Kizilirmak basin where Ankara and Kirikkale cities may use Kizilirmak water as a drinking water supply."

Material and Method

"Of conventional characterization and assessment, samples were taken daily as 24 h composite samples from the influent and the effluent of the treatment plant throughout 2006–2009."

"Average values for each month of the studied years were constructed by analyzing influent and effluent 5-day BOD, COD, total suspended solids (TSS), total nitrogen (TN), and total phosphorus (TP) parameters."

"Three individual measurements were performed and the mean value was reported."

Results and Discussion

"The annual average influent and effluent TSS, COD, BOD_5, TN, and TP concentrations and removal efficiencies are presented."

"According to the average values for the influent wastewater of KWWTP, TSS, COD, $BOD_{5,}$ TN, and TP parameters were found strong."

"The wastewater treatment facility of Kayseri met all permit requirements for TSS, COD, BOD_5, TN, and TP for the years 2006–2009, although some peak values were observed for TSS, TN, and TP."

"Despite the differences in the parameter concentrations in the influent, the results of the conventional characterization study of Kayseri MWWTP wastewaters indicated that the characteristics of the effluent wastewater were stable and did not vary in a wide range except minor changes such as in TN in 2 months in 2009."

"The effluent concentrations for the parameters of TSS, COD, BOD, and TN reached peak values in February of 2006 compared with typical effluent concentrations."

Conclusion

"The performance of Kayseri Wastewater Treatment Plant was evaluated between 2006 and 2009."

"This study is the first for determination of general performance and characterization of wastewater in Kayseri."

"Kayseri Wastewater Treatment Plant was projected for minimum 92% BOD_5 and 94% TSS removal."

"The results emphasize that the wastewater treatment plant has high removal performance and the facility met its respective effluent limitations for BOD, TSS, COD, TN and TP and the flow."

"Taken into consideration of global warming, it would be recommended to use the effluent wastewater of Kayseri Wastewater Treatment Plant as irrigation water."

"Farmers who come across shortage of water for the duration of hot summer months could attempt to use the untreated sewage wastewater for the irrigation."

Tropical and Temperate Wastewater Treatment Plants Assemble Different and Diverse Microbiomes [66]

This is a machine-generated summary of:

Song, Yang; Mhuantong, Wuttichai; Liu, Shuang-Yuan; Pisutpaisal, Nipon; Wongwilaiwalin, Sarunyou; Kanokratana, Pattanop; Wang, Ai-Jie; Jiang, Cheng-Ying; Champreda, Verawat; Qiu, Dong-Ru; Liu, Shuang-Jiang: Tropical and temperate wastewater treatment plants assemble different and diverse microbiomes [66]

Published in: Applied Microbiology and Biotechnology (2021)

Link to original: https://doi.org/https://doi.org/10.1007/s00253-020-11082-0

If you want to cite the papers, please refer to the original.

For technical reasons we could not place the page where the original quote is coming from.

Abstract-Summary

"The diversity and assembly of activated sludge microbiomes play a key role in the performances of municipal wastewater treatment plants (WWTPs), which are the most widely applied biotechnological process systems."

"We investigated the microbiomes of municipal WWTPs in Bangkok, Wuhan, and Beijing that respectively represent tropical, subtropical, and temperate climate regions, and also explored how microbiomes assembled in these municipal WWTPs."

"The assembly of microbiomes in municipal WWTPs followed deterministic and stochastic processes governed by geographical location, temperature, and nutrients."

"We found that both taxonomic and phylogenetic α-diversities of tropical Bangkok municipal WWTPs were the highest and were rich in yet-to-be-identified microbial taxa."

"These results suggest that tropical and temperate municipal WWTPs harbored diverse and unique microbial resources, and the municipal WWTP microbiomes were assembled with different processes."

"Tropical Bangkok WWTPs had more diverse and yet-to-be-identified microbial taxa."

Introduction

"Municipal WWTPs are widely applied at various latitudes of tropical and temperate regions, which makes them ideal for the investigation of microbial biogeography and testing hypothesis of microbial ecology (Daims and others 43; Zhang and others 44)."

"Knowledge is missing of the importance of ranking as well as how these factors are affecting microbiome assembly in WWTPs at a wide latitude scale, which limited the understanding of microbiome assembly and the guidance of WWTP operation."

"Reports from studies on microbiomes of municipal WWTPs at Asia and global scales were inconsistent, especially in deterministic and stochastic processes of microbial assembly (Wu and others 45; Zhang and others 44)."

"The aims of this study were to (1) dissect the microbiomes of each municipal WWTP and to identify the driving force for microbiome assembly and (2) rank the importance of factors that regulate the microbiomes of municipal WWTPs in tropical, subtropical, and temperate regions."

Materials and Methods

"The relative importance and contributions of environmental variables to α-diversity index were calculated by multiple linear regression models with permutation tests using the lmPerm package in R, while the contributions of environmental groups to microbial community structure based on Bray–Curtis (taxonomic diversity) and unweighted Unifrac (phylogenetic diversity) distance matrices were assessed by PERMANOVA."

"To find discriminant microbial OTUs and phylogenetic clades in different sample cities, random forest analysis (randomForest package in R) and LDA effect size (LEfSe) (Segata and others 46) with the criterion of alpha threshold value at 0.05 in non-parametric Kruskal–Wallis rank-sum test and LDA score at 4.0 were applied."

"The OTU tables and representative sequences of the OTUs of the global activated sludge microbial community by GWMC are available on the GWMC website (http://gwmc.ou.edu/data-disclose.html)."

Results

"Temperate Beijing and subtropical Wuhan WWTPs were operated at temperature fluctuations from 14.5 to 27.8 °C, while tropical Bangkok WWTPs were operated at the temperature > 25 °C."

"After quality control, a total of 10,308,162 sequences and 8245 OTUs were obtained, which covered 9 samples from tropical WWTPs (Bangkok), 12 samples from subtropical WWTPs (Wuhan), and 18 samples from temperate WWTPs (Beijing)."

"At the phylum level, > 95% of sequences from tropical and subtropical/temperate WWTPs could be annotated."

"These 19 core OTUs accounted for 4.5%, 9.7%, and 15.1% of total sequences from tropical, subtropical, and temperate WWTPs, respectively."

"The ubiquitous core functional bacterial taxon Casimicrobium was hardly found in tropical Bangkok WWTPs, contrasted by its high abundance in subtropical Wuhan and temperate Beijing WWTPs."

Discussion

"Microbial diversity and community structure correlated with WWTP's performance and would underline how WWTPs respond to environmental changes (Ofiţeru and others 47)."

"We did not find that the WWTP performance such as removal rates of COD and TN was significantly higher in tropical Bangkok WWTPs with higher AS microbial richness than subtropical and temperate WWTPs."

"As stochasticity dominated AS microbial assembly in tropical Bangkok WWTPs, more immigrants and detrimental competitors may suppress community function and WWTP performance because of negative interactions between species (Zhang and others 48)."

"Our results indicated that tropical Bangkok WWTPs had higher microbial diversity as well as more unique microbial taxa than subtropical or temperate WWTPs from Beijing and Wuhan."

"Although higher abundances of functional taxa such as Nitrospira, Denitratisoma, and Thauera were found in tropical Bangkok WWTPs, their performances (COD/N/P removals) were not superior to subtropical Wuhan and temperate Beijing WWTPs."

Municipal Wastewater Treatment Plants: Gap, Challenges, and Opportunities in Environmental Management [67]

This is a machine-generated summary of:

Herrera-Navarrete, Ricardo; Colín-Cruz, Arturo; Arellano-Wences, Hilda Janet; Sampedro-Rosas, María Laura; Rosas-Acevedo, José Luis; Rodríguez-Herrera, América Libertad: Municipal Wastewater Treatment Plants: Gap, Challenges, and Opportunities in Environmental Management [67]

Published in: Environmental Management (2021)

Link to original: https://doi.org/https://doi.org/10.1007/s00267-021-01562-y

Copyright of the summarized publication:

If you want to cite the papers, please refer to the original.

For technical reasons we could not place the page where the original quote is coming from.

Abstract-Summary

"Municipal Wastewater Treatment Plants (MWWTPs) have proven to be sources of adverse environmental impacts; however, integrated management can help improve their efficiency."

"This study aims to evaluate the gap between the current management and another based on an international standard applied to WWTPMs, in order to understand their environmental commitment, and to identify the challenges and opportunities they present for the adoption or certification of an environmental management system (EMS) based on ISO 14001."

"A diagnosis was carried out at the MWWTPs, followed by a SWOT (Strengths, Weaknesses, Opportunities and Threats) analysis to determine internal and external factors until a series of challenges and opportunities was identified."

"Among the challenges, "limited financial resources" were identified followed by "high turnover of managerial staff", while the opportunities with the greatest potential for improvement are related to the factors "candidate for investment" and "environmental policy"."

Introduction

"The demand for efficient management of water resources puts pressure on different governments and their water agencies, mainly derived from the lack of attention to wastewater treatment, a problem that represents a challenge due to the high costs involved in the operation and maintenance of MWWTPs (Guerrini and others 49; Szymański and others 50)."

"Despite the importance of adopting ISO 14001:2015 in the water industry, empirical evidence has shown little research on wastewater treatment systems using the requirements of this standard as a guide for a diagnostic evaluation, in addition to a SWOT (Strengths, Weaknesses, Opportunities and Threats) analysis, a strategic planning tool commonly used in the business field, capable of identifying strengths and weaknesses and converting them into added value in an effective way (Pereira and others 51)."

Materials and Methods

"The wastewater received by each of the treatment units is of domestic origin and at the end of the process they discharge their effluents into natural water bodies."

"This stage is supported by a sand classifier, equipment capable of performing, in parallel, the functions of sand separation, washing, concentration and solids removal, thus ensuring an efficient treatment in the next stage. (2) In the secondary stage, wastewater is treated through biological reactors that operate with a mechanical or extended aeration system."

"Despite the type of treatment these MWWTPs do not use chemicals for phosphorus removal. (3) Finally, the third stage consists of disinfecting the treated water by means of UV radiation or chlorine application, for its subsequent discharge to the receiving bodies."

"In this final stage, Caicedo and others (52) state that it is necessary to perform disinfection processes by applying chlorine or UV radiation before treated wastewater is reused or discharged into natural water bodies."

Results

"This element reflects a critical value (0%), since nonconformity is identified in all the requirements evaluated in both MWWTPs, from the absence of an environmental program to the lack of monitoring of objectives and responsible parties for each activity or process."

"The first element represents zero adherence (0%) linked to actions that become dynamic through planning, by determining risks/opportunities, environmental aspects of their activities, considering legal requirements and other conditions."

"In the performance evaluation, the element related to the monitoring, measurement, analysis, and evaluation of the EMS has an abysmal gap (0%) in both MWWTPs, since it refers to a requirement with a direct link to an EMS, which is absent in this study."

"The results of the clause improvement show a null value (N/A) in the evaluation of the element general (N/A), which is related to stakeholders, risks, and opportunities under an EMS, absent in this case."

Discussion

"The objective of this study was to evaluate the gap between the current management and another based on an international standard applied to WWTPMs, in order to understand how wide or narrow is the gap in environmental commitment, and to identify the challenges and opportunities presented by the municipal level for the adoption or certification of an EMS based on ISO 14001."

"Despite the recommendations to evaluate the environmental performance of an organization (environmental impact evaluation, life cycle analysis or audits) (Iatridis and Kesidou 53; Rodríguez and others 54), it is not clearly specified how an international standard can be adjusted and used as a tool for an environmental evaluation, unless it is executed by consulting companies, as Rukavishnikova and others (55) did in their study, when they considered adopting management systems in the water industry, which implied a high initial cost."

"The study allowed the treatment units to identify challenges and opportunities that would lead them to a consensus to make the best decision regarding the adoption of an international EMS or take actions that will bring them closer to improve critical aspects of their current management from an administrative approach."

Conclusions

"This study applied to municipal treatment plants showed a weak environmental commitment, so the implementation of an EMS is proposed, and the best practices should be replicated in other regions to optimize the operation and management of resources."

"Financial resources under a management system can be a key factor in ensuring the implementation or certification of an environmental initiative in the medium or long term."

"The opportunities identified the factor "candidate for investment" and "environmental policy"."

"From an economic perspective, it would be less costly for those responsible for adopting management systems in MWWTPs than building water transfer systems and desalination plants that intensify the environmental problem or, even worse, pay costly environmental sanctions that further limit financial resources."

Environmental Assessment of Municipal Wastewater Discharges: A Comparative Study of Evaluation Methods [68]

This is a machine-generated summary of:

Teodosiu, Carmen; Barjoveanu, George; Sluser, Brindusa Robu; Popa, Simona Andreea Ene; Trofin, Orest: Environmental assessment of municipal wastewater discharges: a comparative study of evaluation methods [68]

Published in: The International Journal of Life Cycle Assessment (2016)

Link to original: https://doi.org/https://doi.org/10.1007/s11367-016-1029-5

If you want to cite the papers, please refer to the original.

For technical reasons we could not place the page where the original quote is coming from.

Abstract-Summary

"Life cycle assessment (LCA), environmental impact quantification (EIQ) and water footprint (WF) are important instruments for sustainability assessments applied for products, production and consumption evaluations in connection with natural resources depletion and pollution threats."

"This study focuses on the environmental assessment of a municipal wastewater treatment plant (MWWTP) discharges by means of these three evaluation methods with the purpose to understand their (methodological) weak and strong points in capturing the impacts."

"Such a comparative analysis is necessary to understand how the (individual) advantages provided by each of these instruments can be complimentarily used to improve an assessment framework for various stakeholders concerned with water use cycle management (regional water operators, water management authority, public authorities, research entities, societal organizations, etc.)."

"The three assessment methodologies (LCA, EIQ, WF) are presented, implemented and critically analysed based on a unitary set of data concerning the MWWTP of Iasi city (a municipality of approx."

"Although the three methodologies have different principles for environmental impact quantification, the results have shown that most impacts induced to surface waters due to Iasi MWWTP effluents are given by the nutrients (nitrogen and phosphorous compounds), which could induce an eutrophication impact, and to a lesser extent by pollutants responsible for toxicity impacts (such as heavy metals)."

"The strong and weak points that are relevant for each method implementation and their subsequent use by decision-makers and Water Authorities are discussed, in the context of legislative requirements (including the Water Framework Directive), actual development of regional water operators and stakeholders' interests."

Introduction

"In the context of integrated water resources management and more precisely of the wastewater management practices, it is important to develop and use impact evaluation methods that are able to capture (by identification, description and quantification) as easily as possible (from a methodological point of view) the complexity of the water-related impacts associated to wastewater treatment and discharges."

"The main criteria in applying and comparing these methods are related to: (i) the overall data needed and the analysis of resulted impact categories, (ii) the relevance and the easiness of applying these methods for the assessment of actual impacts of municipal wastewater discharges considering the WFD objectives and specific legislation requirements, and (iii) the analysis of combined application of these methods, especially for assessing environmental impacts related to the development of the water system infrastructures of the regional water and wastewater treatment operators and the support for decision making."

Overview of Studies for Environmental Assessment of Wastewater Discharges

"It should be noted that the few research efforts for using the water footprint assessment as a sustainability assessment tool have primarily focused on the blue and green components of the WF targeting more the water use (consumption) impacts and far less the wastewater discharge impacts as indicated by the grey WF."

"Using real data collected from 22 wastewater treatment plants, studies realized by Ene and Teodosiu (56; 57) were the first assessing the grey water footprint for wastewater discharges in Romania, considering the local conditions (natural background concentrations in a receiving water body and the ambient water quality standards for the receiving freshwater body)."

"We only perform the GWF accounting and not water footprint assessment, as our purpose is to better understand how to improve environmental assessment (as a component of sustainability assessment) and to enable methodological comparison with the other instruments."

Case Study and Methodologies

"For these previous studies, an environmental impact index was computed as a function of the impact magnitude (considering the flows of discharged wastewaters) and impact severity (considering the measured concentrations of water quality indicators) and natural quality state of the river (Teodosiu and others 58)."

"The instrument for environmental impact quantification considers the concentrations of water quality indicators and takes into account the local conditions such as: the quality of the receiving water body and the river flow."

"Depending on the reference concentration (c_{ref}), two situations may be distinguished when computing the environmental impact: (a) by considering the flow and quality of the receiving water natural state (water quality indicators determined at the river spring section) and (b) by considering the flow and quality of the receiving water body from just upstream of the discharge point of the MWWTP effluents."

Results and Discussion

"While LCA quantifies environmental impacts as the effects (or damages) that different environmental compartments may suffer (e.g. toxicity or any other type of harm), the environmental impact quantification method and the grey water footprint quantify impacts by comparing the discharged pollution (or load) to a reference pollution level which is related to the quality of the receiving water body and hence to legislative limits."

"This is one of the main differences between these assessment methods: LCA is completely independent in time and space, its impacts depending only on the dimension (pollution load) and effect type, while the other two instruments quantify impacts considering the local conditions like the quality of the receiving water body (upstream of the discharge points or at the river spring) and the corresponding legal requirements."

Conclusions

"The main objective of this paper was to assess the environmental impacts of a municipal wastewater treatment plant from Romania by using three different impact assessment instruments: (1) life cycle assessment (LCA), (2) environmental impact quantification (EIQ) and (3) grey water footprint (GWF) and to critically compare these three methodologies from a practical and applicability point of view in order to support various water management stakeholders and especially the wastewater treatment operators."

"LCA quantifies impact along a causality chain, and it is virtually time and space independent; the EIQ instrument assesses the local wastewater treatment discharges by comparing the actual pollutant contributions (inputs) with the river natural background concentrations, and finally, the grey water footprint measures the virtual water volume needed to dilute pollution to harmless levels, considering the legislative requirements."

"The environmental impact quantification instrument and the grey water footprint accounting methodology could be improved by considering the standardized assessment framework provided by the LCA methodology."

Bibliography

1. Astuti MP, Rangsivek R, Padhye LP (2022) Laboratory and pilot-scale UV, UV/H2O2, and granular activated carbon (GAC) treatments for simultaneous removal of five chemicals of emerging concerns (CECs) in water. J Water Process Eng 47:102730. https://doi.org/10.1016/j.jwpe.2022.102730
2. Kumar R, Sarmah AK, Padhye LP (2019) Fate of pharmaceuticals and personal care products in a wastewater treatment plant with parallel secondary wastewater treatment train. J Environ Manage 233:649–659. https://doi.org/10.1016/j.jenvman.2018.12.062
3. Li Y, Jindal R, Choi K, Kho YL, de Bullen PG (2012) Pharmaceutical residues in wastewater treatment plants and surface waters in Bangkok. J Hazard Toxic Radioact Waste 16:88–91. https://doi.org/10.1061/(ASCE)HZ.2153-5515.0000099
4. Roychowdhury A (2020) (online), Archaeologists find Mughal-era rainwater drain near Delhi Red Fort, Delhi news, The Hindustan Times, Published online 06 January 2020. https://www.hindustantimes.com/cities/delhi-archaeologists-find-mughal-era-rainwater-drain-near-red-fort/story-hutSXzmwADmQlevAyLCegL.html. Accessed 26 May 2021
5. Atanasova N et al (2021) Nature-based solutions and circularity in cities. Circ Econ Sustain 1:319–332. https://doi.org/10.1007/s43615-021-00024-1
6. Oliveira M et al (2021) Circular economy and the transition to a sustainable society: integrated assessment methods for a new paradigm. Circ Econ Sustain 1:99–113. https://doi.org/10.1007/s43615-021-00019-y
7. Morris JC, Georgiou I, Guenther E, Caucci S (2021) Barriers in Implementation of Wastewater Reuse: Identifying the Way Forward in Closing the Loop. Circ Econ Sustain 1:413–433. https://doi.org/10.1007/s43615-021-00018-z
8. Target 6.3, Goal 6, The United Nations, Department of Economic and Social Affairs, Sustainable Development. Targets and Indicators (online), https://sdgs.un.org/goals/goal6

9. Table 1.2-2 Department of Civil Engineering, Indian Institute of Technology, Delhi (2018) Drainage Master Plan of NCT of Delhi, Final Report, Prepared for Department of Irrigation and Floor Control, Government of Delhi, July 2018, pg 4. https://yamuna-revival.nic.in/wp-content/uploads/2020/01/Drainage-Master-Plan-prepared-for-IFC-Department-GNCTD-by-IIT-Delhi-July-2018.pdf. Accessed 26 May 2021
10. Prashad V (2001) Drains, the technology of sanitation in Colonial Delhi, vol 35, No 1. Modern Asian Studies, Cambridge University Press, pp 113–155. https://www.jstor.org/stable/313090
11. R-321 and R-322, Report of the Committee Constituted for the Purpose of Enquiring as to How far the Pollution of Jamuna Water During November 1955, was responsible for the outbreak of Jaundice in Delhi Archives; Proceedings and Reports, (undated), (online); https://archives.delhi.gov.in/delhiarchive/assets/docs/Proceedings_and_Reports.pdf. Accessed 26 May 2021
12. Designated Best Uses of Water, CPCB and Water Quality Standards in India, IS 2296:1992, Annexure-2, Central Water Commission, (undated), www.cwc.gov.in, (online), http://cwc.gov.in/sites/default/files/annexure-2.pdf
13. [ScheduleE – VI], General standards for discharge of environmental pollutants part-A: EFFLUENTS, CPCB Annexure-1,(2000). www.cpcb.ic.in, https://www.cpcb.nic.in/GeneralStandards.pdf
14. Centre for Rural Development and Technology, IIT Delhi (2021) Action Plan on "Alternate Technology for Management of Wastewater in Major Drains Out falling in Yamuna River", South Delhi Municipal Corporation (unpublished)
15. Dalvi V, Chawla P, Malik A (2021) Year-long performance assessment of an on-site pilot scale (100 L) photobioreactor on nutrient recovery and pathogen removal from urban wastewater using native microalgal consortium. Algal Res 55(2021):102228. https://doi.org/10.1016/j.algal.2021.102228
16. Singh A (2018) River turns drains, drowns villages in sorrow, (Published 18 Jan 2018), indiawaterportal.org, India Water Portal. https://www.indiawaterportal.org/articles/river-turns-drain-drowns-villages-sorrow. Accessed 25 May 2021
17. Sharma S (2015) Najafgarh drain contaminates field, Tribune News Service, Gurgaon, Tribune India, Published 25 December 2015, link: https://www.tribuneindia.com/news/archive/haryana/nazafgarh-drain-contaminates-fields-175706. Accessed 25 May 2021
18. Sengupta S (2015) Is sewage farming safe? Down to Earth, Centre for Science and Environment, Published online 25 September 2015. https://www.downtoearth.org.in/coverage/is-sewage-farming-safe-48566. Accessed 25 May 2021
19. Morgan-Sagastume F, Valentino F, Hjort M, Cirne D, Karabegovic L, Gerardin F, Johansson P, Karlsson A, Magnusson P, Alexandersson T, Bengtsson S, Majone M, Werker A (2014) Polyhydroxyalkanoate (PHA) production from sludge and municipal wastewater treatment. Water Sci Technol 69:177–184
20. Santos, M. M. D., Brehm, F. D. A., Filippe, T. C., Knapik, H. G., & Azevedo, J. C. R. D. (2016). Occurrence and risk assessment of parabens and triclosan in surface waters of southern Brazil: a problem of emerging compounds in an emerging country. RBRH, 21(3), 603–617. https://doi.org/10.1590/2318-0331.011616018.
21. Reiss, R., Mackay, N., Habig, C., & Griffin, J. (2002). An ecological risk assessment for triclosan in lotic systems following discharge from wastewater treatment plants in the United States. Environmental Toxicology and Chemistry, 21(11), 2483–2492. https://doi.org/10.1002/etc.5620211130.
22. Lyndall, J., Barber, T., Mahaney, W., Bock, M., & Capdevielle, M. (2017). Evaluation of triclosan in Minnesota lakes and rivers: part I–ecological risk assessment. Ecotoxicology and Environmental Safety, 142, 578–587. https://doi.org/10.1016/j.ecoenv.2017.04.049.
23. Thomaidi, V. S., Matsoukas, C., & Stasinakis, A. S. (2017). Risk assessment of triclosan released from sewage treatment plants in European rivers using a combination of risk quotient methodology and Monte Carlo simulation. Science of the Total Environment, 603, 487–494. https://doi.org/10.1016/j.scitotenv.2017.06.113.
24. Mottaleb, M. A., Usenko, S., O'Donnell, J. G., Ramirez, A. J., Brooks, B. W., & Chambliss, C. K. (2009). Gas chromatography–mass spectrometry screening methods for select UV filters,

synthetic musks, alkylphenols, an antimicrobial agent, and an insect repellent in fish. Journal of Chromatography A, 1216(5), 815–823. https://doi.org/10.1016/j.chroma.2008.11.072.
25. Bound, J., & Voulvoulis, N. (2005). Household disposal of pharmaceuticals as a pathway for aquatic contamination in the United Kingdom. Environmental Health Perspectives, 113(12), 1705–1711.
26. Bowen, E., Comber, S., Makropoulos, C., Rautiu, R., Ross, R., Ruke, K., et al. (2004). Priority hazardous substances, trace organics and diffuse pollution (WFD): urban catchment study and assessment of diffuse inputs. London, UKWIR Report.
27. Rule, K., Comber, S., Ross, D., Tornton, A., Makropoulos, C., & Rautiu, R. (1996). Survey of priority substances entering thirty English wastewater treatment works. Water and Environment Journal, 20(3), 177–184.
28. Rule, K., Comber, S., Ross, D., Thornton, A., Makropoulos, C., & Rautiu, R. (2006a). Diffuse sources of heavy metals entering an urban wastewater catchment. Chemosphere, 63(1), 64–72.
29. Rule, K., Comber, S., Ross, D., Thornton, A., Makropoulos, M., & Rautiu, R. (2006b). Survey of priority substances entering an urban wastewater catchment—Trace organic chemicals. Chemosphere, 63(4), 581–591.
30. EA (2006). Technical guidance note: Monitoring of discharges to water and sewer. Draft Version 2.A ed., Environment Agency Technical Report
31. Thornton, I., Butler, D., Docx, P., Hession, M., Makropoulos, C., Mcmullen, M., et al. (2001). Pollutants in urban wastewater and sewage sludge. Office for Official Publications of the European Communities. Luxembourg, ICON Report
32. Comber, S., & Gunn, A. (1994) Diffuse sources of heavy metals in sewers. Final report to the department of the environment. Department of the environment.
33. Fauser, P., Vikelsoe, J., Sorensen, P., & Carlsen, L. (2003). Phthalates, nonylphenols and LAS in an alternately operated wastewater treatment plant—fate modelling based on measured concentrations in wastewater and sludge. Water Research, 37, 1288–1295.
34. Hollender J, Zimmermann SG, Koepke S, Krauss M, McArdell CS, Ort C et al (2009) Elimination of organic micropollutants in a municipal wastewater treatment plant upgraded with a full-scale post-ozonation followed by sand filtration. Environ Sci Technol 43(20):7862–7869
35. Abegglen C, Escher BI, Hollender J, Koepke S, Ort C, Peter A et al (2009) Ozonation of treated effluent—final report pilot plant Regensdorf. Swiss Federal Office for the Environment 2009, Bern (report in German with English summary)
36. Escher BI, Bramaz N, Ort C (2009) JEM Spotlight: monitoring the treatment efficiency of a full scale ozonation on a sewage treatment plant with a mode-of-action based test battery. J Environ Monit 11(10):1836–1846
37. Stalter D, Magdeburg A, Weil M, Knacker T, Oehlmann J (2010) Toxication or detoxication? In vivo toxicity assessment of ozonation as advanced wastewater treatment with the rainbow trout. Water Res 44(2):439–448
38. Grantham TE, Canedo-Argüelles M, Perrée I, Rieradevall M, Prat N (2012) A mesocosm approach for detecting stream invertebrate community responses to treated wastewater effluent. Environ Pollut 160:95–102
39. BUWAL (2005) Methoden zur untersuchung und Beurteilung der Fliessgewässer. Makrozoobenthos Stufe F (flächendeckend)
40. AquaPlus (2009) Furtbach (ZH): biologische untersuchungen oberhalb und unterhalb der ARA Regensdorf—äusserer aspekt, pflanzlicher bewuchs, kieselalgen und zoobenthos. untersuchungen der jahre 2006, 2007 und 2008. Kurzbericht zürich: AquaPlus, commisioned by Baudirektion Kanton Zürich, AWEL, Abteilung Gewässerschutz
41. Cumming G (2014) The new statistics: why and how. Psychol Sci 25(1):7–29. https://doi.org/10.1177/0956797613504966
42. Halsey LG, Curran-Everett D, Vowler SL, Drummond GB (2015) The fickle P value generates irreproducible results. Nat Methods 12(3):179–185. https://doi.org/10.1038/nmeth.3288
43. Daims H, Taylor MW, Wagner M (2006) Wastewater treatment: a model system for microbial ecology. Trends Biotechnol 24:483–489. https://doi.org/10.1016/j.tibtech.2006.09.002

44. Zhang B, Ning D, Van Nostrand JD, Sun C, Yang Y, Zhou J, Wen X (2020a) Biogeography and assembly of microbial communities in wastewater treatment plants in China. Environ Sci Technol 54:5884–5892. https://doi.org/10.1021/acs.est.9b07950
45. Wu L, Ning D, Zhang B, Li Y, Zhang P, Shan X, Zhang Q, Brown M, Li Z, Van Nostrand JD, Ling F, Xiao N, Zhang Y, Vierheilig J, Wells GF, Yang Y, Deng Y, Tu Q, Wang A, Zhang T, He Z, Keller J, Nielsen PH, Alvarez PJJ, Criddle CS, Wagner M, Tiedje JM, He Q, Curtis TP, Stahl DA, Alvarez-Cohen L, Rittmann BE, Wen X, Zhou J (2019) Global diversity and biogeography of bacterial communities in wastewater treatment plants. Nat Microbiol 4:1183–1195. https://doi.org/10.1038/s41564-019-0426-5
46. Segata N, Izard J, Waldron L, Gevers D, Miropolsky L, Garrett WS, Huttenhower C (2011) Metagenomic biomarker discovery and explanation. Genome Biol 12:R60. https://doi.org/10.1186/gb-2011-12-6-r60
47. Ofiţeru ID, Lunn M, Curtis TP, Wells GF, Criddle CS, Francis CA, Sloan WT (2010) Combined niche and neutral effects in a microbial wastewater treatment community. Proc Natl Acad Sci U S A 107:15345–15350. https://doi.org/10.1073/pnas.1000604107
48. Zhang B, Ning D, Yang Y, Van Nostrand JD, Zhou J, Wen X (2020b) Biodegradability of wastewater determines microbial assembly mechanisms in full-scale wastewater treatment plants. Water Res 169:115276. https://doi.org/10.1016/j.watres.2019.115276
49. Guerrini A, Romano G, Carosi L, Mancuso F (2017) Cost savings in wastewater treatment processes: the role of environmental and operational drivers. Water Resour Manag 31(8):2465–2478. https://doi.org/10.1007/s11269-017-1586-9
50. Szymański K, Morawski AW, Mozia S (2018) Effectiveness of treatment of secondary effluent from a municipal wastewater treatment plant in a photocatalytic membrane reactor and hybrid UV/H2O2—ultrafiltration system. Chem Eng Process Process Intensif 125:318–324. https://doi.org/10.1016/j.cep.2017.11.015
51. Pereira L, Pinto M, Costa R, Dias Á, Goncalves R (2021). The New SWOT for a Sustainable World. J Open Innov 7(1), https://doi.org/10.3390/joitmc7010018
52. Caicedo C, Rosenwinkel KH, Exner M, Verstraete W, Suchenwirth R, Hartemann P et al. (2019) Legionella occurrence in municipal and industrial wastewater treatment plants and risks of reclaimed wastewater reuse: Review. Water Res 149:21–34. https://doi.org/10.1016/j.watres.2018.10.080
53. Iatridis K, Kesidou E (2016) What Drives Substantive Versus Symbolic Implementation of ISO 14001 in a Time of Economic Crisis? Insights from Greek Manufacturing Companies. J Bus Ethics 148(4):859–877. https://doi.org/10.1007/s10551-016-3019-8
54. Rodríguez JP, García CA, Zafra CA (2016) El análisis del ciclo de vida aplicado a las Plantas de Tratamiento de Aguas Residuales. Cienc y Soc 41(3):617–636
55. Rukavishnikova I, Strukova M, Gabova I, Strukova L, Karaeva A (2017) Introduction of iso management systems in municipal enterprises providing water supply, water preparation and wastewater treatment in the major cities of the russian federation. Int J Sustain Dev Plan 12(01):71–78. https://doi.org/10.2495/sdp-v12-n1-71-78
56. Ene SA, Teodosiu C (2011a) Grey water footprint assessment and challenges for its implementation. Environ Eng Manag J 10:333–340
57. Ene SA, Teodosiu C (2011) Grey water footprint assessment of the wastewater treatment plants in the Prut-Barlad catchment. Bul Inst Polit Iasi LVII (LXI):127–143
58. Teodosiu C, Robu B, Cojocariu C, Barjoveanu G (2013) Environmental impact and risk quantification based on selected water quality indicators. Nat Hazards 75:S89–S105
59. Astuti, Maryani Paramita; Notodarmojo, Suprihanto; Priadi, Cindy Rianti; Padhye, Lokesh P. Contaminants of emerging concerns (CECs) in a municipal wastewater treatment plant in Indonesia. Environmental Science and Pollution Research (2022). https://doi.org/10.1007/s11356-022-23567-8
60. Dasgupta, Pinaki; Kumar, Vivek; Malik, Anushree; Kumar, Madan Wastewater Treatment Systems for City-Based Municipal Drains for Achieving Sustainability. Circular Economy and Sustainability (2022). https://doi.org/10.1007/s43615-022-00163-z

61. Valentino, Francesco; Morgan-Sagastume, Fernando; Fraraccio, Serena; Corsi, Giovanna; Zanaroli, Giulio; Werker, Alan; Majone, Mauro Sludge minimization in municipal wastewater treatment by polyhydroxyalkanoate (PHA) production. Environmental Science and Pollution Research (2014). https://doi.org/10.1007/s11356-014-3268-y
62. Mohan, S.; Balakrishnan, P. Triclosan in Treated Wastewater from a City Wastewater Treatment Plant and its Environmental Risk Assessment. Water, Air, & Soil Pollution (2019). https://doi.org/10.1007/s11270-019-4098-9
63. Rowsell, Victoria F.; Tangney, Peter; Hunt, Claire; Voulvoulis, Nikolaos Estimating Levels of Micropollutants in Municipal Wastewater. Water, Air, & Soil Pollution (2009). https://doi.org/10.1007/s11270-009-0112-y
64. Ashauer, Roman Post-ozonation in a municipal wastewater treatment plant improves water quality in the receiving stream. Environmental Sciences Europe (2016). https://doi.org/10.1186/s12302-015-0068-z
65. Özkan, Oktay; Oğuz, Merve; Özdemir, Özgür Characterization and assessment of a large-scale domestic advanced wastewater treatment plant in Turkey. Environmental Monitoring and Assessment (2011). https://doi.org/10.1007/s10661-011-2338-6
66. Song, Yang; Mhuantong, Wuttichai; Liu, Shuang-Yuan; Pisutpaisal, Nipon; Wongwilaiwalin, Sarunyou; Kanokratana, Pattanop; Wang, Ai-Jie; Jiang, Cheng-Ying; Champreda, Verawat; Qiu, Dong-Ru; Liu, Shuang-Jiang Tropical and temperate wastewater treatment plants assemble different and diverse microbiomes. Applied Microbiology and Biotechnology (2021). https://doi.org/10.1007/s00253-020-11082-0
67. Herrera-Navarrete, Ricardo; Colín-Cruz, Arturo; Arellano-Wences, Hilda Janet; Sampedro-Rosas, María Laura; Rosas-Acevedo, José Luis; Rodríguez-Herrera, América Libertad Municipal Wastewater Treatment Plants: Gap, Challenges, and Opportunities in Environmental Management. Environmental Management (2021). https://doi.org/10.1007/s00267-021-01562-y
68. Teodosiu, Carmen; Barjoveanu, George; Sluser, Brindusa Robu; Popa, Simona Andreea Ene; Trofin, Orest Environmental assessment of municipal wastewater discharges: a comparative study of evaluation methods. The International Journal of Life Cycle Assessment (2016). https://doi.org/10.1007/s11367-016-1029-5

Chapter 2
Primary Treatment

Nitin Vitthalrao Thombre

2.1 Introduction by Editor

Primary treatment holds immense importance in municipal sewage treatment plants as it primarily focuses on the removal of solid materials and heavy organic particles from the wastewater. This stage is crucial because untreated wastewater contains a significant amount of suspended solids, debris, and other organic matter, which, if not removed, can result in severe environmental and public health concerns. Primary treatment acts as a preliminary barrier, reducing the organic load and solids content of the wastewater before it undergoes further treatment.

Primary treatment involves several physical and mechanical processes aimed at separating and eliminating solid materials from the wastewater. The primary treatment unit, often referred to as a sedimentation or settling tank, serves as the central element of this stage. The primary treatment processes typically include the following:

Screening: The wastewater passes through screens that effectively remove large debris, such as sticks, rags, plastics, and other solid objects, thereby preventing their entry into the treatment system. This process helps protect pumps, valves, and subsequent treatment units from damage.

Grit Removal: After screening, the wastewater flows into a grit chamber, where heavy inorganic particles like sand, gravel, and grit settle down due to gravity. These abrasive materials can cause wear and damage to equipment if not removed at this stage.

Settling and Sedimentation: In the primary settling tank, the wastewater is allowed to rest undisturbed for a certain period, typically several hours. During this time, the flow velocity decreases, enabling heavier suspended solids and organic matter to

N. V. Thombre (✉)
VA Tech Wabag Ltd., Pune, Maharashtra, India
e-mail: nitin01thombre@gmail.com

N. V. Thombre (ed.), *Municipal Wastewater Reclamation and Reuse*,
https://doi.org/10.1007/978-981-96-9112-8_2

settle to the bottom as sludge, while greases, oils, and lighter particles rise to the surface as scum.

The outputs of primary treatment include two main components: settled sludge and scum. The settled sludge, known as primary sludge, is collected and further processed in subsequent treatment stages, such as secondary treatment or sludge digestion, to extract valuable resources and reduce its volume for disposal. The scum, consisting of fats, oils, and other floating materials, is also removed for proper disposal.

However, it is important to note that primary treatment alone does not achieve the level of wastewater treatment necessary for safe discharge. It primarily focuses on the physical removal of solids but does not address the dissolved organic compounds, nutrients, or pathogens present in the wastewater. Therefore, primary treatment is typically followed by secondary and tertiary treatment processes to achieve a higher degree of purification.

2.2 Machine Generated Summaries

Disclaimer: The summaries in this chapter were generated from Springer Nature publications using extractive AI auto-summarization: An extraction-based summarizer aims to identify the most important sentences of a text using an algorithm and uses those original sentences to create the auto-summary (unlike generative AI). As the constituted sentences are machine selected, they may not fully reflect the body of the work, so we strongly advise that the original content is read and cited. The auto generated summaries were curated by the editor to meet Springer Nature publication standards. To cite this content, please refer to the original papers.

Machine generated keywords: model, settle, primary, primary treatment, sludge, activate sludge, sludge settle, activate, chemically, predict, dynamic, settling, separation, velocity, thickening.

Anaerobic Digestion of Chemically Enhanced Primary Treatment (CEPT) Sludge and the Microbial Community Structure [42]

This is a machine-generated summary of:

Ju, Feng; Wang, Yubo; Lau, Frankie T. K.; Fung, W. C.; Huang, Danping; Xia, Yu; Zhang, Tong: Anaerobic digestion of chemically enhanced primary treatment (CEPT) sludge and the microbial community structure [42]

Published in: Applied Microbiology and Biotechnology (2016)

Link to original: https://doi.org/10.1007/s00253-016-7730-2

If you want to cite the papers, please refer to the original.

For technical reasons we could not place the page where the original quote is coming from.

Abstract-Summary

"The effectiveness and treatment conditions of $FeCl_3$- and $AlCl_3$-coagulated municipal sewage sludge from chemically enhanced primary treatment (CEPT) using anaerobic digestion (AD) and the structure of microbial community were investigated."

"The results based on 297 measurements under different operational conditions demonstrate good average AD performance of CEPT sludge, that is, percent volatile solid reduction of 58%, specific biogas production (or biogas yield) of 0.92 m^3/kg volatile solids (VS) destroyed, and methane content of 65.4%."

"Methanomicrobia methanogens were better enriched in the anaerobic digesters of CEPT sludge than in the combined sludge."

"Combined, this study supports AD as an efficient technology for CEPT sludge treatment and poses first insights into the microbial community structure."

Introduction

"Whether CEPT sludge from iron and aluminum coagulation can be anaerobically digested efficiently to give at least the same benefits as anaerobic digestion (AD) of sewage sludge in conventional secondary treatment works, namely sludge reduction and biogas production as a renewable energy, is unknown."

"Although microbial community structure in anaerobic digesters of combined primary and/or secondary sewage sludge has been well described (Regueiro and others 1; Sundberg and others 2; Yu and others 3; Yu and others 4), whether the different chemical property and biodegradability of CEPT sludge, compared with combined sewage sludge, may accommodate microbial community with a different structure in anaerobic digesters remains to be explored."

"This work focuses on the AD performance of $FeCl_3$- and $AlCl_3$-coagulated CPET sludge and the microbial community structure and addresses the following issues: (1) the effects of operational parameters, including $FeCl_3$ dosing, organic loading rate (OLR), temperature, hydraulic retention time (HRT), and salinity, on the AD performance of CEPT sludge, and (2) the structure of microbial community that contributes to AD of CEPT sludge and its comparison with that of combined sludge digesters."

Materials and Methods

"The effect of $FeCl_3$ dosing (specific gravity = 1.45 g/cm^3; concentration = 40% w/w), OLR, temperature, HRT, and sewage salinity on AD performance of sulfate-rich saline CEPT sludge of SCI-STWs and Siu Ho Wan STWs (SHW-STWs) were evaluated in four 0.8-L semi-continuous digesters, namely, M1, M2, M3, and M4, with mild magnetic stirring and pH control at 6.8–7.2 by adding 1 mol/L $NaHCO_3$ solution."

"For microbial structure analysis, sludge samples were taken from R1 and R2 on days 76 and 90 when both digesters were under steady operation with feeding CEPT sludge of SCI-STWs (R1) and SmT-STWs (R2)."

"Due to unavoidable differences between digesters of this and previous studies, such as feed sludge type (e.g., CEPT sludge, TPS, and TSAS) and digester configurations (e.g., size and shape), it is unfair to compare their biogas production (V_G) or volatile solid reduction (VSR)."

Results

"For the CEPT sludge of SCI-STWs, lower OLRs led to significantly (P values < 0.02) higher VSR% (59–60%) than higher OLRs (53%)."

"There was no significant difference (P values > 0.05) in CH_4% between lower and higher OLR conditions for CEPT sludge of both STWs."

"AD performance of CEPT sludge of SCI-STWs and SHW-STWs showed no significant (all P values > 0.05) difference under each HRT, revealing negligible impact of tested salinity difference on CEPT sludge digestion under test conditions."

"AD of CEPT sludge of SCI-STWs or SHW-STWs both showed significantly (P values < 0.003) higher VSR, VSR (%), and CH_4% at an HRT of 7 days, whereas significantly higher SMP and SBP at an HRT of 12 days, revealing higher solid reduction efficiencies but much lower specific methane/biogas production at a shorter HRT of 7 days."

Discussion

"Batch operations over 90 days suggest that $FeCl_3$ dose efficiently reduces H_2S production and significantly improves AD performance of sulfate-rich saline CEPT sludge of SCI-STWs, including the specific methane production and methane content in biogas."

"Further studies are required to differentiate and resolve the effects of different chemical ingredients of CEPT sludge (e.g., coagulant type, salinity, organic compositions) on the variations of microbial community structure in different anaerobic digesters."

"While these results may implicate the influences of organic compositions (primary sludge vs. secondary biological sludge) and chemical properties (e.g., salinity and coagulants) of feed sludge on the anaerobic digester microbial community structure, microbiological data from more AD reactors under comparable operational conditions are required to completely resolve the effects of various chemical characteristics of feed sludge on the structure and performance of methanogenic microbial communities."

Combined Ultrasound-Ozone Treatment for Reutilization of Primary Effluent—A Preliminary Study [43]

This is a machine-generated summary of:

Rossi, Giada; Mainardis, Matia; Aneggi, Eleonora; Weavers, Linda K.; Goi, Daniele: Combined ultrasound-ozone treatment for reutilization of primary effluent—a preliminary study [43]

Published in: Environmental Science and Pollution Research (2020)

Link to original: https://doi.org/10.1007/s11356-020-10467-y

If you want to cite the papers, please refer to the original.

For technical reasons we could not place the page where the original quote is coming from.

Abstract-Summary

"The present work is a preliminary study on the potential of low-frequency ultrasound irradiation coupled with O_3 process for the disinfection of a primary effluent from a municipal wastewater treatment plant preserving nutrient levels (in particular nitrogen and phosphorous), for its possible reuse in civil, industrial, and agricultural sectors."

"The results showed that the combined US-O_3 process produces a good removal efficiency regarding soluble Chemical Oxygen Demand, sCOD (ca."

"The process also reached high disinfection performances, obtaining 4 logs for E. coli and 5 log abatement for Total Coliforms."

"The high removal efficiency is matched by an outstanding retention of nutrients (total nitrogen and orthophosphate) highlighting a high potential value for agricultural reuse of the treated primary effluent, with possible significant saving of chemical fertilizers."

"It was concluded that low-frequency ultrasound pre-treatment, combined with ozonation, could be a useful process for primary effluent recovery for several purposes."

Introduction

"Primary effluents (prEFFs) from wastewater treatment plants (WWTPs) have a good agricultural reuse potential, because most of the nutrients are conserved throughout the primary treatment, while only high-density particulate matter is removed (Abdessemed and Nezzal 6)."

"Ozone disinfection of wastewater from primary treatment could reduce the pathogenic risk and the microbial toxicity of the effluents, degrading electrophilic organics and maintaining, at the same time, nitrogen and phosphorus."

"The potential benefit of a combined sonozone treatment relates to sonication-induced pressure waves that, besides stimulating the production of hydroxyl radicals, might augment ozone transfer rate into water (Yargeau and Danylo 7)."

"The combined ultrasound and ozonation process has proved to be more efficient than ultrasound and ozonation alone in the treatment of real abattoir wastewater, showing significant biochemical oxygen demand (BOD) and COD removal, coupled with a good effluent disinfection, due to the complete inactivation of Total Coliforms (Alfonso-Muniozguren and others 8)."

Materials and Methods

"The wastewater flowed into the reactor by means of a peristaltic pump (Cellai Perinox SF3, 230 V, 80 W, 1.6x1.6 3R) with flowrate ranging from 7 to 23 L/h. The selected ultrasonic contact times were 6, 12, and 20 s, corresponding respectively to 10 (US-1), 20 (US-2), and 50 (US-3) kJ US energy/L. US power was verified by calorimetric measurements (Wei and others 9), and the sonochemical efficiency in the whole reaction system was quantified using the Weissler reaction (Kimura and others 10)."

"All the other chemical parameters were measured following Standard Methods (APHA 11); in particular, sCOD was evaluated using method 5220D, orthophosphate using method 4500-P C, and MBAS using method 5540 C. Microbiological characterization was performed using membrane filtration method 9222, utilizing ECD MUG agar (Fluka) for the Total Coliforms and E. coli."

"The specific energy consumption of US and ozone apparatus was calculated for the different operating conditions from the instrument powers, the results of the energy meter measurements and the treated wastewater volume (see the "Bench-top ultrasonic and ozonation apparatus" section)."

Results and Discussion

"Orthophosphate concentration did not substantially change throughout the tests, both when ultrasound and ozonation were applied alone and in the combined US-O_3 prEFF treatment, again underlining a favorable condition for agricultural reuse of the treated effluent."

"As for ozonation treatment, they observed a low removal for an applied O_3 dose of 0.05 mg O_3/L, while a complete abatement was obtained increasing ozone dosage to 0.16 mg O_3/L (O_3 dose was significantly lower than the concentration used in the present study)."

"Barzegar and others (12), instead, studied a combined electrocoagulation and ozone treatment for greywater reuse, having COD concentration of 460 mg/L (higher than the analyzed prEFF), Total Coliforms in the range of 2–5 $\times$ 10^7 CFU/100 mL (comparable with that of actual wastewater) and E. coli concentration of 2–2.8 $\times$ 10^3 CFU/100 mL (lower than actual prEFF)."

Conclusions

"A series of combined low-frequency ultrasound and ozonation tests were planned and executed to evaluate the potential reuse of a primary effluent from municipal wastewater treatment plant for civil, industrial, and agricultural applications."

"During the US-O_3 combined treatment, the nutrient charge of the mixture, measured as total nitrogen and orthophosphate, was well-maintained: consequently, a high potential value for agricultural reuse of the treated primary effluent was highlighted, with possible significant saving of chemical fertilizers."

"Primary effluent disinfection was more effective when a combined US-O_3 treatment was applied, reaching maximum values of 5 logs on the Total Coliform parameter."

"It was concluded that low-frequency ultrasound pre-treatment, combined with ozonation, could be a useful process for primary effluent recovery for several purposes."

Effect of Ferrate Pretreatment on Anaerobic Digestibility of Primary Sludge Spiked with Resin Acids [44]

This is a machine-generated summary of:

Das, Sreejon; Manoli, Kyriakos; Sharma, Virender K.; Dagnew, Martha; Ray, Madhumita B.: Effect of ferrate pretreatment on anaerobic digestibility of primary sludge spiked with resin acids [44]

Published in: Environmental Science and Pollution Research (2022)

Link to original: https://doi.org/10.1007/s11356-022-21599-8

If you want to cite the papers, please refer to the original.

For technical reasons we could not place the page where the original quote is coming from.

Abstract-Summary

"Anaerobic digestion (AD) is a common practice stabilizing sludge; however, high concentration of resin acids affects the AD process."

"The aim of this research was mainly to determine the impact of ferrate (Fe (VI)) oxidation on selected resin acids and anaerobic digestibility of ferrate-treated primary sludge (PS) spiked with the resin acids."

"Anaerobic treatability studies of primary sludge spiked with resin acids both under control condition and optimum ferrate pretreatment were conducted."

"Up to 97% oxidation of resin acids occurred in pure water, while only 44%–62% oxidation of resin acids occurred in PS with an increasing Fe (VI) dosage from 0.034 to 0.137 mg Fe (VI)/mg $tCOD_{fed}$."

"The pretreatment did not affect the anaerobic biodegradability of resin acids; however, it lowered their negative influences on the PS digestibility."

"The potential benefits of ferrate pretreatment of P&P primary sludge include resin acids oxidation and subsequent toxicity reduction, higher sludge solubilization enhancing methane production and enabling anaerobic digestion at higher COD loading."

Introduction

"The reported lethal concentration is much lower than the resin acids concentration range (10 to 10,000 mg/L) in P&P wastewater (Meyer and Edwards 13); thus, it is important to understand the fate of the resin acids in P&P wastewater prior to disposal to freshwater bodies."

"Resin acids above 100 mg/L inhibited the AD of P&P sludge, as demonstrated by a decrease in gas production during batch anaerobic digestion studies conducted by Das and others (14)."

"In a study by Das and others (14), pretreatment of resin acids in primary sludge showed a reduction of 50–70% spiked resin acids and improved methane gas production by about 42% with a substantially high ozone dose of 0.52 mg O_3/mg $tCOD_{fed}$."

"Fe (VI) oxidation was conducted in Milli-Q water spiked with model resin acids to determine the degradation kinetics and suitable oxidation conditions to be applied for sludge pretreatment."

Materials and Methods

"A combination of resin acids containing 22% DhA and 78% AbA/isoP was used to assess the oxidative efficiency of Fe (VI) on resin acids and their subsequent anaerobic digestibility of primary sludge."

"The AD was conducted using unpretreated and pretreated (0.076 mg Fe (VI)/mg $tCOD_{fed}$) resin acids spiked PS."

"Untreated resin acids in PS were incubated with anaerobic seed sludge to evaluate AD performance in the presence of resin acids (untreated digester)."

"Treated resin acids in primary sludge (pretreated digester) were also incubated to measure the impact of Fe (VI) pretreatment on the anaerobic digestibility of resin acids."

"The Fe (VI) concentrations were reasonably higher (1052–3692 mg/L) for the primary sludge spiked with resin acids, but the normalized dosage varied from 0.034 to 0.119 mg Fe (VI)/mg $tCOD_{fed}$, very similar to the dosages used in control oxidation experiments conducted in Milli-Q water."

Results and Discussions

"The impact of pH (6.0, 7.0 and 8.0) on the oxidation performance of resin acids in phosphate buffer solution was accompanied using 40 mg/L of Fe (VI)."

"The process variables such as Fe (VI) dosage (mg/L), pH and reaction time (min) were optimized based on the oxidation efficiency (response) of tested resin acids using Minitab (version 17) software."

"The optimization study showed that maximum oxidation of resin acids was achieved with a Fe (VI) dosage, pH and reaction time of 40 mg/L (0.137 mg Fe (VI)/mg $tCOD_{fed}$), 7 and 20 min, respectively."

"In comparing the treatment of resin acids spiked in Milli-Q water and raw PS, the degradation efficiency of tested resin acids decreased by 54–70% in primary sludge at 0.034 mg Fe (VI) dosage/mg $tCOD_{fed}$ because of the background organics in PS."

Conclusions

"The performance of Fe (VI) pretreatment of tested resin acids spiked in municipal PS on AD was evaluated in this study."

"Fe (VI) oxidation increased the removal efficiency of resin acids and sludge solubilization."

"Pretreatment with the dosage of 0.076 mg Fe (VI)/mg tCODfed at pH 7.0 and 20 min reaction time reduced the initial mass of resin acids by 29–48% and increased sCOD by around 18–21% in the digester."

"Biodegradation study of both untreated and Fe (VI)-pretreated resin acids in municipal PS showed negligible biodegradation of resin acids over 20 days of AD."

Correlation Between Operating Parameters and Removal Efficiency for Chemically Enhanced Primary Treatment System of Wastewater [45]

This is a machine-generated summary of:

Al Bazedi, Ghada A.; Abdel-Fatah, Mona A.: Correlation between operating parameters and removal efficiency for chemically enhanced primary treatment system of wastewater [45]

Published in: Bulletin of the National Research Centre (2020)

Link to original: https://doi.org/10.1186/s42269-020-00368-y

If you want to cite the papers, please refer to the original.

For technical reasons we could not place the page where the original quote is coming from.

Abstract-Summary

"“Chemically enhanced primary treatment” (CEPT) is an approach to wastewater treatment."

"The aim of this study is to create an empirical model of separation efficiency for wastewater chemically enhanced primary treatment."

"The present work provides an approach for using chemically enhanced primary treatment of wastewater."

"The obtained results showed that the empirical model can predict removal efficiencies with $R^2 = 0.973$, and 0.978 for COD and TSS."

"The advantage of this model is that it would allow better process control and treatment efficiency."

"The results show that chemically enhanced primary treatment method can be used as an efficient method in conventional municipal wastewater treatment plants to reduce the organic load of biological treatment and enhance nutrients removal."

Background

"CEPT can be used to achieve many different goals in wastewater treatment facilities: (i) to increase the TSS elimination performance of the primary settling process based on substantial treatment, (ii) to limit organic loading rates thereby reducing the demand for oxygen in aerobic biological treatment facilities, (iii) to achieve excessive phosphorus removal through precipitation and ultimately to achieve excessive phosphorus removal through precipitation, and ultimately (iv) to allow increased hydraulic charging quotes to an existing PST, thereby enhancing plant life with excessive moist weather flows (Ayoub & Abdel-fattah, 15; Ayoub and others, 16; Neupane and others, 17)."

"Different studies showed that CEPT process has effectively removed 85% and 57% of total suspended solids (TSS) and biological oxygen demand (BOD), respectively."

"Different researchers have assessed the use of coagulants and the effectiveness of combining chemical treatment with conventional processes."

"The current work investigates combining different coagulants together to reach acceptable removal efficiency to reduce the organic load of biological treatment and enhance nutrients removal."

Methodology

"The main objective of the analysis was to investigate the relation between the operating parameters including COD, pH, TSS, TDS, and the removal efficiency of both COD and TSS using enhanced primary treatment technology in Gitis and Hankins (Gitis & Hankins, 18)."

"The effect of "pH" on the efficiency of the process has been assessed for each coagulant type, e.g., the removal percentage of "COD", turbidity, and "TSS"."

"The effect of coagulant type and dosage, as well as pH, was investigated."

"Coagulant aids (a) Magnafloc155, Magnafloc1011 A mathematical correlation was formulated using a multivariate regression model for experimental results between operating parameters and removal efficiency, and operating parameters and suspended solids level."

Results

"Experiments with alum show higher COD removals than those with $FeCl_3$ and lime."

"The correlation between TSS and operating parameters (COD, pH, and TDS) using multiple regression analysis with coefficient of determination $R^2 = 0.997$."

"The correlation between TSS and operating parameters (COD, pH, settling time, temperature, flow rate, TSS, and TDS) using multiple regression analysis with coefficient of determination $R^2 = 0.973$."

"The correlation between SS and operating parameters (COD, pH, settling time, temperature, flow rate, SS, and TDS) using multiple regression analysis with Coefficient of determination $R^2 = 0.978$."

Discussion

"Investigating the turbidity removal efficiency for various coagulant dosages indicates that $FeCl_3$ and Alum are eminently effective coagulants for turbidity, with removal efficiency above 90%."

"Lime shows to be less adequate and mediocre, where the removal efficiency reaches 80% when using a lime with a dose of 400 mg/l. There is no appreciable effect of differing lime concentrations on turbidity removal efficiency."

"With $FeCl_3$, it was noticed in dosages of 40 mg/l and higher, while for Lime, the optimum dosage was determined to be 400 mg/l. The experiments revealed that it is better to use each coagulant alone due to interference between coagulant mixtures, which leads to undesired changes in pH. Treatment with the use of Alum was proved to be more efficient than when using $FeCl_3$ and lime, with the advantage that higher removal rates are attained at normal pH levels."

Conclusion

"Multivariate regression model empirical model was developed using the simulation data obtained from the pilot plant and the experimental studies using different types of coagulant ($FeCl_3$, alum, lime, and Magna-floc 155)."

"The removal efficiencies of COD and TSS can be predicted to be ($R^2 = 0.973$ and 0.978, respectively)."

"The advantage of this model is that it would allow a better process control."

Dynamic Modelling of Solids in a Full-Scale Activated Sludge Plant Preceded by CEPT as a Preliminary Step for Micropollutant Removal Modelling [46]

This is a machine-generated summary of:

Baalbaki, Zeina; Torfs, Elena; Maere, Thomas; Yargeau, Viviane; Vanrolleghem, Peter A.: Dynamic modelling of solids in a full-scale activated sludge plant preceded by CEPT as a preliminary step for micropollutant removal modelling [46]

Published in: Bioprocess and Biosystems Engineering (2016)

Link to original: https://doi.org/10.1007/s00449-016-1715-5

If you want to cite the papers, please refer to the original.

For technical reasons we could not place the page where the original quote is coming from.

Abstract-Summary

"The presence of micropollutants in the environment has triggered research on quantifying and predicting their fate in wastewater treatment plants (WWTPs)."

"Since the removal of micropollutants is highly related to conventional pollutant removal and affected by hydraulics, aeration, biomass composition and solids concentration, the fate of these conventional pollutants and characteristics must be well predicted before tackling models to predict the fate of micropollutants."

"Results also demonstrated that particular attention must be given to characterizing incoming TSS to obtain a representative solids balance in the presence of a chemically enhanced primary treatment, which is key to predict the fate of micropollutants."

Introduction

"Such modelling activity involves purpose-dependent calibration of models representing the bioreactor and the settler, which are the two essential constituents of an activated sludge plant."

"To the bioreactor model, a representative secondary settler model is essential for accurate modelling of the bioprocesses in the bioreactors as the return sludge from the settler influences the solids balance in the whole system [21]."

"The emphasis of the conventional pollutants calibration exercise should be adapted to its final goal (i.e. supporting proper modelling of the fate of micropollutants in this context)."

"This paper presents a procedure for the calibration of a conventional activated sludge model based on standard data available from WWTPs (i.e. with a minimal additional sampling effort) for the specific objective of a subsequent development and calibration of a micropollutant model that is tackled in another study."

Methodology

"In the final model structure, it was necessary to separately consider the two parallel trains of aeration tanks (i.e. two parallel trains of three tanks in series) to capture the distinct hydraulic behaviour of each train, since the incoming wastewater and return sludge flow rates are observed to be different for each train."

"The used influent TSS/COD ratios assume a certain fixed amount of inorganic suspended solids (i.e. iSS, such as minerals) that are not measured as COD [20]."

"Both TSS characterization methods (i.e. based on particulate COD and based on TSS measurements) should give the same results, but differences can occur when not properly accounting for colloids and inorganics."

"The second step involved achieving a match between the simulated and the measured TSS concentrations at different locations in the activated sludge process, such as the waste sludge (i.e. WAS TSS), the aeration tank (i.e. MLSS) and the secondary effluent (i.e. effluent TSS)."

Results and Discussion

"Reaching a proper solids balance by achieving a match between the measured and simulated TSS in aeration tank (i.e. MLSS), and more importantly, TSS in the waste sludge (i.e. WAS TSS) is an essential requirement for predicting the fate of micropollutants."

"To remediate the discrepancies in the solids balance and to achieve an agreement between the simulated and the measured MLSS and WAS TSS concentrations, the characterization of incoming solids was analysed in more detail."

"This obtained ratio between the calibrated and the measured TSS in the input to activated sludge treatment is slightly lower than that reported by Henze and others [22] for an ASM2d model (i.e. 1.5)."

"This indicated that a less extensive calibration was required for the Bürger–Diehl model (i.e. only one parameter) compared to the Takács model in order to match the effluent TSS measurements."

Conclusion

"An activated sludge model was calibrated with 1-year, full-scale conventional pollutants measurements and a relatively small, dedicated 3-day detailed measurement campaign as a preparatory step for the subsequent fate modelling of micropollutants."

"Careful calibration of the solids balance through detailed influent characterization and appropriate selection of the settler model allowed the measured nutrient profiles in the effluent to be captured by the model using default ASM parameters."

"The routine measurements available from the WWTP based on 24-h composite samples were proven to be sufficient to successfully model the general yearly profiles of WAS TSS, effluent TSS, DO and nutrients."

"Since these variables are linked to the main removal processes of micropollutants (i.e. through SRT, aeration and solids balance), it is shown that the measurement effort and resources required for the conventional modelling that is performed as a preliminary step for the calibration of the micropollutant model can remain limited."

2.3 Screening and Grit Removal

Introduction by Editor

Screening is a preliminary step in wastewater treatment that serves as the first line of defense against unwanted materials entering the treatment plant. Its primary purpose is to remove large solid objects, including sticks, rags, plastics, and other debris that could disrupt operations, damage equipment, hinder subsequent treatment processes, or pose a threat to downstream environments. Effective screening prevents clogging, minimizes wear and tear on pumps and equipment, and ensures the efficient operation of the entire treatment plant.

Municipal wastewater treatment plants employ various types of screens to achieve effective solid separation. The selection of the appropriate screen type depends on factors such as wastewater characteristics, expected solids content, flow rate, and available space. Common types of screens used in municipal wastewater treatment plants include:

1. Bar Screens: Bar screens consist of vertical or inclined bars placed at regular intervals. They allow wastewater to pass through while capturing and removing solids for subsequent disposal.
2. Drum Screens: Drum screens feature rotating cylindrical drums with perforations or mesh screens. As wastewater flows through the drum, solids are retained and lifted out of the water for disposal.
3. Fine Screens: Fine screens utilize smaller openings to remove smaller debris and solids. These screens are often used as a secondary treatment step to complement primary screening and achieve a higher level of solids removal.
4. Static Screens: Static screens employ fixed plates or bars to capture solids as wastewater flows through them. They are commonly used in smaller treatment plants or as a preliminary step before further screening.
5. Step Screens: Step screens utilize a series of inclined steps or bars to capture and lift solids out of the wastewater. They offer increased screening surface area and are effective in handling larger flows and higher debris loads.

The screening process in municipal wastewater treatment plants typically involves multiple steps. Initially, raw wastewater enters the screening chamber containing the screens. As the wastewater passes through the screens, larger solids are retained on the surface or within the screen openings. Mechanical cleaning mechanisms such as rakes or brushes are often employed to prevent clogging and maintain operational efficiency by removing the accumulated solids from the screens. These solids are then transported to a separate disposal unit or a subsequent treatment process.

Grit consists of inorganic particles such as sand, gravel, silt, and other heavy materials that enter wastewater from sources like storm-water runoff and industrial discharges. If not efficiently removed, grit can cause operational issues and damage downstream treatment processes and equipment. It can lead to abrasion and wear on pumps, valves, and mechanical components, reducing their efficiency and lifespan. Additionally, grit accumulation in sedimentation tanks can reduce their capacity and interfere with the settling of organic solids. Therefore, effective grit removal is vital for the smooth operation and longevity of municipal wastewater treatment plants.

Municipal wastewater treatment plants employ various techniques for efficient grit removal, broadly categorized as gravity-based and mechanically assisted methods.

1. Gravity-based Grit Removal:

    - Grit Chambers: Grit chambers, also known as detritus tanks, utilize gravity to decrease wastewater flow, allowing the sedimentation of heavy particles. The settled grit is then removed through controlled flushing or a grit pump.
    - Aerated Grit Chambers: Aerated grit chambers introduce air to create turbulence, facilitating the separation of grit from organic matter. Lighter organic particles float, while heavier grit settles for subsequent removal.

2. Mechanically Assisted Grit Removal:

- Grit Cyclones: Grit cyclones use centrifugal force to separate and remove grit. Wastewater enters the cyclone tangentially, causing rotational motion. Heavy grit particles are forced towards the outer wall, while cleaner water is discharged from the center.
- Grit Classifiers: Grit classifiers combine sedimentation and mechanical agitation to separate and dewater grit. Wastewater enters a settling tank, where grit settles at the bottom. Mechanical agitators remove the settled grit, transferring it to a dewatering system.
- Grit Screw Conveyors: Grit screw conveyors feature a rotating screw that transports settled grit from the bottom of a tank or chamber to a dewatering unit or disposal system.

Proper disposal of grit is essential once it is removed from the wastewater. Municipal wastewater treatment plants may employ various methods, including dewatering, washing, and landfill disposal. Dewatering processes, such as grit screw conveyors, reduce the moisture content of grit, facilitating its handling and transportation. Some treatment plants may also wash the grit to remove organic matter before disposal. Final disposal of grit typically occurs in designated landfill facilities.

2.4 Settling

Introduction by Editor

It explores the underlying principles governing settling and sedimentation, describes the design and operation of sedimentation tanks, discusses the factors influencing settling efficiency, and highlights the importance of proper sludge management.

The settling and sedimentation process involves the gravitational settling of suspended solids, organic matter, and other particulate pollutants present in wastewater. Stokes' Law, a fundamental principle governing this process, describes the settling velocity of particles in a fluid medium. According to Stokes' Law, particle settling velocity is directly proportional to particle size and density while inversely proportional to the dynamic viscosity of the fluid. This law forms the foundation for understanding particle behavior during sedimentation.

Sedimentation tanks, also referred to as clarifiers, are meticulously designed to create an optimal environment for the settling and separation of particles from wastewater. These tanks typically exhibit rectangular or circular shapes and comprise multiple compartments. Incoming wastewater enters the tank, undergoing quiescent or low-velocity flow, which allows ample time for particles to settle.

The design parameters of sedimentation tanks, such as surface area, depth, and detention time, are critical in ensuring efficient settling and sedimentation. The tank's geometry and hydraulic characteristics must be thoughtfully considered to maximize particle removal while minimizing issues such as short-circuiting or turbulence, which can disrupt the settling process.

Several factors influence the efficiency of settling and sedimentation in wastewater treatment plants. Particle characteristics, including size, density, shape, and surface properties, significantly impact settling rates. Smaller particles generally exhibit lower settling velocities and may necessitate longer detention times for effective removal. Moreover, the aggregation of particles into flocs, facilitated by the addition of coagulants or flocculants, can enhance settling efficiency by increasing particle size and density.

Hydraulic conditions within the sedimentation tank, such as flow velocity, turbulence, and tank loading rate, also influence settling performance. Proper flow distribution and the implementation of hydraulic control mechanisms are vital to minimize short-circuiting and ensure uniform particle settling throughout the tank.

During settling, a layer of sludge accumulates at the bottom of the sedimentation tank. Efficient sludge management is crucial to ensure the long-term performance of wastewater treatment plants. Regular removal and treatment of the collected sludge are necessary to prevent excessive buildup, odors, and adverse impacts on downstream treatment processes and the receiving environment. Various techniques, including sludge scraping, sludge pumping, and sludge thickening, are employed for effective sludge management.

Machine generated keywords: settle, sludge settle, model, settling, velocity, thickening, activate sludge, activate, bulking, sludge, density, plastic, zone, thicken, separation.

Assessment of Microbial Community Structure and Function in Serially Passaged Wastewater Electro-Bioreactor Sludge: An Approach to Enhance Sludge Settleability [47]

This is a machine-generated summary of:

ElNaker, Nancy A.; Elektorowicz, Maria; Naddeo, Vincenzo; Hasan, Shadi W.; Yousef, Ahmed F.: Assessment of Microbial Community Structure and Function in Serially Passaged Wastewater Electro-Bioreactor Sludge: An Approach to Enhance Sludge Settleability [47]

Published in: Scientific Reports (2018)

Link to original: https://doi.org/10.1038/s41598-018-25509-2

If you want to cite the papers, please refer to the original.

For technical reasons we could not place the page where the original quote is coming from.

Abstract-Summary

"The main objective of this study was to carry out functional characterization of microbial community structure of wastewater electro-bioreactor sludge as it undergoes serial passaging in the presence or absence of a current density over 15 days."

"Illumina MiSeq sequencing and QIIME were used to assess sludge microbial community shifts over time. (α) and (β) diversity analysis were conducted to assess the microbial diversity in electro-bioreactors."

"Phylogenetic analysis revealed stark differences in the development of sludge microbial communities in the control and treatment reactor."

"There was no mention of any studies aimed at characterizing functional microbial communities under electric field and the results communicated here are the first, to our knowledge, that address this gap in the literature."

Introduction

"Filamentous bacteria are a necessary component of the activated sludge, but their excessive growth results in sludge bulking and foaming, two of the most common technical problems of wastewater treatment plant bioreactor operation [23–25]."

"Our data is the first to investigate the impact of serial passaging on the performance and microbial community structure of wastewater electro-bioreactors."

"The main objectives of this study were (a) to investigate the effect of CD on the performance of serially passaged electro-bioreactors; (b) to illustrate the effect of serial passaging on microbial community structure and function; (c) to differentiate between the effects of CD and serial passaging linking the performance and functional bacterial groups associated with sludge settling problems in both bioreactors and electro-bioreactors."

"This research is the first to study the integration of serial passaging technique with bio-electrochemical stimulation on the microbial communities and reactor performance efficiency and stability in wastewater electro-bioreactors."

Results and Discussion

"Results showed that SVI was reduced from 325 to 236.6, 141.5, and 54.7 mLg^{-1} in serially passaged control bioreactor and electro-bioreactors operated at CD of 3 and 7 Am^{-2}, respectively reflecting the importance of applying electric field in the sludge cohesion and density."

"It is noticed that when comparing the relative abundances of bacterial communities present in serially passaged electro-bioreactors and control bioreactor, there was an inflection point at certain passaging days where it reached its highest abundance then slightly decreased which could be due to serial passaging effect."

"Our results showed that Enterobacteriaceae, Pseudomonadaceae, Streptococcaceae, Acetobacteraceae and Rhodocyclaceae had the highest average relative abundance across serially passaged control bioreactor samples than electro-bioreactors operated at CD of 3 and 7 Am^{-2} (18.8, 4.6 and 11.7%; 15, 1.2 and 2.7%; 8, 0.1 and 0.1%; 4.5, 0 and 0%; 2.1, 1.6 and 1.5%, respectively)."

Methods

"The system used for this study was aerobic batch electro-bioreactors containing sludge and synthetic wastewater prepared in the laboratory."

"The total effective volume of the reactor is 1500 mL of which 300 mL of raw sludge was mixed with 1200 mL of synthetic wastewater."

"The sludge was serially passaged and fed with synthetic wastewater every 24 h for a two-week experiment."

"The sludge was then fed with synthetic wastewater and left to operate for 24 h and so on."

"The effective surface area of the anode was calculated in each operating condition depending on the effective volume of the sample, by multiplying the width by the immersed length of the anode in the bioreactor."

"Synthetic wastewater and sludge dissolved oxygen (DO in mgL^{-1}), pH, temperature (T in °C) and electrical conductivity (EC in μScm^{-1}) were analyzed using a HACH HQ40d single-input multi-meter probe (Hach Company, Loveland, CO, USA)."

Fractional Order-Based Hierarchical Controller Design and Evaluation with Bürger-Diehl Settler Model in a Total Nitrogen Removal Wastewater Treatment Process [48]

This is a machine-generated summary of:

Eagalapati, Sree Santoshi Tejaswini; Sheik, Abdul Gaffar; Ambati, Seshagiri Rao: Fractional order-based hierarchical controller design and evaluation with Bürger-Diehl settler model in a total nitrogen removal wastewater treatment process [48]

Published in: Environmental Science and Pollution Research (2022)

Link to original: https://doi.org/10.1007/s11356-022-20473-x

If you want to cite the papers, please refer to the original.

For technical reasons we could not place the page where the original quote is coming from.

Abstract-Summary

"The primary objective of this study was to establish two-level structured control techniques based on the globally known benchmark simulation model no."

"1 (BSM1) and the Bürger-Diehl settler model in order to improve effluent quality."

"A two-level hierarchical control structure was considered to maintain the dissolved oxygen concentration basing on the ammonia levels and also a nitrate controller was considered to improve nitrogen removal efficiency."

"Fractional order PI controllers were considered at the secondary level and advanced control techniques, namely, MPC and fuzzy, were implemented at the primary level."

"Two advance control schemes, being, FPI-MPC and FPI-fuzzy were designed in the present work."

"The controllers were designed based on the plant model which was identified using prediction-error minimization method."

Introduction

"A variety of wastewater treatment plant layouts and situations may be employed to apply the suggested control strategy, making it difficult to compare controller performance."

"COST (Cooperation in Science and Technology) 624, a European initiative for evaluating control techniques in wastewater treatment plants (WWTPs), produced a benchmark called as benchmark simulation model no."

"Numerous studies have been published in the literature that demonstrate the use of a variety of control mechanisms, ranging from basic PI structures to more advanced tactics such as model predictive control (MPC)."

"An energy-efficient control system with better plant performance using SCADA-based hierarchical techniques is developed (Crisan and others 26)."

"To the best of the authors' knowledge, no advanced control strategies have been rigorously tested using BSM1 integrated with the Bürger-Diehl settler model."

"The goal of this research is to determine the effect of fractional order-based hierarchical control techniques on the modified BSM1 plant."

Methodology

"BSM1 is emerged from a collection of several sub-models that have been used to represent different processes."

"To the usual structure of BSM1, the secondary settler in this study is represented using the Bürger-Diehl settler model (Bürger and others 27)."

"The biological reactor is composed of five tanks, two of which are anoxic and three of which are aerobic, and the settler is modeled with 10 layers."

"The ASM1 model is used to represent the physical and chemical processes happening in an activated sludge reactor."

"The BD model is flexible and includes functions that describe the different phenomena taking place, i.e., settling velocity, compression, and dispersion."

"The corresponding model is represented by 1-D PDE as follows: Here, C is the concentration of biomass at depth z and time t. The convective transport and gravity settling are represented by the first term."

Control Structure

"BSM1's core control technique exhibits control over the denitrification and nitrification processes and measures the closed loop's EQI and OCI."

"FPI is designed as the secondary controller for the two control loops in this framework."

"Fuzzy and MPC controllers are recommended as primary controllers in combination with the secondary controller."

"The controllers, $C_{NO,2}$ and $C_{SO,5}$, are designed accordingly applying the tuning relations on the FOPTD models identified using the prediction error minimization (PEM) approach (Ljung 28)."

"This MPC provides set-points to the secondary controller, FPI."

"By controlling the DO set-points, the fuzzy control method removes $S_{NH,5}$ peaks."

"The rules are defined using steady state data obtained through simulations and the fuzzy rules are taken as a direct proportionality relation which exists between the concentrations of ammonia and DO, i.e., If the concentration of ammonia is "low," the concentration of dissolved oxygen is "low.""

Results and Discussion

"The lower level FPI is combined with the higher level fuzzy controller for DO control loop."

"The higher level MPC provides a dynamic set-point of DO to the FPI controller at the lower level in this configuration."

"The tuning parameters selected to design the higher level MPC are as follows: Control horizon = 5 Prediction horizon = 20 Sampling time = 0.02 days. The purpose of this arrangement is to maintain an ammonia level of 1.5 g $N{\cdot}m^{-3}$ at a predetermined set-point."

"FPI-MPC produces the least violations in ammonia and on contrary, FPI-fuzzy produces the least violations in total nitrogen."

"There is also a significant reduction in EQI and effluent violations when fuzzy is used as the primary controller and FPI is used as the secondary controller."

"FPI-fuzzy resulted in noticeable 60% and 53% reduction in total nitrogen violations for dry and storm climatic conditions, respectively."

Conclusion

"This research focusses on the improvement of effluent quality from a total nitrogen removal plant by implementing two-level advanced control strategies on BSM1 merged with Bürger-Diehl settler model."

"In order to meet the objectives of a higher plant performance, the default benchmark control loops are considered and advanced control approaches are used, which incorporate sophisticated control strategies like fractional order control, fuzzy, and MPC."

"Of the Bürger-Diehl settler model being more efficient and sophisticated hierarchical control mechanisms being used, a considerable reduction in effluent quality index (EQI) and effluent violations is obtained."

Settling Velocity of Irregularly Shaped Microplastics Under Steady and Dynamic Flow Conditions [49]

This is a machine-generated summary of:

Wang, Zhen; Dou, Ming; Ren, Pengju; Sun, Bin; Jia, Ruipeng; Zhou, Yuze: Settling velocity of irregularly shaped microplastics under steady and dynamic flow conditions [49]

Published in: Environmental Science and Pollution Research (2021)

Link to original: https://doi.org/10.1007/s11356-021-14654-3

If you want to cite the papers, please refer to the original.

For technical reasons we could not place the page where the original quote is coming from.

Abstract-Summary

"To predict the settling trajectory of MPs in aquatic environments, this study investigates the settlement law of MPs under static and dynamic conditions."

"The experimental results show that the main factors affecting the settling velocity of MPs were shape irregularity, density, and particle size."

"The settling velocity of irregular MPs was significantly lower than that of perfectly spherical MPs."

"We proposed a model that predicts the correlation between the settling velocity of MPs and their shape, density, particle size, and water density."

Introduction

"Khatmullina and Isachenko (29) mainly studied the settling velocity of regular-shaped spheres, cylinders, and fishing lines, and suggested that the particle shapes can significantly affect their settlement."

"Khatmullina and Isachenko (29) studied the influence of regular shapes such as spheres and cylinders on the settlement of MPs and provided a prediction equation for regular shapes."

"Kaiser and others (30) studied the settlement rule of MPs with irregular shapes and sizes less than 1 mm in a laminar flow state, and presented an equation for the relationship between settling velocity, particle size, and excess density."

"The main objective is to understand the settlement behavior of MPs of different materials, diameters, and shapes in static and dynamic aquatic environments with different salinity levels under laboratory conditions."

"Based on a force analysis, we derived a theoretical equation for the settling velocity of MPs, which changes according to their material, shape, particle size, and water density."

Materials and Methods

"A preliminary experiment showed that even if the MP density was greater than that of the aqueous solution, it was often difficult for the particles to settle smoothly because of their irregular shapes and high surface roughness."

"Where w is the particle's bounded settling velocity, m/s; w^{∞} is the particle's unbounded velocity, m/s (the settling velocity in the following is the unbounded settling velocity); d is the particle diameter, where ESD, m; and L is the diameter of the water column container, m. Changes in temperature can change the water density, thereby affecting the settling process (Kaiser and others 30)."

"A near-sphere with a CSF index of 0.9–1 and a PS MP with an ESD range of 0.371–1.195 mm were selected, and the settling experiment was performed for salinity of 0‰. where Re is the particle Reynolds number, 0–1 is laminar flow, 1–10^3 is transitional flow, and 10^3–10^5 is turbulent flow; w is particle settling velocity, m/s; υ is the kinematic viscosity, m^2/s; and others are consistent with the above formula."

Results

"The PA, PET, and PVC MPs were selected for settling experiments under static water conditions with salinities of 0‰, 15‰, and 36‰. A total of nine groups with approximately 115 MP particles each were investigated, with a total of approximately 1040 particles."

"An increase in salinity decreased the settling velocity of PVC MPs, but the impact was not significant."

"The settlement law for salinity of 15‰ and 36‰ was similar to that for 0‰. In the dynamic water conditions, the settling velocity of approximately 230 MPs was measured, including that of PS and PET particles, and the salinity was set to 0‰. The two types of MPs were subjected to vibrate at 50, 100, and 150 rpm, for a total of six groups."

"There was a significant linear relationship between the settling velocity of near-spherical and polygonal ellipsoid particles and ESD, but fragments presented a poor correlation."

Discussion

"The settling velocity of MPs was stable, and most MPs were freed from settlement."

"The relative average error between the measured and predicted data was only 0.25, and R^2 was 0.8145, indicating that the interpretation rate of the model for the MP settling velocity was 81.45%."

"The model developed in this study was used to calculate the settling velocity under dynamic water conditions."

"For PS, the measured settlement velocity under dynamic conditions was smaller than the predicted velocity."

"This occurred because under dynamic water conditions, in addition to gravity, buoyancy, and resistance, MPs were also affected by drag forces that decreased their settling velocity."

"The measured and predicted settling velocities under dynamic conditions presented a good fit."

Conclusions

"This study mainly investigated the settlement of MPs under static and dynamic water conditions, and comprehensively considered the influence of MP density, particle size, shape, and water salinity on the settlement process."

"Based on a large amount of experimental data, the main factors affecting the settling velocity were studied, including size, density, and shape of MPs, and the salinity and turbulence of the water body."

"The salinity of the water body also influenced the settling velocity, especially for MPs with density of 1000–1250 kg/m^3, for which higher salinity values led to slower settling velocities."

"For MPs larger than 1250 kg/m^3, the dynamic water condition slightly reduced their settling velocity, but higher speeds did not significantly change the settling velocity."

"Considering the particle size, density, shape, and water salinity of the MPs, a static water settlement equation was fitted, and adjustments were made to obtain the settlement equation of low-density MPs under dynamic water conditions."

Simple Model of Sludge Thickening Process in Secondary Settlers [50]

This is a machine-generated summary of:

Zhang, Yuankai; Wang, Hongchen; Qi, Lu; Liu, Guohua; He, Zhijiang; Jiang, Songzhu: Simple model of sludge thickening process in secondary settlers [50]

Published in: Frontiers of Environmental Science & Engineering (2014)

Link to original: https://doi.org/10.1007/s11783-014-0758-6

If you want to cite the papers, please refer to the original.

For technical reasons we could not place the page where the original quote is coming from.

Abstract-Summary

"In wastewater treatment plants (WWTPs), a secondary settler acts as a clarifier, sludge thickener, and sludge storage tank during peak flows and therefore plays an important role in the performance of the activated sludge process."

"By detecting the hindered zone from the complete thickening process of activated sludge, a simple model for the sludge thickening velocity, $u_s = aX^b$ ($a = 0.9925SSVI_{3.5}$, $b = 3.541\ln(SSVI_{3.5}) + 12.973$), describing the potential and performance of activated sludge thickening in the hindered zone was developed."

"Sludge thickening in the compression zone was not studied because sludge in the compression zone showed limited thickening."

"This empirical model was developed using batch settling data obtained from four WWTPs and validated using measured data from a fifth WWTP to better study sludge thickening."

Monitoring of Activated Sludge Settling Ability Through Image Analysis: Validation on Full-Scale Wastewater Treatment Plants [51]

This is a machine-generated summary of:

Mesquita, D. P.; Dias, O.; Amaral, A. L.; Ferreira, E. C.: Monitoring of activated sludge settling ability through image analysis: validation on full-scale wastewater treatment plants [51]

Published in: Bioprocess and Biosystems Engineering (2008)

Link to original: https://doi.org/10.1007/s00449-008-0255-z

If you want to cite the papers, please refer to the original.

For technical reasons we could not place the page where the original quote is coming from.

Abstract-Summary

"The monitoring of the activated sludge contents allowed for the detection of bulking events proving that the developed image analysis methodology is adequate for a continuous examination of the morphological changes in microbial aggregates and subsequent estimation of the sludge volume index."

"The obtained results proved that the developed image analysis methodology is a feasible method for the continuous monitoring of activated sludge systems and identification of disturbances."

Introduction

"The settling phase is considered a critical stage of the process in which filamentous bulking and deflocculation processes are the most common problems, causing the decrease of the sludge settling ability and effluent quality deterioration [31, 32]."

"It comes as no surprise that microscopic image analysis procedures have been used, in recent years, to complement the wastewater treatment processes mainly on the morphological characterization and evaluation of aggregated and filamentous bacteria contents."

"As the activated sludge process efficiency is dependent on the aggregates physical properties, some basic parameters have been used to study the settling ability of the microbial aggregates such as the filamentous bacteria contents, aggregates contents, size, shape, density, porosity and settling velocity, as well as consistency and specific surface area [33, 34]."

"Image analysis procedures are, nowadays, considered to be a feasible method to characterize quantitatively aggregates and protruding filamentous bacteria, and subsequently used to relate to the settling abilities [35, 36], assess biomass morphology changes and monitor bulking events in pilot and full-scale plants [37–39]."

Materials and Methods

"Samples were taken to perform physical measurements, on one hand, and microscopic observations, on the other, in order to estimate the contents and morphology of the microbial aggregates and protruding filamentous bacteria by image acquisition and analysis."

"The image acquisition of the Frossos2 WWTP, had already been performed for an earlier study and, relied on 20 images per sample acquired in bright field microscopy in a SZ 4045TR-CTV Olympus stereo microscope (Olympus, Tokyo), with 40× magnification, coupled to a Sony CCD AVC–D5CE (Sony, Tokyo) grey scale video camera."

"The image processing step determined the binary images from the aggregated biomass and the protruding filamentous bacteria and thereafter, the morphological parameters were determined."

"Supported on the previous study of Amaral and Ferreira [39], 36 parameters were determined either directly from the image analysis program either in association with the sludge physical properties for a total around 400 000 aggregates."

Results and Discussion

"It was also found that the difference between the values is smaller for the lower SVI values, around 24 mL/g for the set of SVI values below 200 mL/g, and larger for higher SVI values, around 49 mL/g for the set of SVI values above 200 mL/g. Once again this behavior may be explained by the fact the majority of the data points above 200 mL/g rely on an average of 20 images per sample instead of the average 200 images for the data points below 200 mL/g. In most cases for the SVI below the bulking limit (150 mL/g according to Jenkins and others [40]) there is a correspondence to the predicted SVI (presenting also values below 150 mL/g)."

"Only 5 days (days 28, 49, 56, 63 and 70) presented SVI values higher than 150 mL/g (bulking phenomena), ranging from 160 to 265 mL/g. The predicted values for this WWTP were, for the most, in accordance with the observed SVI values with larger deviations occurring on days 28, 56 and 105 (all above 50 mL/g)."

"The predicted values for this WWTP were, for the most, in accordance with the observed SVI values with larger deviations occurring on days 14 and 128 (above 50 mL/g)."

Conclusions

"The PLS analysis of the image analysis data allowed to predict the SVI values with a reasonable correspondence as it can be inferred by the 0.947 multiple correlation coefficient obtained between the observed and predicted SVI values."

"It should be kept in mind, however, that although the obtained correlation holds true for the SVI range presented in the surveyed WWTPs (up to 700 mL/g) no direct extrapolations can be performed for higher SVI values regarding severe bulking phenomena."

"It was also found that, for the current methodology, the observed and predicted SVI values distanced from each other an average 24 mL/g for the set of SVI values below 200 mL/g, which comprises the bulking threshold limit."

"The obtained results showed that the developed image analysis and PLS methodology proved to be adequate for continuous examination of sludge settling properties, in terms of the SVI."

Settleability and Dewaterability of Sewage Sludge with Modified Diatomite [52]

This is a machine-generated summary of:

Wang, Ling-Yu; Li, Xue; Luo, Kun; Song, Yu-Hao; Liu, Ren-Guo: Settleability and dewaterability of sewage sludge with modified diatomite [52]

Published in: Environmental Science and Pollution Research (2022)

Link to original: https://doi.org/10.1007/s11356-022-20368-x

If you want to cite the papers, please refer to the original.

For technical reasons we could not place the page where the original quote is coming from.

Abstract-Summary

"Cationic polyacrylamide (CPAM)-modified diatomite and cetyl trimethyl ammonium bromide (CTAB)-modified diatomite were synthesized and used as conditioners for sewage sludge dewatering."

"The effects of these two types of modified diatomite on the dewaterability and settling performance of the activated sludge were studied."

"The mechanisms of the two modified diatomite types in the activated sludge system were elucidated."

"The efficiency of the CPAM-modified diatomite was better than that of the CTAB-modified diatomite in improving the settleability and dewaterability of sludge."

Introduction

"Natural diatomite used directly as a conditioner cannot fully meet the demands of sludge dewatering and the environment."

"Surfactants and polymers are effective diatomite modifiers that can improve the pore structure and surface charge, while the subsidence performance of the entire system is enhanced in water samples (Yang and others 41)."

"Few studies have investigated the role of modified diatomite as a conditioner for sludge dewatering."

"This study aims to investigate the role of modified diatomite particles in the dewatering performance of an activated sludge system."

"The sedimentation and dewatering efficiency of the sludge was evaluated by taking the dosage of these two types of modified diatomite and pH as control factors, from the sedimentation volume (SV), specific resistance to filtration (SRF), cake moisture content, and EPS concentration."

"Differences between natural diatomite and modified diatomite and the morphology of dewatering sludge were investigated by SEM."

Materials and Methods

"The roasted diatomite (40 g) and 5–10% HCl (200 ml) were mixed in a 500 ml beaker and then placed in a water bath at 50–65 °C and was stirred for 2 h at a stirring speed of 150–300 r/min."

"20 g acidifying diatomite into a beaker, added 100 ml of 2% CPAM/CTAB solution, and then stirred for 2 h at the speed of 100 r/min at room temperature, followed by suction filtration."

"A specific dosage of the modified diatomite and 150 ml activated sludge mixture was added to a 200 ml beaker and blended violently with a glass rod."

"SRF was measured using a vacuum extraction and filtration device; 100 ml of sludge was poured into a standard 9 cm Buchner funnel with pre-wetting quantitative filter paper at a constant vacuum pressure of 0.03 MPa for 10–15 min."

Results and Discussion

"When the dosage of CPAM-modified diatomite was below 0.4%wt, the more added amount, the better sludge settling was obtained."

"It was clear that the sludge settling performance by adding CPAM-modified diatomite was better than that of natural diatomite and CPAM."

"The CPAM-modified diatomite was better than the CTAB-modified diatomite in improving the settling performance of the sludge."

"A few clusters of sludge flocs floating on the surface of the liquid supernatant with 0.8%wt CTAB-modified diatomite addition was much less than that of 0.8%wt CPAM-modified diatomite."

"Adding the modified diatomite, EPS on the surface of sludge particles can quickly separate from sludge and dissolve in water under a surfactant/coagulant, thus reducing the interstitial water, thereby reducing SRF and moisture content."

"With an increase in modified diatomite, the sludge surface released more EPS into the water, resulting in the obstruction of filtration."

Conclusion

"Two types of organic-modified diatomite (CPAM/CTAB-modified diatomite) were synthesized by roasting and acid modification."

"The settleability and dewaterability of sewage sludge can be improved significantly after conditioning with CPAM/CTAB-modified diatomite by polymer absorption, charge neutralization, particle bridging, decreasing cake blinding, and increasing cake permeability."

"The experimental data indicated that the efficiency of CPAM-modified diatomite was better than that of CTAB-modified diatomite in improving the settleability and dewaterability of sludge."

"This research proves the potential of CPAM/CTAB-modified diatomite for sewage sludge pretreatment in mechanical dewatering."

Bibliography

1. Regueiro L, Veiga P, Figueroa M, Alonso-Gutierrez J, Stams AJ, Lema JM, Carballa M (2012) Relationship between microbial activity and microbial community structure in six full-scale anaerobic digesters. Microbiol Res 167(10):581–589
2. Sundberg C, Al-Soud WA, Larsson M, Alm E, Yekta SS, Svensson BH, Sørensen SJ, Karlsson A (2013) 454 pyrosequencing analyses of bacterial and archaeal richness in 21 full-scale biogas digesters. FEMS Microbiol Ecol 85(3):612–626
3. Yu B, Zhang D, Shan A, Yuan H, Huang X, Yuan W, Dai X, Zhu N (2015) Methane-rich biogas production from waste activated sludge with the addition of ferric chloride under thermophilic anaerobic digestion system. RSC Adv 5:38538–38546
4. Yu H, Wang Q, Wang Z, Sahinkaya E, Li Y, Ma J, Wu Z (2014) Start-up of an anaerobic dynamic membrane digester for waste activated sludge digestion: temporal variations in microbial communities. PLoS One 9(4):e93710
5. Guo WQ, Yin RL, Zhou XJ, Du JS, Cao HO, Yang SS, Ren NQ (2015) Sulfamethoxazole degradation by ultrasound/ozone oxidation process in water: kinetics, mechanisms, and pathways. Ultrason Sonochem 22:182–187
6. Abdessemed D, Nezzal G (2003) Treatment of primary effluent by coagulation-adsorption-ultrafiltration for reuse. Desalination 152(1–3):367–373
7. Yargeau V, Danylo F (2015) Removal and transformation products of ibuprofen obtained during ozone - and ultrasound-based oxidative treatment. Water Sci Technol 72(3):491–500
8. Alfonso-Muniozguren P, Hazzwan Bohari M, Sicilia A, Avignone-Rossa C, Bussemaker M, Saroj D, Lee J (2020) Tertiary treatment of real abattoir wastewater using combined acoustic cavitation and ozonation. Ultrason Sonochem 64:104986
9. Wei Z, Kosterman JA, Xiao R, Pee GY, Cai M, Weavers LK (2015) Designing and characterizing a multi-stepped ultrasonic horn for enhanced sonochemical performance. Ultrason Sonochem 27:325–333
10. Kimura T, Sakamoto T, Leveque JM, Sohmiya H, Fujita M, Ikeda S, Ando T (1996) Standardization of ultrasonic power for sonochemical reaction. Ultrason Sonochem 3(3):S157–S161
11. APHA (2012) Standard methods for the examination of water and wastewater, 22nd ed. American Public Health Association, America Water Works Association, Water Environment Federation, Washington DC.
12. Barzegar G, Wu J, Ghanbari F (2019) Enhanced treatment of greywater using electrocoagulation/ozonation: investigation of process parameters. Process Saf Environ Prot 121:125–132
13. Meyer T, Edwards EA (2014) Anaerobic digestion of pulp and paper mill wastewater and sludge. Water Res 65:321–349. https://doi.org/10.1016/j.watres.2014.07.022
14. Das S, Dagnew M, Ray MB (2021) Anaerobic digestibility of resin acids in primary sludge: Effect of ozone pretreatment. Chemosphere 262:128316. https://doi.org/10.1016/j.chemosphere.2020.128316
15. Ayoub M, Abdel-fattah A (2016) A parametric study of alum recovery from water treatment sludge. Water Science and Technology 74(2):516–523. https://doi.org/10.2166/wst.2016.241
16. Ayoub M, Afify H, Abdel-fattah A (2017) Chemically enhanced primary treatment of sewage using the recovered alum from water treatment sludge in a model of hydraulic clariflocculator. Journal of Water Process Engineering, 19, 133–138. https://doi.org/10.1016/j.jwpe.2017.07.014
17. Neupane DR, Riffat R, Murthy SN, Peric MR, Wilson TE (2008) Influence of source characteristics, chemicals, and flocculation on chemically enhanced primary treatment. Water Environment Research 80(4):331–338. https://doi.org/10.2175/106143007x221355
18. Gitis V, Hankins N (2018) Water treatment chemicals: trends and challenges. Journal of Water Process Engineering 25:34–38. https://doi.org/10.1016/j.jwpe.2018.06.003
19. Tik S, Vanrolleghem PA (2012) Gestion des eaux usées en temps de pluie—De l'alun en périodes critiques à la Ville de Québec (in French). Vecteur Environn:32–33

20. Henze M, Gujer W, Mino T, Matsuo T, Wentzel MC, Marais GVR, Van Loosdrecht MCM (1999) Activated sludge model No. 2D, ASM2D. Water Sci Technol 39(1):165–182. https://doi.org/10.1016/S0273-1223(98)00829-4
21. Smith M, Dudley J (1998) Dynamic process modelling of activated-sludge plants. J Chart Inst Water Environ Manag 12(5):346–356. https://doi.org/10.1111/j.1747-6593.1998.tb00196.x
22. Henze M, Gujer W, Mino T, Van Loosdrecht MCM (2000) Activated sludge models ASM1, ASM2, ASM2d and ASM3. IWA Publishing, London
23. Cydzik-Kwiatkowska, A. & Zielińska, M. Bacterial communities in full-scale wastewater treatment systems. World J. Microbiol. Biotechnol. 32, 66 (2016).
24. Ma, Q. et al. Bacterial community compositions of coking wastewater treatment plants in steel industry revealed by Illumina high-throughput sequencing. Bioresour. Technol. 179, 436–443 (2015).
25. Martins, A. M., Pagilla, K., Heijnen, J. J. & van Loosdrecht, M. C. Filamentous bulking sludge—a critical review. Water Res. 38, 793–817 (2004).
26. Crisan R, Harja G, Nascu I, Nicoara I (2018) Hierarchical control system for energy savings in wastewater treatment plant. In: 2018 IEEE international conference on automation, quality and testing, robotics, AQTR 2018 - THETA 21st edition, proceedings. pp 1–6
27. Bürger R, Diehl S, Farås S, Nopens I, Torfs E (2013) A consistent modelling methodology for secondary settling tanks: a reliable numerical method. Water Sci Technol 68:192–208. https://doi.org/10.2166/wst.2013.239
28. Ljung L (1999) System identification - a theory for the user. Prentice Hall International, Upper Saddle River, NJ, USA
29. Khatmullina L, Isachenko I (2016) Settling velocity of microplastic particles of regular shapes. Mar Pollut Bull 114(2):871–880. https://doi.org/10.1016/j.marpolbul.2016.11.024
30. Kaiser D, Estelmann A, Kowalski N, Glockzin M, Waniek JJ (2019) Sinking velocity of sub-millimeter microplastic. Mar Pollut Bull 139:214–220. https://doi.org/10.1016/j.marpolbul.2018.12.035
31. Sezgin M, Jenkins D, Parker D (1978) A unified theory of filamentous activated sludge bulking. J Water Pollut Control Fed 50:362–381
32. Schuler AJ, Jang H (2007) Density effects on activated sludge zone settling velocities. Water Res 41(8):1814–1822
33. Cenens C, Jenné R, Van Impe JF (2002) Evaluation of different shape parameters to distinguish between flocs and filaments in activated sludge images. Water Sci Technol 45(4–5):85–91
34. Jenné R, Banadda EN, Gins G, Deurinck J, Smets IY, Geeraerd AH, Van Impe JF (2006) Use of image analysis for sludge characterisation: studying the relation between floc shape and sludge settleability. Water Sci Technol 54(1):167–174
35. Grijspeerdt K, Verstraete W (1997) Image analysis to estimate the settleability and concentration of activated sludge. Water Res 31:1126–1134
36. Jenné R, Banadda EN, Smets IY, Van Impe JF (2004) Monitoring activated sludge settling properties using image analysis. Water Sci Technol 50(7):281–285
37. da Motta M, Amaral AL, Casellas M, Pons MN, Dagot C, Roche N, Ferreira EC, Vivier H (2001a) Characterisation of activated sludge by automated image analysis: validation on full-scale plants. In: Proceedings of the IFAC computer applications in biotechnology, Québec City, pp 427–431
38. da Motta M, Pons MN, Roche N (2001) Automated monitoring of activated sludge in a pilot plant using image analysis. Water Sci Technol 43(7):91–96
39. Amaral AL, Ferreira EC (2005) Activated sludge monitoring of a wastewater treatment plant using image analysis and partial least squares regression. Anal Chim Acta 544:246–253
40. Jenkins D, Richard MG, Daigger G (2003) Manual on the causes and control of activated sludge bulking, foaming and other solids separation problems. Lewis publishing, Boca Raton
41. Yang W, Wang P, Luo WL, Zhu J, Zhang Y (2011) The diatomite modified by PAM and applied to adsorb Pb(II) in the simulated wastewater. Adv Mater Res 233–235:382–389. https://doi.org/10.4028/www.scientific.net/AMR.233-235.382

42. Ju, Feng; Wang, Yubo; Lau, Frankie T. K.; Fung, W. C.; Huang, Danping; Xia, Yu; Zhang, Tong Anaerobic digestion of chemically enhanced primary treatment (CEPT) sludge and the microbial community structure. Applied Microbiology and Biotechnology (2016). https://doi.org/10.1007/s00253-016-7730-2
43. Rossi, Giada; Mainardis, Matia; Aneggi, Eleonora; Weavers, Linda K.; Goi, Daniele Combined ultrasound-ozone treatment for reutilization of primary effluent—a preliminary study. Environmental Science and Pollution Research (2020). https://doi.org/10.1007/s11356-020-10467-y
44. Das, Sreejon; Manoli, Kyriakos; Sharma, Virender K.; Dagnew, Martha; Ray, Madhumita B. Effect of ferrate pretreatment on anaerobic digestibility of primary sludge spiked with resin acids. Environmental Science and Pollution Research (2022). https://doi.org/10.1007/s11356-022-21599-8
45. Al Bazedi, Ghada A.; Abdel-Fatah, Mona A. Correlation between operating parameters and removal efficiency for chemically enhanced primary treatment system of wastewater. Bulletin of the National Research Centre (2020). https://doi.org/10.1186/s42269-020-00368-y
46. Baalbaki, Zeina; Torfs, Elena; Maere, Thomas; Yargeau, Viviane; Vanrolleghem, Peter A. Dynamic modelling of solids in a full-scale activated sludge plant preceded by CEPT as a preliminary step for micropollutant removal modelling. Bioprocess and Biosystems Engineering (2016). https://doi.org/10.1007/s00449-016-1715-5
47. ElNaker, Nancy A.; Elektorowicz, Maria; Naddeo, Vincenzo; Hasan, Shadi W.; Yousef, Ahmed F. Assessment of Microbial Community Structure and Function in Serially Passaged Wastewater Electro-Bioreactor Sludge: An Approach to Enhance Sludge Settleability. Scientific Reports (2018). https://doi.org/10.1038/s41598-018-25509-2
48. Eagalapati, Sree Santoshi Tejaswini; Sheik, Abdul Gaffar; Ambati, Seshagiri Rao Fractional order-based hierarchical controller design and evaluation with Bürger-Diehl settler model in a total nitrogen removal wastewater treatment process. Environmental Science and Pollution Research (2022). https://doi.org/10.1007/s11356-022-20473-x
49. Wang, Zhen; Dou, Ming; Ren, Pengju; Sun, Bin; Jia, Ruipeng; Zhou, Yuze Settling velocity of irregularly shaped microplastics under steady and dynamic flow conditions. Environmental Science and Pollution Research (2021). https://doi.org/10.1007/s11356-021-14654-3
50. Zhang, Yuankai; Wang, Hongchen; Qi, Lu; Liu, Guohua; He, Zhijiang; Jiang, Songzhu Simple model of sludge thickening process in secondary settlers. Frontiers of Environmental Science & Engineering (2014). https://doi.org/10.1007/s11783-014-0758-6
51. Mesquita, D. P.; Dias, O.; Amaral, A. L.; Ferreira, E. C. Monitoring of activated sludge settling ability through image analysis: validation on full-scale wastewater treatment plants. Bioprocess and Biosystems Engineering (2008). https://doi.org/10.1007/s00449-008-0255-z
52. Wang, Ling-Yu; Li, Xue; Luo, Kun; Song, Yu-Hao; Liu, Ren-Guo Settleability and dewaterability of sewage sludge with modified diatomite. Environmental Science and Pollution Research (2022). https://doi.org/10.1007/s11356-022-20368-x

Chapter 3
Secondary Treatment

Nitin Vitthalrao Thombre

3.1 Introduction by Editor

Secondary treatment is a vital stage in municipal wastewater treatment plants that focuses on the removal of persistent organic matter and dissolved pollutants, including nitrogen and phosphorus compounds, that remain after primary treatment. Various methods are commonly employed to remove organic matter, such as the activated sludge process, trickling filter process, and rotating biological contactor process. These processes harness the metabolic activities of microorganisms to facilitate the biological decomposition of organic matter.

Apart from organic matter, the removal of nutrients, particularly nitrogen and phosphorus, is crucial to prevent eutrophication and preserve water quality. Specialized microorganisms are involved in processes such as biological nitrogen removal, nitrification, denitrification, and enhanced biological phosphorus removal to target the reduction of nitrogen and phosphorus compounds.

Efficient operation and design considerations significantly impact the performance of secondary treatment processes. Operational parameters, including hydraulic retention time (HRT), mixed liquor suspended solids concentration (MLSS), dissolved oxygen levels (DO), and sludge age, must be meticulously monitored and controlled to maintain optimal conditions for microbial activity. Effective aeration and mixing systems ensure proper oxygen transfer and homogeneous distribution of microorganisms throughout the treatment system.

Design parameters, such as tank sizing, hydraulic loading rates, and process configuration, are critical for achieving desired treatment objectives. Adequate clarifier design, proper hydraulic and solids management, and the incorporation of advanced monitoring and control systems are essential to maximize treatment efficiency while minimizing operational issues. These factors collectively contribute to

N. V. Thombre (✉)
VA Tech Wabag Ltd., Pune, Maharashtra, India
e-mail: nitin01thombre@gmail.com

N. V. Thombre (ed.), *Municipal Wastewater Reclamation and Reuse*,
https://doi.org/10.1007/978-981-96-9112-8_3

the optimization of secondary treatment processes in municipal wastewater treatment plants.

3.2 Machine Generated Summaries

Disclaimer: The summaries in this chapter were generated from Springer Nature publications using extractive AI auto-summarization: An extraction-based summarizer aims to identify the most important sentences of a text using an algorithm and uses those original sentences to create the auto-summary (unlike generative AI). As the constituted sentences are machine selected, they may not fully reflect the body of the work, so we strongly advise that the original content is read and cited. The auto generated summaries were curated by the editor to meet Springer Nature publication standards. To cite this content, please refer to the original papers.

Machine generated keywords: sludge, compost, anaerobic, dewatere, sewage sludge, dewaterability, reactor, sludge dewaterability, granule, soil, aerobic, digestion, granular, anaerobic digestion, granular sludge.

Assessment of Casuarina glauca as Biofiltration Model of Secondary Treated Urban Wastewater: Effect on Growth Performances and Heavy Metals Tolerance [272]

This is a machine-generated summary of:

Slaimi, Ridha; Abassi, Mejda; Béjaoui, Zoubeir: Assessment of Casuarina glauca as biofiltration model of secondary treated urban wastewater: effect on growth performances and heavy metals tolerance [272]

Published in: Environmental Monitoring and Assessment (2021)

Link to original: https://doi.org/10.1007/s10661-021-09438-8

If you want to cite the papers, please refer to the original.

For technical reasons we could not place the page where the original quote is coming from.

Abstract-Summary

"A hydroponically experiment was conducted to evaluate C. glauca ability to remove heavy metals from secondary treated urban wastewater (SWW)."

"The effect of the SWW on plant biomass, some physiological parameters, heavy metals (Cd, Pb, Ni and Zn) bioaccumulation and removal from wastewater was evaluated."

"Casuarina glauca plants were able to remove heavy metal ions Cd, Pb, Ni and Zn from SWW and the removal efficiency was 92%, 77%, 83% and 73%, respectively."

"Casuarina glauca plants accumulated concentrations of heavy metals (Cd, Pb, Ni and Zn) in their roots higher than the shoots."

"The results indicated that C. glauca can act as scavengers of heavy metal ions from polluted water and confirms their ability for wastewater treatment."

Introduction

"The reuse of treated wastewater for crop irrigation could decrease water shortage, support the agriculture sector, reduced demand for soil fertilizer, and protect water resources (Bedbabis and others, 2)."

"Long-term irrigation with secondary treated wastewater proved to have different effects on physicochemical characteristics of soil and groundwater quality and pose a toxicity threat to human health, animals, and crop plants (Khan and others, 3; Sachan and others, 4). It is well known that the heavy metal pollution has severe negative consequences for human health and negative impact on the environment."

"This work was aimed to evaluate (i) the potential of C. glauca to improve the quality of secondary treated wastewater (ii) the impact of the secondary wastewater charged with metals on plant growth and biomass organs (roots and shoots) (iii) finally, evaluated the heavy metals accumulation strategies of the C. glauca using bioconcentration and translocation factors indices."

Materials and Methods

"The electrolyte leakage (EL) was defined as follows: To determine proline concentration, 100 mg of dry powder needle and root samples were homogenized in 2 ml of 40% methanol following the method described by Monneveux and Nemmar (5)."

"To determine soluble sugars concentration, 100 mg of the dry needle and root samples were homogenized with 10 ml of 80% ethanol following the method described by Albouchi and others (6)."

"0.5 ml of extract was mixed with 3 ml of anthrone solution (200 mg of anthrone dissolved in 100 ml concentrated sulphuric acid, W/W) and immediately placed in a boiling water bath for 10 min."

"The total phenolic content was expressed as gallic acid mg equivalents per gram of dry mass (mg GAE/g)."

"500 μl of 1 M NaOH were added and incubated for 5 min at room temperature and the final volume was made up to 2.5 ml with distilled water."

Results

"The BOD5, COD, and SM of the TWW were found reduced by 91.8%, 82.8%, and 31.4% respectively."

"All heavy metal concentrations were reduced tremendously in the TWW confirming the effectiveness of C. glauca in removing heavy metals from wastewater."

"The removal efficiency (RE)of heavy metals from the SWW were about 92.3%, 95.3%, 94.7%, 85.5%, 76.7%, 82.8% and 73% for Cd, Mn, Cu, Fe, Pb, Ni and Zn respectively."

Growth Parameters

"The shoot/root ratio was significantly higher in the treated plants compared with controls."

"The concentrations of Zn, and Cd in treated plants were significantly higher ($P \leq 0.05$) than in control plants."

"In the shoots of C. glauca the total content of polyphenol, flavonoids and tannin were significantly higher ($P < 0.05$) in plants grown in SWW than the control respectively with 113.61%, 277.5% and 203.8%."

"Whereas in roots there was no significant difference between the polyphenol content of treated and control plant."

"In the shoots the total concentration of proline and soluble sugar were significantly higher ($P < 0.05$) in plants grown in SWW than the control one with 511% and 32% respectively."

"In roots the concentration of proline and soluble sugar were significantly higher ($P < 0.05$) in treated plants compared to their controls with respectively 367% and 57%."

Discussion

"Several papers reported the use of rhizofiltration systems for the purification of water contaminated with heavy metals such as Cd, Zn, Pb and Ni involving different metals accumulator plants species such as Hydrocotyle verticillata Thunb, Myriophyllum spicatum L., Hydrocotyle umbellate L., water hyacinth, Phragmites australis and Typha latifolia, Eichhornia crassipes, Pistia stratiotes, Lemna minor, Azolla pinnata, Spirodela polyrhiza and Colocasia esculenta (L.) (Elias and others, 7; Kumari & Tripathi, 8; Ning and others, 9; Rana & Maiti, 10; Upadhyay and others, 11). These plants as C. glauca can uptake heavy metals from wastewater but with different potential for heavy metal removal."

"Similar findings were observed on Vigna radiata chlorophyll production irrigated with lower effluent concentrations of sugar mill effluent (10%) (Baskaran and others,

12). The higher growth and optimal photosynthesis of the treated plants is associated with the ability of C. glauca to tolerate wastewater pollutants, particularly heavy metals."

Conclusion

"The results of this experiment revealed that the microbiological and physicochemical quality of secondary treated wastewater was improved by C. glauca."

"The secondary treated STW has no detrimental effect on this plant species; on the contrary, it promotes its growth."

"These data experimental in hydroponic confirm that C. glauca can be used as an alternative to tertiary treatment with low cost of operation and low energy to improve the quality effluents of wastewater treatment plants used for crop irrigation."

17α-Ethinylestradiol and 17β-Estradiol Removal from a Secondary Urban Wastewater Using an RBC Treatment System [273]

This is a machine-generated summary of:

Maurício, R.; Dias, R.; Ribeiro, V.; Fernandes, S.; Vicente, A. C.; Pinto, M. I.; Noronha, J. P.; Amaral, L.; Coelho, P.; Mano, A. P.: 17α-Ethinylestradiol and 17β-estradiol removal from a secondary urban wastewater using an RBC treatment system [273]

Published in: Environmental Monitoring and Assessment (2018)

Link to original: https://doi.org/10.1007/s10661-018-6701-8

If you want to cite the papers, please refer to the original.

For technical reasons we could not place the page where the original quote is coming from.

Abstract-Summary

"Wastewater treatment plants are the main point source of E2 and EE2 into aquatic environments, but conventional wastewater treatment systems are not specifically designed for steroid removal."

"Biological tertiary treatment may be a solution for the removal of emergent pollutants such as E2 and EE2."

"The main purpose of the present study is to provide a solution based on the optimization of a rotating biological contactor system to remove estrogens, specifically E2 and EE2, and to quantify their removal efficiency on different matrices, namely real wastewater and different synthetic wastewaters."

"All assays presented viable removal efficiencies for compound E2 with values always above 50%; real wastewater yielded the highest removal efficiencies."

"EE2 removal had better removal efficiencies with synthetic wastewater as feed solution, with removals above 15%, whereas the removal efficiency with real wastewater was inexistent."

Introduction

"Wastewater treatment plants (WWTP) are known as the main point source of E2 and EE2 into aquatic environments (Esperanza and others 13; Hamid and Eskicioglu 14)."

"Their removal efficiency depends on several factors, such as the type of biological treatment, the E2 and EE2 concentrations in the influent flow, temperature, and operational conditions such as hydraulic and solid retention times (Liu and others 15)."

"Biological tertiary treatment may become the solution to remove emergent pollutants such as E2 and EE2."

"Considering the risks that E2 and EE2 can pose to humans and wildlife, it is crucial to study and optimize a solution to remove these compounds from wastewater."

"The main purpose of the present paper is to provide a solution based on the optimization of a pilot-scale RBC system to remove estrogens, specifically E2 and EE2, and to quantify the removal efficiency on different matrices with a known input concentration of these compounds."

Materials and Methods

"The SBSE bar was extracted and placed in a vial with 3 mL of acetonitrile (CH_3CN, Carlo Erba – Heagenis, molecular weight (MW), 41.05 g/mol), and stirred for 1 h at 300 rpm."

"Three experimental phases were performed to study E2 and EE2 biodegradation in advanced wastewater treatment logic using an RBC treatment system (very low effluent COD concentration, simulating the conditions observed in a secondary effluent)."

"In this phase, the two COD concentrations (measured as O_2) for the feed solution were 125 and 250 mg/L, simulating the two final discharged COD concentrations."

"The main goal in this phase was to study the biodegradation process of each compound individually and to observe if the feed solution concentration and the type and amount of microorganisms present in the RBC consequentially affected the removal efficiency from the liquid phase."

"After the 7 days for biofilm growth, each solution was spiked with 100 μg/L of the studied compounds (E2 and EE2)."

Results and Discussion

"Compared with phase 1 results, the phase 2 system performance to remove E2 was not affected by the presence of both compounds (in a mixture) (i.e., registered no inhibition or catalytic phenomena)."

"In phase 2, the EE2 removal efficiency varied between 6.3 and 27.4%, with an average of 20%."

"Comparing phases 1 and 2 with the same methodology and conditions but with only EE2 present and not in a mixture, the removal efficiencies obtained varied between 0 and 64%, with an average removal of 37%."

"Comparing the results obtained in phases 1 and 3 verifies that the WW matrix presented better results (i.e., higher removal efficiencies)."

"Comparing the removal efficiency of E2 in phase 3 with the previous phases, a microorganism preference seems to exist for this compound when both compounds are in the WW matrix."

Conclusions

"The main goal of this work was to study the removal of E2 and EE2 from treated wastewaters (secondary effluent) using an RBC treatment system, operated under a continuous mode."

"All phases presented promising E2 removal efficiencies with values always above 50%, with the real wastewater matrix having the best removal efficiencies considering secondary effluent characteristics."

"The removal efficiencies for both compounds in the synthetic wastewater matrix were promising, with the tested concentration higher and above the natural concentrations (reported monitoring studies) for E2 and EE2 feed concentration (100 μg/L)."

"The results obtained with real wastewater for E2 removal were promising but inconclusive for EE2."

"RBC systems were shown to be viable for E2 removal in a tertiary wastewater treatment, offering removal efficiencies equivalent to other biological systems but much more competitive when incorporated into an already existent WWTP because they are more compact and introduce much lower head losses."

Amplicon-Based Profiling of Bacteria in Raw and Secondary Treated Wastewater from Treatment Plants Across Australia [274]

This is a machine-generated summary of:

Ahmed, Warish; Staley, Christopher; Sidhu, Jatinder; Sadowsky, Michael; Toze, Simon: Amplicon-based profiling of bacteria in raw and secondary treated wastewater from treatment plants across Australia [274]

Published in: Applied Microbiology and Biotechnology (2016)

Link to original: https://doi.org/10.1007/s00253-016-7959-9

If you want to cite the papers, please refer to the original.

For technical reasons we could not place the page where the original quote is coming from.

Abstract-Summary

"We investigated the use of Illumina high-throughput sequencing of 16S ribosomal RNA (rRNA) amplicons to explore microbial diversity and community structure in raw and secondary treated wastewater (WW) samples from four municipal wastewater treatment plants (WWTPs A–D) across Australia."

"Sequence reads were analyzed to determine the abundance and diversity of bacterial communities in raw and secondary treated WW samples across the four WWTPs."

"Sequence reads were also characterized to phenotypic features and to estimate the abundance of potential pathogenic bacterial genera and antibiotic-resistant genes in total bacterial communities."

"Raw and secondary treated WW samples were dominated by members of the genera Pseudomonas, Arcobacter, and Bacteroides."

"The abundance of potentially pathogenic genera in raw WW samples was higher than that found in secondary treated WW samples."

"These findings provided a comprehensive profile of bacterial communities, including potential bacterial pathogens and antibiotic-resistant genes, in raw and secondary treated WW samples from four WWTPs across Australia and demonstrated that Illumina high-throughput sequencing can be an alternative approach for monitoring WW quality in order to protect environmental and human health."

Introduction

"Treated effluent supplied to growers should not contain >1 E. coli, or other FIB, per 100 mL. It has been demonstrated that monitoring FIB does not reflect the occurrence of pathogens, especially protozoa and viruses in treated WW, possibly due to their high susceptibility to chlorination (Miescier and Cabelli 17; Harwood and others 18)."

"The aims of this study were to (i) determine the abundance and diversity of bacterial communities in raw and secondary treated WW samples from four facilities across Australia using an Illumina-based next-generation sequencing approach, (ii) investigate whether the bacterial community structure differs in raw and secondary treated WW samples across the four WWTPs, (iii) determine the fraction of the bacterial population in secondary WW that can be attributed to raw WW in order to determine the percent and composition of the bacterial community removed by treatment, and (iv) estimate the abundance of potential pathogenic bacterial genera and antibiotic-resistant genes in bacterial communities from raw and secondary treated WW samples."

Materials and Methods

"Raw (approximately 100 mL each) and secondary treated WW samples (approximately 10 L each) were collected from each WWTP in sterile containers."

"Raw and secondary treated WW samples were grouped by WWTPs and corresponding states [WWTP A (Brisbane), WWTP B (Melbourne), WWTP C (Perth) and WWTP D (Hobart)]."

"The Bayesian classifier software program SourceTracker version 0.9.8 (Knights and others 19) was used to determine the proportion of bacterial communities in secondary treated WW samples that could be attributed to raw WW."

"Input files (consensus taxonomy, shared file, and metadata) were created in mothur using all raw and secondary treated WW samples."

"Phylogenetic Investigation of Communities by Reconstruction of Unobserved States (PICRUSt) was used to infer the prevalence of genes encoding selected antibiotic resistance features in each raw and secondary treated WW sample (Langille and others 20)."

Results

"A total of 26,183 and 20,074 OTUs were observed, at 97% similarity, in raw and secondary treated WW samples from the four WWTPs."

"Differences in sample coverage and richness did not differ significantly ($P > 0.05$) between raw and secondary treated WW samples among the four WWTPs."

"Alpha diversity, measured by either the Shannon or ACE indices, did not differ significantly ($P > 0.05$) between raw and secondary treated WW samples among four WWTPs."

"Secondary treated WW samples from WWTP A were predominated by OTUs that could not be classified to any genera, while WWTP C had greater abundances of Paludibacterium and Conchiformibius than other WWTPs examined."

"Raw and secondary treated WW samples collected from WWTPs A and C had significantly different beta diversity from each other ($P < 0.05$) and other WWTPs (B and D) ($P < 0.05$)."

Discussion

"We determined the abundance and diversity of bacterial communities in raw and corresponding secondary treated WW samples from four WWTPs across Australia."

"Among the genera detected, Pseudomonas and Arcobacter were highly abundant in both raw and secondary treated WW samples in WWTPs B, C, and D. WW treatment systems appear to supply a unique niche where Arcobacter can flourish, and as a result, members of this genus are often found in great abundances (4–11%) in association with raw WW (Collado and others 21; Cai and others 22; Shchgolkova and others 23)."

"25 potential pathogenic genera were identified among raw and secondary treated WW samples from various geographical locations."

"Among the pathogenic genera detected, Acinetobacter, Aeromonas, Arcobacter, and Pseudomonas were most abundant in both raw and secondary treated WW samples, indicating similar potential pathogenic bacterial communities."

"25 potential pathogenic genera were identified among raw and secondary treated WW samples with varying abundance."

Utilization of Secondary-Treated Wastewater for the Production of Freshwater Microalgae [275]

This is a machine-generated summary of:

Gómez-Serrano, C.; Morales-Amaral, M. M.; Acién, F. G.; Escudero, R.; Fernández-Sevilla, J. M.; Molina-Grima, E.: Utilization of secondary-treated wastewater for the production of freshwater microalgae [275]

Published in: Applied Microbiology and Biotechnology (2015)

Link to original: https://doi.org/10.1007/s00253-015-6694-y

If you want to cite the papers, please refer to the original.

For technical reasons we could not place the page where the original quote is coming from.

Abstract-Summary

"We studied the potential use of secondary-treated wastewater as nutrient source in the production of freshwater microalgae strains."

"We demonstrated that all the tested strains can be produced by using only secondary-treated wastewater as the nutrient source."

"The utilization of secondary-treated wastewater imposes nutrient-limiting conditions, with maximal biomass productivity dropping to 0.5 g l^{-1} day^{-1} and modifies the biochemical composition of the biomass by increasing the amount of lipids and carbohydrates while reducing the biomass protein content."

"We demonstrated that all the tested strains were capable of completely removing the nitrogen and phosphorus contained in the secondary-treated wastewater, and while the use of this effluent reduced the cells' photosynthetic efficiency, the nitrogen and phosphorus coefficient yield increased."

"and S. subpicatus were selected as the most promising strains for outdoor production using secondary-treated wastewater as the culture medium; this was not only because of their high productivity but also their photosynthetic efficiency, of up to 2.5%, along with nutrient coefficient yields of up to 96 $g_{biomass}$ $g_N{}^{-1}$ and 166 $g_{biomass}$ $g_P{}^{-1}$."

Introduction

"Conventional activated sludge processes transform wastewater contaminants into non-valuable products such as N_2 and CO_2, and in the process consume large amounts of energy (0.5–1.0 kWh m^{-3}), whereas the utilization of microalgae-bacteria consortiums allows one to recover the nitrogen and phosphorus contained in the wastewater yet consume far less power (0.1–0.2 kWh m^{-3}) (Aqualia, personal communication)."

"Life cycle analyses (LCA) concluded that the environmental burden associated with microalgae production for biofuels could be considerably reduced by coupling biomass cultivation and processing to wastewater treatment and industrial CO_2 capture (Collet and others 24; Lardon and others 25)."

"The coupling of microalgae production processes to tertiary treatment in wastewater treatment plants allows one to recover the nutrients contained in the wastewater and to produce valuable biomass, especially where nutrient removal is required prior to wastewater discharge."

Materials and Methods

"Experiments were performed simultaneously in all reactors, which were inoculated with 10% of the culture volume from the same standard inoculum."

"The dry weight biomass concentration (C_b) was measured by filtering 50 ml of culture through 0.45-μm filters and drying it in an oven at 80 °C for 24 h. Optimal quantum yield measurements were taken with a Junior-PAM fluorimeter (Walz, Germany)."

"The average irradiance at which cells are exposed inside a culture (I_{av}) is a function of irradiance in the absence of cells (I_o), the biomass extinction coefficient (K_a), the biomass concentration (C_b), and the light path inside the reactor (p)."

"To determine the biochemical composition of the biomass, we quantified the lipids, proteins, carbohydrates, and ash content, in addition to the fatty acid content."

Results

"Comparing these values with the expected ratio in the biomass, of 100:5:1, it was concluded that both culture mediums were carbon limited; Arnon medium contained a significant excess of phosphorus with respect to the nitrogen content."

"Using Arnon medium, the biomass concentration was higher than 1.5 g l^{-1} for all strains, but using STWW, only cultures performed using Muriellopsis sp."

"Comparing the results from both culture mediums, we observed that the electron transfer rate (and so the performance of the photosynthetic apparatus) reduced in all the tested strains when STWW was used compared to values obtained using Arnon medium."

"Results showed that for all strains, the lipid and carbohydrate content of the biomass increased when using secondary-treated wastewater compared to values obtained using Arnon medium."

"These results agree with the expected behavior of nitrogen-limited cultures (when using STWW) compared to nutrient-replete conditions (using Arnon medium)."

Discussion

"The results reported here show that all the strains tested were capable of growing in secondary-treated wastewater (STWW) without additional nutrients although the productivity was lower compared to using standard culture medium caused by nutrient limitation in the STWW."

"Under nutrient-limited conditions, using STWW as the culture medium, the light utilization efficiency of all the strains reduced, with Muriellopsis sp."

"Results show that under nutrient-limited conditions, using STWW as the culture medium, the nitrogen and phosphorus coefficient yields significantly increased with respect to those measured under nutrient-sufficient conditions, as when using Arnon medium."

"Of the low nitrogen and phosphorus concentration found in STWW, of 17 and 10 mg l^{-1} respectively, the cultures become nutrient limited, thus reducing productivity to 0.5 g l^{-1} day^{-1} compared to that obtained under nutrient-sufficient conditions, of 0.9 g l^{-1} day^{-1}."

Performance of Horizontal Flow Constructed Wetland for Secondary Treatment of Domestic Wastewater in a Remote Tribal Area of Central India [276]

This is a machine-generated summary of:

Shukla, Reetika; Gupta, Deepak; Singh, Gurudatta; Mishra, Virendra Kumar: Performance of horizontal flow constructed wetland for secondary treatment of domestic wastewater in a remote tribal area of Central India [276]

Published in: Sustainable Environment Research (2021)

Link to original: https://doi.org/10.1186/s42834-021-00087-7

If you want to cite the papers, please refer to the original.

For technical reasons we could not place the page where the original quote is coming from.

Abstract-Summary

"The purification of the primary treated domestic sewage was performed in the present study through the horizontal sub-surface flow constructed wetland (CW) of 10 × 3.5 m dimension."

"The CW was operated in continuous mode at an average hydraulic loading rate of 250 L h^{-1} and the treated effluent was analysed twice every week at four different sampling points having hydraulic retention times (HRT) of 12, 24, 36 and 48 h for important sewage quality parameters All the three setups of CW were able to clean the primary treated sewage significantly."

"Among the three sets of wetlands used, CW 3 was the best performer removing 79, 77, 79, 79, and 78% of biochemical oxygen demand, chemical oxygen demand, nitrate, ammonia, and phosphate respectively in 48 h HRT."

"Among the three sets of wetlands, the CW 3 removed the highest percent of total coliforms, fecal coliforms, and E. coli as 64, 61 and 52% respectively."

Introduction

"The performance and efficiency of CWs concerning pollutant removal from the wastewater are governed by its components like growth media, plant, microbes, and pattern of water flow in the wetland system [26]."

"Several studies dealing with the application of CWs for the treatments of municipal wastewater and subsequent reuse of treated effluents have been performed in India [27]."

"Rana and Maiti [10], performed the treatment of municipal wastewater through CW planted with Colocasia esculenta and T. latifolia, in a mesocosm study with findings able to remove several important parameters like chemical oxygen demand (COD), by 71%; total Kjeldahl nitrogen by 64–72% and some of the heavy metals."

"Most of the studies have been conducted on a lab-scale or at a mesocosm scale; therefore, more studies are required at field scale to establish CW technology as a sustainable approach for wastewater treatment in India [27]."

Materials and Methods

"CW 3 (CW planted with two different species of macrophyte, T. latifolia and C. benghalensis) in which both the species were given equal cover, i.e., 50% area of the wetland to grow."

"To achieve the further treatment, primary treated sewage from the sewage treatment plant (STP) of the IGNTU campus was passed through the experimental CWs in the continuous operation mode."

"After the collection, plants were washed properly for the removal of soil and debris attached to them, followed by this the macrophytes were planted in the gravel media, in a nursery established at IGNTU campus, after 1 month's adaptation period macrophytes were transferred to the experimental CW."

"From the collection tank, the primary treated sewage was managed to feed the CW at an average HLR of 250 L h^{-1}."

Results and Discussion

"The removal of pollutants by three different types of CWs (CW 1, CW 2 and CW 3) at different HRT (i.e., 0, 12, 24, 36, and 48 h) has been evaluated for this primary treated sewage."

"The present study has demonstrated a substantial removal of TC, FC, and E. coli from primary treated sewage during 48 h HRT."

"Treatment of various parameters of primary treated sewage such as BOD, COD, nitrogen, phosphate, and pathogens in a CW was facilitated by a combination of various natural processes including physical, chemical, and biological processes."

"The process of phosphate removal in CW depends upon redox chemistry, pH, and temperature of the wetland."

"Due to the limited contact opportunity between the wetland and phosphate, its removal in most of the CWs is not very efficient, however, our study removed almost 60% of phosphate by CW 3 which can be treated as an efficient performance."

Conclusions

"Horizontal sub-surface flow constructed wetland under present investigation treated strong strength primary treated sewage after HRT of 48 h. Three different setups of CWs (CW 1; CW 2; and CW 3) have achieved good removal efficiency for the removal of various physico-chemical and biological parameters."

"The performance of planted CW was found better in comparison to the unplanted wetland and the removal of various parameters increased with increasing HRT (up to 48 h)."

"During approximately 1 year of operation, the performance of the different CWs has continuously enhanced for the treatment of primary treated sewage."

"Various factors such as easy availability of land, availability of wetland filter media (gravel and sand) in the local area, and availability large number of native macrophytes in the study area makes this technology an ideal option for the sewage treatment in such areas of India."

Removal of Nitrogen and Phosphorus from Effluent of a Secondary Wastewater Treatment Plant Using a Pond–Marsh Wetland System [277]

This is a machine-generated summary of:

Kim, Dong-Gill; Park, Joohyun; Lee, Dowon; Kang, Hojeong: Removal of Nitrogen and Phosphorus from Effluent of a Secondary Wastewater Treatment Plant Using a Pond–Marsh Wetland System [277]

Published in: Water, Air, & Soil Pollution (2010)

Link to original: https://doi.org/10.1007/s11270-010-0399-8

If you want to cite the papers, please refer to the original.

For technical reasons we could not place the page where the original quote is coming from.

Abstract-Summary

"A constructed wetland composed of a pond- and a marsh-type wetland was employed to remove nitrogen (N) and phosphorus (P) from effluent of a secondary wastewater treatment plant in Korea."

"Results from mesocosm experiments indicated that NH was removed similarly in both pond- and marsh-type mesocosms within 1 day, but that NO was removed more efficiently in marsh-type mesocosms, which required a longer retention time (2–4 days)."

"Phosphorus was significantly removed similarly in both pond- and marsh-type mesocosms within 1 day."

"We infer that wetland system composed of a pond- and a marsh-type wetland consecutively can enhance nutrient removal efficiency compared with mono-type wetland."

"The reason is that removal of NH and P can be maximized in the pond while NO requiring longer retention time can be removed through both pond and marsh."

"Results of this study suggest that a constructed wetland composed of a pond- and a marsh-type wetland is highly effective for the removal of N and P from effluents of a secondary wastewater treatment plant."

Introduction

"Treatment wetlands, which are less costly and more easily maintained than conventional systems, have been proposed and constructed as an effective alternative for nutrient removal from wastewater (e.g., Gschlößl and others 28; Horne 29; Masi and others 30)."

"One of the wastewater treatment facilities employed experimental wetlands to remove N and P from some of the plant's effluent which had been previously subjected to secondary wastewater treatment processes."

"Most of the previous studies focused mainly on filtering effects for storm runoff or drainage systems, rather than application of constructed wetlands as a complementary treatment system for N and P removal in effluent of a secondary wastewater treatment plant."

"We aimed to address (1) the characteristics and effectiveness of constructed wetlands for removal of inorganic N and P from effluent of a secondary wastewater treatment plant, and (2) the main environmental variables affecting the removal efficiencies."

Materials and Methods

"Water samples were collected by a manual sampling method (APHA 31) at three points: inflow to the pond (i.e., effluents from the Nanji wastewater treatment plant), outflow from the pond, and outflow from the marsh."

"Five hundred milliliters of water samples was collected and transported immediately on ice to a laboratory to analyze BOD, NH, NO, total N (TN), and total P (TP)."

"Pond mesocosms (soil layer 0.2–0.3 m; water depth 0.6 m) had no macrophyte, but reeds (P. communis) were planted in marsh mesocosms (soil layer 0.2–0.3 m; water depth 0.3 m)."

"The first experiment was conducted after exchanging water with newly collected effluents of Nanji wastewater treatment plant in August 23–29, 2001 (mesocosm experiment A)."

"Water samples were collected twice a day during the operation periods to analyze NO, TP, soluble reactive P (SRP), total organic carbon (TOC), and dissolved organic carbon (DOC)."

Results

"In experiment A, NO concentration changed from 19.0 ± 0.0 to 13.4 ± 0.3 mg L^{-1} during 2 days in the pond-type mesocosms but the change was not significant ($P = 0.1$)."

"NO concentration significantly decreased by 60.9% during 2 days in the marsh-type mesocosms ($P < 0.001$)."

"NO concentration significantly decreased by 86.7% (from 13.6 ± 0.7 to 6.0 ± 0.9 mg L^{-1}) during 2 days in the marsh-type mesocosms ($P = 0.002$)."

"Concentrations of NO and TN significantly decreased by 24.4% and 25.4%, respectively, through the pond and the marsh ($P < 0.001$) and concentrations of NO and TN in the marsh was significantly lower than the one in input and the pond."

Discussion

"The pond–marsh wetland system removed 24–78% (concentration based) and 72–75% (loading based) of NH, NO, TN, and TP in secondary treated effluent of conventional wastewater treatment systems."

"If N or P concentration in the inflows does not exceed the range of the wetland background limit, the treatment wetland may not effectively remove N and P. These results suggest that nutrient concentrations of typical secondary treated effluent from conventional wastewater treatment systems including ours are higher than the wetland background limit, and hence secondary treated effluent can be further treated in treatment wetlands."

"The results from this study indicate that constructed wetlands have applicability for the removal of N and P from secondary treated effluent of conventional wastewater treatment systems."

"The results from our study indicate that constructed wetlands, even of a simple design, are highly effective in the removal of N and P in effluent from a conventional wastewater treatment plant."

Advanced Treatment of Municipal Secondary Effluent by Catalytic Ozonation Using Fe_3O_4-CeO_2/MWCNTs as Efficient Catalyst [278]

This is a machine-generated summary of:

Bai, Zhiyong; Wang, Jianlong; Yang, Qi: Advanced treatment of municipal secondary effluent by catalytic ozonation using Fe_3O_4-CeO_2/MWCNTs as efficient catalyst [278]

Published in: Environmental Science and Pollution Research (2017)

Link to original: https://doi.org/10.1007/s11356-017-8619-z

If you want to cite the papers, please refer to the original.

For technical reasons we could not place the page where the original quote is coming from.

Abstract-Summary

"The advanced treatment of municipal secondary effluent was performed by catalytic ozonation using Fe_3O_4-CeO_2/MWCNTs as catalyst."

"The experimental results showed that in catalytic ozonation system, the removal efficiency of soluble COD was more than 46% after 30 min reaction, and about 36%

of effluent organic matters (EfOMs) were mineralized, which was four times higher than that in single ozonation system."

"In catalytic ozonation system, the ozone utilization increased, which is favorable for the degradation of EfOM."

"The catalyst could be stable after five-time reuse for catalytic ozonation of effluent."

Introduction

"The advanced treatment of the secondary effluent to remove the effluent organic matters, especially the toxic micropollutants, is crucial to guarantee the safety of the reclaimed water (Giannakis and others 32; Cai and Lin 33)."

"The ozonation has recently been considered as a competitive technique for advanced treatment of municipal secondary effluents."

"Heterogeneous catalytic ozonation have proved its high efficiency for the removal of refractory compounds (Bing and others 34; Goncalves and others 35; Huang and others 36; Ikhlaq and others 37; Bai and others 38), natural organic matter (Park and others 39; Park and others 40), and tertiary treatment of industrial wastewater, such as bio-treated dyeing and finishing wastewater (Wu and others 41), heavy oil refining wastewater (Chen and others 42), and Lurgi coal gasification wastewater (Zhuang and others 43)."

"A few studies focused on the catalytic ozonation of municipal secondary effluents."

"Fe_3O_4-CeO_2/MWCNTs can be used as ozonation catalyst for the advanced treatment of municipal secondary effluents."

"The objective of this study was to investigate the catalytic ozonation of municipal secondary effluents using Fe_3O_4-CeO_2/MWCNTs as catalyst."

"We focused on the removal efficiency of EfOM in municipal secondary effluents and the utilization efficiency of ozone in catalytic ozonation system."

Materials and Methods

"A certain ratio of $FeSO_4 \cdot 7H_2O$, $Fe_2(SO_4)_3$, H_2SO_4, and MWCNTs was dissolved or dispersed in water under ultrasound for 10 min, and then the mixed solution was added dropwise into four-necked flask 80 °C under Ar gas protection."

"1 L secondary effluent and 0.3 g catalyst were added into the reactor, and then ozone was continuously added to the solution through a porous diffuser device."

"p-HBA (10 mg L^{-1}) was dissolved in deionized water (DW), MSEF, and MSEF with low alkalinity, respectively."

"With the same conditions described above, samples were collected and the concentration of p-HBA was detected."

"The usage of SMT and CBZ in the experiments is 20 and 10 mg L^{-1} under the same reaction conditions, respectively."

"The concentration of p-HBA, SMT, and CBZ was measured using high-performance liquid chromatography (HPLC, Agilent 1200) (Bai and others 44, 45, Wang and Wang 46)."

Results and Discussion

"The addition of the catalyst significantly improved the SCOD removal efficiency, which was 51% in catalytic ozonation system."

"The polysaccharide concentration decreased weakly; their concentration was 5.56 and 5.27 mg L^{-1} after ozonation and catalytic ozonation, respectively; and the removal efficiency was only 5.64 and 9.97%."

"In catalytic ozonation system, only after 5 min, the value of UV_{254} decreased to 0.055; the corresponding removal efficiency was 53.04%."

"The results were consistent with the decreasing tendency of SCOD and TOC removal; as discussed above, the organic pollutants in effluent can be oxidized by ozone, but hydroxyl radical is more powerful for their mineralization."

"Catalytic ozonation is effective for degrading micropollutants in the municipal secondary effluents by using Fe_3O_4-CeO_2/MWCNTs as efficient catalyst."

Conclusions

"Municipal secondary effluent was further treated by catalytic ozonation system using Fe_3O_4-CeO_2/MWCNTs as catalyst."

"The catalytic ozonation system exhibited high degradation efficiency for organic pollutants in the effluent."

"The addition of catalyst increased ozone utilization obviously."

Removal Processes for Tributyltin During Municipal Wastewater Treatment [279]

This is a machine-generated summary of:

Scrimshaw, M. D.; Hala, D.; Okiemute, A. E.; Cartmell, E.; Lester, J. N.: Removal Processes for Tributyltin During Municipal Wastewater Treatment [279]

Published in: Water, Air, & Soil Pollution (2013)

Link to original: https://doi.org/10.1007/s11270-012-1400-5

If you want to cite the papers, please refer to the original.

For technical reasons we could not place the page where the original quote is coming from.

Abstract-Summary

"Both sites had two inlet streams, and each utilised high rate biological filters (biofilters) on one the streams, before treatment of the combined flows on trickling filters, with one having additional tertiary processes, installed to remove ammonia and solids."

"At the treatment works with tertiary processes, the mass flux showed the overall removal of TBT was 68%, predominantly due to removal with solids in the primary settlement processes."

"The two works both removed TBT, but at different treatment stages and by different processes."

"Differences in the form (solubility) of TBT in the influent may have attributed to this, although further understanding of factors controlling degradation would allow for a more complete assessment of the potential of biological processes to remove hazardous compounds from wastewaters."

Introduction

"The fate, behaviour and environmental impact of tributyltin (TBT) in the aquatic environment as the result of its use as an antifoulant on the hulls of boats and ships is well understood (Clark and others 47; Dowson and others 48)."

"There are a limited number of detailed studies of the fate and behaviour of TBT and its breakdown products DBT and MBT in WWTWs and include those by Fent and Muller (49) and Voulvoulis and others (50). The study by Fent and Muller (49) was undertaken at Zurich WWTWs, where treatment included primary clarification, activated sludge treatment and tertiary treatment by sand filtration."

"Fent and Muller (49) concluded that TBT was efficiently removed mainly by sedimentation in the primary clarifier."

"Given the concentrations of TBT in both primary sludge and waste activated sludge (WAS), they concluded that adsorption was the paramount process in their removal."

"This study was undertaken to further elaborate on the fate and behaviour of TBT during conventional wastewater treatment at two WWTWs."

Materials and Methods

"The 'low level' crude sewage following primary sedimentation underwent high-rate biological treatment in biotowers, the effluent from which underwent further settlement."

"Liquor from the GBT were returned to the head of the works, and liquors from the plate press were combined with brewery crude sewage for treatment in the high-rate biological filters."

"The humus sludge and return liquor were sampled once every 24 h for 96 h. There were eight return/sludge streams for the tertiary treatment works, each sampled five times, yielding 40 samples."

"There were six return liquor sludge streams at the conventional trickling filter works which were each sampled five times yielding 30 samples."

"Samples with high solids (sludges and return flows at the site) were transferred into a 50-ml centrifuge tube with 2 ml of concentrated trace metal analysis grade hydrochloric acid (Fisher Scientific)."

Results

"After the primary settling stage, the 'low level' flow was treated in the high-rate biological filter and underwent subsequent sedimentation (the biotower process)."

"After the biotower processes, the 'low level' flow combined with the flow from the 'high level' primary tanks and this combined flow (55.2 Ml day^{-1} referred to as filter feed) was sampled before the secondary biological trickling filters."

"However, the trickling filters, BAFF and RGF removed 0.7 g day^{-1}, or 10% of the total removal of TBT (6.8 g day^{-1}), highlighting the significance of the primary sedimentation processes in accounting for TBT removal at the tertiary treatment works."

"The concentration of TBT in the settled sewage (primary tank effluent) flowing to the trickling filters was 0.663 μg l^{-1}."

Discussion

"In the biological stage of the conventional works in this study, removal of TBT was greater than 90%."

"It is possible that the solubility of TBT that reduced removal during primary sedimentation was responsible for facilitating uptake and subsequent biodegradation during secondary biological treatment in the conventional trickling filter works."

"The critical role of bacteria the biodegradation removal mechanism appears to be clearly established by the studies of Vossler and others 51 using azide to eliminate biological activity and with it TBT breakdown."

"Not only do high concentrations of bacteria appear to enhance TBT breakdown but the presence of soluble organic matter, which is abundant in sewage, also appears to enhance removal (Kawai and others 52)."

"Utilising a laboratory-scale batch activated sludge process for biological wastewater treatment supplied with synthetic sewage, Stasinakis and others (53) observed half lives for TBT, DBT and MBT of 1.4, 3.6 and 9.8 days, respectively."

Conclusions

"The removal of TBT showed a significant correlation with suspended solids removal at one works; however, at the other site, this relationship was not present."

"At the conventional treatment works, degradation of TBT was observed in two fixed film biological processes; however, due to the nature of the growth in these systems, no inference on degradation rates could be made."

"Differences in TBT speciation (dissolved or associated with solids in the sewage) at these two wastewater works resulted in different removal mechanisms."

"Further understanding of the organisms responsible for and the mechanisms of biodegradation may allow for more effective biological treatment processes for the removal of hazardous chemicals."

3.3 Aerobic Stabilization

Introduction by Editor

In the secondary treatment phase of municipal wastewater treatment plants, aerobic stabilization plays a pivotal role in the effective removal of organic matter, nutrients, and pollutants. Aerobic stabilization utilizes aerobic microorganisms to decompose organic matter present in wastewater. These microorganisms metabolize complex organic compounds, breaking them down into simpler forms such as carbon dioxide and water. The process of aerobic stabilization reduces biochemical oxygen demand (BOD) and effectively removes potential pathogens, thereby improving water quality.

Successful implementation of aerobic stabilization requires careful attention to operational considerations. Parameters such as aeration rate, dissolved oxygen levels, temperature, pH, and nutrient availability must be controlled to optimize microbial activity and organic matter degradation. Aeration systems, such as diffusers or mechanical aerators, supply oxygen and promote the growth and activity of aerobic microorganisms. Monitoring and controlling dissolved oxygen levels within the aerobic reactor are critical to maintaining optimal conditions for microbial metabolism.

Several process configurations are employed for aerobic stabilization in secondary treatment. The activated sludge process is the most commonly used configuration, involving the continuous cultivation of a mixed population of aerobic microorganisms in a suspended growth reactor. This process offers high treatment efficiency, flexibility, and the ability to handle variations in influent wastewater characteristics.

Other configurations include sequencing batch reactors (SBRs) and oxidation ditches. SBRs operate in a batch mode, allowing greater control over the treatment process and improved nutrient removal. Oxidation ditches are circular or oval channels where wastewater and activated sludge are continuously mixed, promoting aerobic degradation. The choice of configuration depends on specific site requirements and treatment objectives.

Aerobic stabilization is vital for achieving high-quality effluent in municipal wastewater treatment plants. It significantly reduces the concentration of organic matter, pathogens, and pollutants, thereby improving water quality and protecting the receiving environment. Additionally, aerobic stabilization contributes to the removal of nitrogen and phosphorus compounds, helping prevent eutrophication and maintaining ecological balance in receiving water bodies.

Machine generated keywords: granule, granular, granular sludge, aerobic, aerobic granule, bacteria, reactor, formation, microbial, liu, emission, biomass, community, microbial community, diameter.

The Influence of Solid-Liquid Coefficient in the Fate of Pharmaceuticals and Personal Care Products in Aerobic Wastewater Treatment [280]

This is a machine-generated summary of:

Kramer, Rafael D.; Filippe, Tais C.; Prado, Marcelo R.; de Azevedo, Júlio César R.: The influence of solid-liquid coefficient in the fate of pharmaceuticals and personal care products in aerobic wastewater treatment [280]

Published in: Environmental Science and Pollution Research (2018)

Link to original: https://doi.org/10.1007/s11356-018-2609-7

If you want to cite the papers, please refer to the original.

For technical reasons we could not place the page where the original quote is coming from.

Abstract-Summary

"Wastewater treatment plants (WWTPs) are considered to be a source of environmental contamination by micropollutants, especially from pharmaceuticals and personal care products (PCPs)."

"The pathway of those compounds during sewage treatment has been investigated, but data from real-scale WWTPs is still missing (for example, the values of the solid-liquid coefficient (K_d) during treatment)."

"This paper uses the K_d values for some pharmaceuticals and PCPs (fenofibrate, gemfibrozil, propranolol, metoprolol, salicylic acid, acetylsalicylic acid, ibuprofen, diclofenac, naproxen, fenoprofen, caffeine, triclosan, methylparaben, ethylparaben, propylparaben, butylparaben, and benzylparaben) to describe the micropollutants' behavior in the treatment process."

"Each micropollutant had a specific way to be removed during wastewater treatment."

Introduction

"Some of these pollutants, when found in the environment, have been called micropollutants or emerging contaminants."

"The term emerging contaminants is commonly used to describe compounds that were not originally detected or found in the environment but are currently the most frequently studied and widely found in water matrixes, effluents, soil, or air."

"Municipal WWTPs are considered point sources of emerging contaminants in aquatic environments for the following reasons: by receiving a large influent load, by applying conventional treatment systems that cannot deal with them, and by not being designed to remove these complex molecules (Kasprzyk-Hordern and others 54; Narumiya and others 55; Pomiès and others 56; Ratola and others 57; Verlicchi and others 58)."

"The presence of micropollutants in sludge cannot be considered as adequate removal by WWTPs but just as a phase transfer (liquid to solid) since sludge needs treatment before being disposed of or reused (Semblante and others 59)."

Materials and Methods

"Some characteristics (flow rate, hydraulic retention time, sludge age, and biological mass) of the WWTP during the sampling period was obtained with the operating system of the WWTP and can also be seen in the Supplementary Materials."

"With mass values at the input (dissolved phase and adsorbed in SS) and at the output (dissolved phase, adsorbed in SS, and in sludge), it was feasible to calculate mass balance and, particularly, the pathway of contaminants during the treatment process."

"Some considerations and assumptions had to be made for the development of the calculation: (1) even with chemical and physical differences between influent and effluent solids, the same K_d values were applied to both; (2) there is no influence of biodegradation in the K_d values; and (3) the single input considered was the influent of the WWTP and the outputs considered were the effluent and the sludge, which went through the thickener."

Results and Discussions

"The $\log K_{oc}$ (organic carbon adsorption coefficient) and $\log K_{ow}$ (octanol-water coefficient) values indicate that some compounds have a higher capacity of being adsorbed in a solid phase; this is the case with triclosan and gemfibrozil."

"Because of ibuprofen's higher tendency to be adsorbed, and considering all the analyzed drugs, its mass in the sludge had one of the highest values (49.1 mg day^{-1} median), followed only by gemfibrozil which had the highest median (99.1 mg day^{-1})."

"Triclosan, known as a compound with more interaction with the solid phase (Huang and others 60), had a median load at the input of 1893 mg day^{-1}; 2.4% was adsorbed in the SS (a major proportion of PCPs analyzed)."

"Ibuprofen, salicylic acid, and caffeine were the most removed compounds, and this could be related with sludge age (Verlicchi and others 58) because removal was higher in the sampling with the highest sludge age."

Conclusion

"The observation of different types of micropollutant characteristics can be important for predicting their behavior during treatment."

"Calculation of solid-liquid coefficient (K_d) was an important way to understand the behavior of micropollutants in WWTPs."

"Other kinds of compounds (e.g., triclosan, gemfibrozil, and ibuprofen) showed that they can be removed by adsorption, especially in treatment where sludge is fluidized."

"The development of mass balance allowed observation of the behavior of pharmaceuticals and PCPs during sewage treatment—in this case, in an aerobic treatment."

Effect of Carbon Source on Nitrous Oxide Emission Characteristics and Sludge Properties During Anoxic/Aerobic Wastewater Treatment Process [281]

This is a machine-generated summary of:

Yan, Xu; Yang, Jie; Guo, Dongli; Ma, Jiahui; Su, Xianfa; Sun, Jianhui: Effect of carbon source on nitrous oxide emission characteristics and sludge properties during anoxic/aerobic wastewater treatment process [281]

Published in: Environmental Science and Pollution Research (2021)

Link to original: https://doi.org/10.1007/s11356-021-14713-9

If you want to cite the papers, please refer to the original.

For technical reasons we could not place the page where the original quote is coming from.

Abstract-Summary

"Three parallel sequencing batch reactors (SBRs) were processed with acetic acid, propionic acid, and a 1:1 mixture of both acids (calculated in COD) to study the effect of carbon sources on N_2O generation and sludge properties (including intracellular polymer content, extracellular polymeric substance (EPS) composition, particle size distribution, settleability, and microbial community structure)."

"The results showed that the highest COD, NH_4^+-N, and TP removal efficiencies (92.2%, 100%, and 82.3%, respectively) were achieved by the reactor with mixed acid as the carbon source, whereas the reactor using acetic acid had the highest TN removal rate (82.6%) and the lowest N_2O–N conversion rate (1.4%, based on TN removal)."

"The reactor with the carbon source of mixed acid produced the highest polyhydroxyalkanoate (PHA) content, which led to an increase in N_2O generation from the aerobic denitrification pathway."

"The SBR with mixed acid carbon source also had the highest concentration of EPS, which resulted in the largest particle size and the lowest settleability of sludge flocs among the SBRs."

"The carbon source of acetic acid promoted the growth of denitrifying bacteria (DNB), which led to the highest TN removal rate."

Introduction

"The carbon source can significantly impact the nitrification and denitrification processes and is closely linked to N_2O generation."

"The activated sludge process is the most widely used treatment method in WWTPs owing to its low operating costs, high removal efficiency, and facile operation (Maddela and others 61)."

"Carbon sources are important for microbial metabolism and the formation of activated sludge flocs during the treatment processes."

"In biological wastewater treatment processes, the N_2O generation characteristics, sludge properties, and their internal relationships under different carbon sources have not been sufficiently studied."

"The effects of the carbon source on N_2O generation characteristics and activated sludge properties, including particle size distribution, PHA content, EPS composition, settleability, and microbial community structure, were explored."

"The objective of this work was to comprehensively understand the effect of carbon sources on the activated sludge treatment system and provide a scientific reference for optimizing WWTPs."

Materials and Methods

"The mixed liquor suspended solids (MLSS) of the reactors were maintained at 3000–3500 mg/L. The concentration of mixed liquor volatile suspended solids (MLVSS) was 2200–2700 mg/L. The components of influent were as follows: 155 mg/L NH_4Cl, 4.89 mg/L $MgSO_4$, 200 mg/L $NaHCO_3$, 14 mg/L $CaCl_2$, 16 mg/L KH_2PO_4, and 0.3 mL/L trace element solution (150 mg/L $FeCl_3 \cdot 6H_2O$, 60 mg/L $MnCl_2 \cdot 4H_2O$, 150 mg/L H_3BO_3, 120 mg/L $Na_2MoO_4 \cdot 2H_2O$, 180 mg/L $CuSO_4 \cdot 5H_2O$, 150 mg/L $ZnSO_4 \cdot 7H_2O$, 120 mg/L KI, and 100 mg/L EDTA)."

"After 60 days of operation, when the SBRs reached a stable state, three sequential sample collections of typical cycles according to time points were conducted."

"The formaldehyde-NaOH method was employed for extracting EPS from sludge (Liu and Fang 62)."

"The gas sampling was performed as described by Yan and others (63)."

"The effluent gas was collected using 50 mL sampling bags (Delin, China)."

"Sludge samples were obtained following sample collection of typical recycles."

Results and Discussion

"This result revealed that the carbon source primarily affected the production of the functional groups of LB-EPS in the activated sludge."

"Ye and others (64) also observed variations in the LB-EPS content of sludge with different carbon sources for a similar reason."

"During the nitrogen removal process, nitrification was primarily conducted by AOB and NOB to convert NH_4^+-N to NO_2^--N, followed by NO_2^--N to NO_3^--N. In this study, variation in the AOB content in the presence of different carbon sources was observed."

"The type of carbon source was closely related to the microbial metabolic pathway, leading to a difference in the composition and content of intracellular polymers and EPS, and the microbial community (as well as the functional microorganisms of AOB, NOB, and DNB)."

"As the carbon source, the high PHA content in the sludge was advantageous for nitrifier denitrification and N_2O generation."

Conclusion

"The performance, N_2O emission, sludge properties, and microbial community structure of an SBR process with carbon sources of acetic acid, propionic acid, and mixed acid were analyzed and compared in this study."

"The highest COD, NH_4^+-N, and TP removal efficiencies occurred in the reactor with a carbon source of mixed acid, whereas acetic acid had the best TN removal rate and the lowest N_2O generation."

"The mixed acid produced the highest PHA content, resulting in the promotion of N_2O generation from the aerobic denitrification pathway."

"The carbon source of acetic acid promoted the growth of DNB, which led to the highest TN removal rate."

Aerobic Biological Treatment of Low-Strength Synthetic Wastewater in Membrane-Coupled Bioreactors: The Structure and Function of Bacterial Enrichment Cultures as the Net Growth Rate Approaches Zero [282]

This is a machine-generated summary of:

Chen, Ruoyu; LaPara, Timothy M.: Aerobic Biological Treatment of Low-Strength Synthetic Wastewater in Membrane-Coupled Bioreactors: The Structure and Function of Bacterial Enrichment Cultures as the Net Growth Rate Approaches Zero [282]

Published in: Microbial Ecology (2006)

Link to original: https://doi.org/10.1007/s00248-005-0081-7

If you want to cite the papers, please refer to the original.

For technical reasons we could not place the page where the original quote is coming from.

Abstract-Summary

"The goal of the current research was to determine if the stringent nutrient limitation imposed by membrane-coupled bioreactors (MBRs) could be used to force mixed bacterial communities to exhibit a zero net growth rate over an extended time period."

"This zero net growth rate could be achieved when the amount of energy available for growth is balanced by the maintenance requirements of the bacterial community."

"Leucine aminopeptidase activities concomitantly increased by at least 10-fold, suggesting that bacterial catabolic activity remained high even while growth rates became negligible."

"A-glucosidase and heptanoate esterase activities decreased, indicating that the bacterial community specifically adapted to the carbon source in the feed medium."

"Bacterial community analysis by denaturing gradient gel electrophoresis of PCR-amplified 16S rRNA gene fragments (PCR-DGGE) suggested that the bacterial community structure completely changed from the beginning to the end of each MBR."

Introduction

"Particulate material is physically removed and then the activated sludge process is used to biodegrade the remaining soluble wastewater pollutants."

"The principal limitation of the activated sludge process is the efficiency by which biomass is separated from the treated effluent [65]."

"From an ecological perspective, MBRs impose a stringent nutrient limitation on the biomass, which results in lower cellular levels of RNA and ATP [66], lower respiratory potentials [66, 67], less rapid responses to nutrient pulses [68], and bacterial growth rates similar to those observed in nature [67]."

"The goal of the current research was to determine if the stringent nutrient limitation imposed by MBRs could be used to force mixed bacterial communities to exhibit a zero net growth rate over an extended time period."

"This situation is possible in MBRs treating low-strength wastewater because the nutrient flux is relatively low, whereas the biomass density can be very high."

"Bench-scale MBRs were fed a constant supply of synthetic feed medium while cells were allowed to accumulate, thus imposing an increasingly stringent nutrient limitation on the biomass."

Materials and Methods

"Laboratory-scale MBRs were inoculated with 1 mL of biomass from a 100-mL batch culture containing a gelatin-based synthetic wastewater that had been incubated overnight at 22°C."

"These batch cultures had been inoculated with 1 mL of cryopreserved activated sludge (quick-frozen in 15% glycerol) collected from the aeration tanks of the Metropolitan Wastewater Treatment Plant (St. Paul, MN, USA)."

"In these analyses, a unit of enzyme activity corresponded to the production of 1 μmol product min^{-1}; enzymatic activities were normalized to biomass concentrations measured as total particulate protein."

"These PCR products were purified using the Geneclean II kit (Qbiogene) prior to nucleotide sequence determination."

"All nucleotide sequences were determined at the Advanced Genetic Analysis Center at the University of Minnesota using an ABI 3100 Genetic Analyzer (Applied Biosystems, Foster City, CA, USA)."

"Nucleotide sequences were determined fully in both directions for each purified PCR-DGGE band using 338F and 518R as sequencing primers."

Results

"Leucine aminopeptidase activity either decreased or remained relatively constant for the first 15–20 days, after which substantial increases (~10-fold) in both V_m and a^0 were observed."

"Substantial shifts were observed in the community structures from the beginning until the end of each MBR."

"Cluster analysis of the community fingerprints demonstrated a complete shift in community structure from the beginning until the end of each MBR (analyses not shown)."

"The adaptation in community structure was initially rapid; several of the bands that were originally detectable disappeared in the first 20 days, whereas other bands concomitantly appeared that were not initially detectable."

"The shifts in community structure continued throughout each experiment, with previously undetectable bands appearing as late as 100 days following the initiation of the MBR."

Discussion

"The primary objective of this research was to operate laboratory-scale MBRs at very long cell residence times and at high biomass densities to determine if the bacterial community would exhibit zero net growth while metabolizing the organic constituents of a synthetic wastewater."

"In some of the MBRs, however, substantial increases inleucine aminopeptidase activity were observed during the last 20–40 days of the experiments, during which time less substantial adaptations in the bacterial community structures were detected by PCR-DGGE."

"Previous work has demonstrated that MBRs are an excellent model system to study the ecology of nutrient-limited bacterial communities [67]."

"As with previous studies, PCR-DGGE suggested that the bacterial community consisted of multiple bacterial populations [67, 69, 70]; the current study was unique, however, in that the complexity of the bacterial communities observed herein exceeded the complexity of the nutrient medium (principally gelatin)."

"We also demonstrated that the bacterial community actively adapted its structure and function in response to the increasingly stringent nutrient limitation imposed by the MBR."

The Role of Stress Agents as Operating Factors in Formation and Functioning of Granular Aerobic Activated Sludge at Model Domestic Wastewater Treatment [283]

This is a machine-generated summary of:

Khokhlachev, Nikolay S.; Kalenov, Sergei V.; Zanina, Olga S.; Tyupa, Dmitry V.; Baurina, Marina M.; Kuznetsov, Alexander Ye.: The role of stress agents as operating factors in formation and functioning of granular aerobic activated sludge at model domestic wastewater treatment [283]

Published in: Bioprocess and Biosystems Engineering (2014)

Link to original: https://doi.org/10.1007/s00449-014-1149-x

If you want to cite the papers, please refer to the original.

For technical reasons we could not place the page where the original quote is coming from.

Abstract-Summary

"Maintenance of the wastewater treatment plants and increasing the efficiency of existing aerobic biological reactors depend on the stability of activated sludge characteristics under varying wastewater parameters within significant limits and/or influence of some environmental factors."

"To improve aerobic wastewater treatment characteristics, a possibility of using impact of stress conditions upon activated sludge has been studied."

"Under conditions of fractional hydrogen peroxide addition at diffused lighting, the granular aerobic activated sludge adapted to hydrogen peroxide has been obtained."

"It also endures an impact of one-time hydrogen peroxide addition up to 1.2–1.5 g H_2O_2/l. The conditions under which the steady aerobic granules of the diameter from 2 to 5 mm were formed with high treatment ability have been chosen."

"The granules were being stabilized at passages with hydrogen peroxide treatment and they endured up to 2.4–3.0 g/l of one-time H_2O_2 addition."

Introduction

"The problems of the aerobic stage are solved partially: bioavailability of organic pollution increases, the amount of excess activated sludge decreases [71] but the standards on wastewater treatment remain difficult to reach even at combination of modern anaerobic reactors with aerobic reactors of various designs."

"The study of borderline states of aerobic activated sludge in which some changes of the microbial community structure take place can be of a considerable interest."

"Obtaining such stable structures under conditions of changing loading could solve a number of the problems connected with instability of aerobic activated sludge communities."

"Fragmentary information on the factors influencing stability of aerobic granules of activated sludge allows assuming that different types of stress and the borderline states caused by them play a crucial role in the community regulation [72]."

Materials and Methods

"Activated sludge was cultivated in the conical flasks of 250 ml volume with the sludge suspension volume of 100 ml and initial activated sludge (in the form of flakes) concentration of 2–4 g/l. At passages for adaptation to impact of hydrogen peroxide the concentration of H_2O_2 was of 0.3–1.2 g/l. The typical characteristics of the model wastewater obtained by mixing the broth of aqueous extracts from foodstuff measured were: COD = 400 ± 50 mg O_2/l, $PO_4^{3-} = 5 \pm 1$ mg/l, $N_{total} = 25 \pm 3$ mg/l and they were varied depending on the purpose of the experiment."

"After a cycle of cultivation the contents of the conical flasks were transferred quantitatively into the measured cylinder of 100 ml volume and allowed to settle within 5–10 min."

"The sample was transferred back into the conical flask, and the cultivation of developing biocenose was carried out in a shaker flask at 200 rpm (without oxygen limitation)."

Results

"After stabilization of domestic wastewater treatment characteristics in control variants the study of the cultivation of activated sludge was continued with a set of passages at the addition of H_2O_2."

"During the first several passages the decrease of biocenose activity of every line with hydrogen peroxide addition was observed in comparison with the control variants C_7 and T_7."

"At the first passages with hydrogen peroxide the biocenose of activated sludge was in suppression (an initial stage of adaptation) and this activated sludge did not reach the characteristics of pollution removal in contrast with the control variant."

"With flocculi of the activated sludge, the granules adapted for hydrogen peroxide endured a larger concentration of H_2O_2 in 4–5 times; and in comparison with the granules obtained by a batch operation at peak loading and the subsequent starvation, they endured a larger concentration of H_2O_2 in 2–3 times that can be a sign of stability to adverse environmental factors."

Discussion

"The agglomerates of activated sludge found in some cases in a batch operation in the control variants and also at adapting to the stressor, hydrogen peroxide, had not been steady for a long-term time."

"These granules were being stabilized at passages with hydrogen peroxide and subsequently they endured one-time addition of hydrogen peroxide up to 2.4–3.0 g/l. The change of specific composition of activated sludge in the form of flakes and granules at adaptation to peroxide is of great interest for further investigations."

"The increased mechanical strength of the adapted granules of aerobic activated sludge, the possibility of regulation of the sizes of these granules at addition of the stressor in increased doses, and improved treatment characteristics allow saying about possibility of creation of new technologies of aerobic treatment and modernization of existing ones."

Conclusion

"The findings of our investigation have shown a possibility to ameliorate the stability and operability of flocs and granules of aerobic activated sludge by optimal action of some stress factors like hydrogen peroxide."

"The positive effect is observed at the adaptation of activated sludge to optimal doses of hydrogen peroxide and concomitant illumination of medium by diffused visible light of low intensity (about 40 mWt/l)."

"There must be periodic increase in loading with the subsequent long-term activated sludge starvation and hemicontinuous process for formation of granules."

"At the addition of H_2O_2 the granular formation is followed by reduction of a granular species' diversity that testifies to sufficiency of minimum variety of species for functioning and stability of these granules."

Performance and Microbial Characteristics of Biomass in a Full-Scale Aerobic Granular Sludge Wastewater Treatment Plant [284]

This is a machine-generated summary of:

Świątczak, Piotr; Cydzik-Kwiatkowska, Agnieszka: Performance and microbial characteristics of biomass in a full-scale aerobic granular sludge wastewater treatment plant [284]

Published in: Environmental Science and Pollution Research (2017)

Link to original: https://doi.org/10.1007/s11356-017-0615-9

If you want to cite the papers, please refer to the original.

For technical reasons we could not place the page where the original quote is coming from.

Abstract-Summary

"By modification of the operational conditions of batch reactors, a municipal wastewater treatment plant was upgraded from activated sludge to aerobic granular sludge (AGS) technology."

"The concentration of biomass in the reactors increased twofold; the average biomass yield was 0.6 g MLVSS/g COD, and excess granular sludge was efficiently stabilized in aerobic conditions."

"Bacteria belonging to these genera produce extracellular polymeric substances, which stabilize granule structure and accumulate phosphorus."

"The results of this study will be useful for designers of AGS wastewater treatment plants, and molecular data given here provide insight into the ecology of mature aerobic granules from a full-scale facility."

Introduction

"The settling properties of granular sludge are better than those of conventional activated sludge, and granular sludge is more resistant to changes in environmental conditions, such as temperature, organic load, or pH. The formation of aerobic granules and the morphology of the resulting granules are influenced by many operational factors such as the applied shear stress, settling time of biomass, organic loading rate, volumetric exchange rate, or hydraulic retention time (Liu and Tay 73)."

"According to Pronk and others (74), the energy use in a full-scale WWTP with aerobic granules treating domestic wastewater was 58–68% lower than in a typical conventional activated sludge plant."

"Reports on aerobic granulation in WWTPs have mainly focused on technological research, and little is known about the microbial bases of wastewater purification in full-scale facilities with granular sludge."

"To meet this need, here we report on the process efficiency and on the microbiological and morphological structure of aerobic granules during long-term operation of a municipal WWTP that was upgraded from activated sludge to aerobic granular technology by modifying the technological line and operational parameters."

Materials and Methods

"Before upgrading, the reactor cycle had separate phases for reactor feeding and discharge of treated wastewater."

"The operational mode of the reactors was modified to favor granulation (shortened settling, shorter filling, precise removal of excess sludge from aeration tanks)."

"The superficial gas velocity in the reactors was 0.18 cm/s; the sludge retention time (SRT) was about 30 days; hydraulic retention time was about 1 day, and the minimum settling velocity in the aerobic granular sludge reactors was about 1.6 m/h. In the aerobic stabilization tank, the biomass concentration reached 15–16 g mixed liquor suspended solids (MLSS)/L, and the hydraulic retention time was 17 days."

"At the end of the study, pollutant concentrations were also measured in duplicate during the reactor cycle to examine the kinetics of removal of organics, nitrogen, and phosphorus by the aerobic granular sludge."

"For molecular analyses, sludge from the two biological reactors was analyzed four times over 7 months."

Results and Discussion

"Betaproteobacteria is usually the most abundant class of bacteria in activated sludge, and the fact that their abundance in the aerobic granules at the end of the study was more than three times as much as in the activated sludge is favorable because they are largely responsible for organic and nutrient removal (Nielsen and others 75; Hu and others 76; Wang and others 77)."

"The results of the present study indicate a transition in microbial structure during granule maturation from backbone-forming filamentous bacteria in activated sludge to EPS-producing bacteria in aerobic granules."

"The abundance of Actinobacteria decreased from 32.7% in activated sludge to 14.3 ± 3.7% in mature granules at the end of the experiment."

"The abundance of bacteria belonging to the order Sphingomonadales increased during the process of biomass adaptation and reached 4.8 ± 0.0% in the mature granules, in comparison to 0.6% in the activated sludge."

Conclusions

"Application of AGS allowed for more compact wastewater treatment than with activated sludge; after upgrading, the volume of the reactors was reduced by 30%, but the quality of the effluent was improved."

"AGS was easily thickened and dewatered, which allowed both the volumes of the tanks in the sludge treatment line and the volume of excess sludge to be decreased."

"The abundances of Betaproteobacteria, Deltaproteobacteria, Flavobacteria, and Cytophagia, largely responsible for organic and nutrient removal, increased more than three times in biomass after upgrading, which significantly increased the treatment efficiency."

Potential of Aerobic Membrane Bioreactor for Recycling and Reuse of Domestic Wastewater for Irrigation [285]

This is a machine-generated summary of:

Yılmaz, F.; Otuzaltı, M. M.; Perendeci, N. A.; Karatay, M.; Ünşar, E. Kökdemir; Ateş, M.; Akin, R.; Yıldız, O.; Orhon, D.: Potential of aerobic membrane bioreactor for recycling and reuse of domestic wastewater for irrigation [285]

Published in: Environmental Earth Sciences (2020)

Link to original: https://doi.org/10.1007/s12665-020-09006-2

If you want to cite the papers, please refer to the original.

For technical reasons we could not place the page where the original quote is coming from.

Abstract-Summary

"Performances of two different full-scale treatment plants, one designed as aerobic membrane bioreactor (AMBR) equipped with a hollow fiber membrane with a nominal pore size of 0.03 μm, and the other, as a conventional nitrogen removal activated sludge plant with pre-denitrification (A^2O) were evaluated and compared."

"The effluent of the A^2O system included an average of 109 mg L^{-1} chemical oxygen demand (COD), 32 mg L^{-1} total suspended solids (TSS) and 17 mg L^{-1} of total nitrogen."

"The AMBR system was able to reduce the effluent COD level to 24 mg L^{-1}, obviously with lower TSS and turbidity."

"The effluent quality of the AMBR was in compliance with irrigation water standards; its nitrogen content should be evaluated as an added nutritional value for most plants."

Introduction

"Municipal wastewater, which is basically wasted in large quantity, should be regarded as a sustainable resource for water recovery and reuse, especially in agriculture."

"In 2016, the Ministry of Environment and Urbanization (MoEU) completed a significant project to determine the level of wastewater treatment plant (WWTP) in Turkey including the efficiency and operational issues facing municipal WWTP across the country."

"The total solids retention potential of MBR process also offered a much useful perspective—stable system operation at much lower SRTs and biomass levels compared with conventional activated sludge flow scheme."

"This study's goal was to evaluate and compare the performances of two full-scale treatment plants, one designed and operated as a nutrient removal (A^2O) process, and the other as an aerobic MBR, both treating municipal sewage of the same region in Antalya city, Turkey."

Materials and Methods

"The design wastewater flow rate of Elmalı A^2O type WWTP is 2400 $m^3\ d^{-1}$ with an average COD content of 600 mg L^{-1}."

"AMBR type WWTP located in Hurma WWTP of Antalya city—Turkey is a compact system, consists of the fine screen (size 1 mm) and aerobic tank (133 m^3) with the membrane tank (17 m^3) having 660 m^2 membrane surface area."

"The raw wastewater fed to AMBR WWTP was obtained from the urban Hurma WWTP pretreatment system composed of 30 mm and 10 mm coarse and fine screens, aerated grid chamber and primary clarifier to remove rubbish, sand, oils and suspended solids."

"Full-scale AMBR designed with a permeate flow rate of 16.5 $m^3\ h^{-1}$ with the recycle flow rate of 5Q. A constant sludge withdrawal was performed from the aerobic tank to provide 18 d of SRT, with the MLSS concentration of 10,000 mg L^{-1}."

"Wastewater flow rates are quite different for Hurma AMBR and Elmalı district A^2O WWTPs."

Results and Discussion

"The Elmalı A^2O treatment plant designed for an influent flow rate of 2400 $m^3\ d^{-1}$ and a COD value of 600 mg L^{-1}, corresponding to a daily COD load of 1440 kgCOD d^{-1}."

"The observed difference essentially relates to the size distribution of the residual COD (Noyan and others 78) (i) Particulate (biomass) COD escape from the AMBR system using a membrane with a 30 Nm nominal pore size is not possible, while the average effluent TSS of the Elmalı plant operating with a traditional gravity settling is 32 mgTSS L^{-1}, which is basically responsible for more than 40 mg L^{-1} of the COD in the effluent."

"When the energy cost was considered, the use of AMBR in wastewater treatment seems like a sustainable alternative process in water-scarce areas due to its advantages as modularity, less qualified employee needs and high quality permeates."

Conclusion

"Observations on the effluent quality of the full-size treatment plants provided conclusive proof for the irrefutable advantage of the activated sludge system with membrane filtration, constituting a new era for the century-old activated sludge process discarding the traditional gravity settling."

"This system offers the best option for effluent recovery and reuse, with complete retention of particulate matter (biomass); bacteria and capture of the residual COD except a fraction of smaller size than the effective filtration size of the membrane."

"The desired membrane module can be coupled with any activated sludge configuration to also secure additional treatment such as nutrient removal or low sludge generation."

"In effluent reuse for irrigation, the decisive factor is definitely the quality of the water supply and especially the total dissolved solids level, which cannot be reduced with conventional MBR systems."

Microalgal-Bacterial Granular Sludge Process Outperformed Aerobic Granular Sludge Process in Municipal Wastewater Treatment with Less Carbon Dioxide Emissions [286]

This is a machine-generated summary of:

Guo, Dabin; Zhang, Xuechun; Shi, Yuting; Cui, Baihui; Fan, Jie; Ji, Bin; Yuan, Julin: Microalgal-bacterial granular sludge process outperformed aerobic granular sludge process in municipal wastewater treatment with less carbon dioxide emissions [286]

Published in: Environmental Science and Pollution Research (2020)

Link to original: https://doi.org/10.1007/s11356-020-11565-7

If you want to cite the papers, please refer to the original.

For technical reasons we could not place the page where the original quote is coming from.

Abstract-Summary

"The aerobic granular sludge (AGS) process and microalgal-bacterial granular sludge (MBGS) process were comparably applied for municipal wastewater treatment in sequencing batch reactors with a height to diameter ratio of eight."

"The nutrients removal performances by MBGS process were generally slightly better than AGS process."

"Mass balance calculation estimated that MBGS process appeared to achieve much less CO_2 emission (5.8%) compared with AGS process (44.4%)."

"It proved that MBGS process, with the merits of potentially low energy cost, limited CO_2 emission, and excellent performance, showed more prospects in municipal wastewater treatment than AGS process."

Introduction

"The development of urbanization and civilization was bound to produce a large amount of wastewater, so the nutrients removal and recovery (e.g., soluble organic matter, nitrogen, and phosphorus) in wastewater treatment and its byproducts of activated sludge have always been the focus and research direction in the wastewater treatment area (Azam and others 85; Pang and others 86)."

"The microalgal-bacterial granular sludge (MBGS) processes have been paid attention to wastewater treatment (Zhang and others 87)."

"It remains elusive the releasing amount of the main greenhouse gas of CO_2 that can be saved by using MBGS process compared with AGS process for wastewater treatment."

"This study aimed to compare AGS process with MBGS process for municipal wastewater treatment in a bioreactor with a height to diameter ratio of 8, which could reduce the footprint effectively."

"Results of this study may gain more knowledge on the emerging process in municipal wastewater treatment."

Material and Methods

"Chemical oxygen demand (COD), ammonia nitrogen (NH_4^+-N), and total phosphorus (TP) were about 320, 35, and 9 mg/L, supplied by CH_3COONa, NH_4Cl, and KH_2PO_4, respectively."

"The aerobic granules were divided into the two SBRs (R1 and R2) with the mixed liquid suspended solids (MLSS) were about 3.9 g/L. Dissolved oxygen (DO) concentrations were 4.0–5.0 mg/L and lower than 0.4 mg/L in aerobic and anaerobic stages, respectively."

"The attributes of aerobic granules and microalgal-bacterial granules on day 40 were measured by standard methods (APHA, 88), including 5 min sludge volume index (SVI_5), 2 min sludge volume index (SVI_2), MLSS, and MLVSS."

"The bacterial and algal communities were studied by Illumina MiSeq sequencing and microscopic examination, respectively."

"For bacterial community analysis, granule samples were collected on day 58."

Results and Discussion

"It was noteworthy that, unlike previous regular spherical granular sludge, the shape of the aerobic granules and microalgal-bacterial granules were strip and elliptical, respectively."

"The statistical results indicated that the average length and width of aerobic granules were 4.0 and 0.62 mm, which were 2.05 and 0.75 mm for microalgal-bacterial granules, respectively."

"The class Chloroflexi of aerobic granules was higher than that of microalgal-bacterial granules (10.3, 1.9%)."

"Acidobacteria contributed about 8.0% and Bacteroidetes accounted for about 6.2% of the bacterial abundance, respectively, which likely caused the unique strip-shaped aerobic granules (4.0 mm $\times$ 0.62 mm on average)."

"Future studies should focus on novel processes based on microalgal-bacterial granules for various kinds of wastewater treatment with low or no aeration towards sustainable wastewater treatment (Ji and others 89; Ji and others 90)."

Conclusions

"The AGS process and MBGS process were comparably studied for municipal wastewater treatment in two sequencing batch reactors."

"It proved that the MBGS process had more advantages in terms with abundant microbial community, better nutrients removal, potential less energy cost, and less CO_2 emission than AGS process."

"It implied that the MBGS process showed much more prospects than AGS process towards environmentally sustainable wastewater treatment and reclamation."

Bioaugmentation and Enhanced Formation of Microbial Granules Used in Aerobic Wastewater Treatment [287]

This is a machine-generated summary of:

Ivanov, Volodymyr; Wang, Xiao-Hui; Tay, Stephen Tiong-Lee; Tay, Joo-Hwa: Bioaugmentation and enhanced formation of microbial granules used in aerobic wastewater treatment [287]

Published in: Applied Microbiology and Biotechnology (2005)

Link to original: https://doi.org/10.1007/s00253-005-0088-5

If you want to cite the papers, please refer to the original.

For technical reasons we could not place the page where the original quote is coming from.

Abstract-Summary

"Microbial aggregates of an aerobic granular sludge can be used for the treatment of industrial or municipal wastewater, but their formation from a microbial activated sludge requires several weeks."

"The aim of this research was the selection of microbial cultures to shorten the granule-forming period from several weeks to a few days."

"An enrichment culture with the ability to accelerate granulation was obtained by repeating the selection and batch cultivation of fast-settling microbial aggregates isolated from the aerobic granular sludge."

"A mixture of these strains with the activated sludge was used as an inoculum in an experimental sequencing batch reactor to start up an aerobic granulation process."

Introduction

"Since Liu and Tay (73) study on the application of an aerobic granular sludge, research has focused on the performance of an aerobic granular sludge wastewater treatment and the microbiology of microbial granules."

"The primary aim of the formation of strong microbial granules and their application in industrial and municipal wastewater treatment is to avoid the construction of secondary settling tanks or to diminish their size."

"The formation of aerobic granules is very crucial for their applicability in wastewater treatment while this process takes several weeks to start up an aerobic granular system from a conventional activated sludge (Morgenroth and others 91; Beun and others 92; Tay and others 93; Malik and others 94)."

"By an analogy with other wastewater treatment systems, the formation of granules can be enhanced by selected microbial cultures (Beun and others 95)."

Materials and Methods

"The control process in reactor R1 was started with 100 ml of the activated sludge with a mixed liquor volatile suspended solid (MLVSS) concentration of 3,000 mg l^{-1}."

"The microbial granules were collected from different laboratory SBRs and were dispersed by beating 200 mg (wet weight) of granules in a 2-ml tube without glass beads for 3 min at 420 strokes min^{-1} in a Mini-Beadbeater (Biospec, Bartlesville, OK, USA)."

"1 ml of cell suspension and aggregates from the bottom of the conical tube was transferred into a fresh medium and incubated for 48 h at 25°C on a shaker operating at 150 rpm."

"Starting from the second transfer, the microbial aggregates were collected and washed five times in PBS using 15-ml conical tubes."

"The microbial aggregates of the enrichment culture after the 13th transfer were dispersed in PBS by beating 200 mg (wet weight) of aggregates in a 2-ml tube for 3 min at 420 strokes min^{-1}."

Results

"Microbial strains B and F had AIs of 65 and 51%, respectively."

"It is shown that the addition of strains B and F with higher aggregation abilities than that of the activate sludge diminished the duration of the granulation process from several weeks to 8 days."

"The cultivation of cells from the dispersed granules on a solid medium showed that $12 \pm 3\%$ of colonies from the granules formed during the 10 days of cultivation in the experimental reactor (R2) were colonies of Klebsiella pneumoniae strain B and that $40 \pm 6\%$ colonies were colonies of Pseudomonas veronii strain F. The concentrations of K. pneumoniae and P. veronii were both 36 mg l^{-1} at the beginning of cultivation."

"It indicated that the added strains B and F were dominant cultures in formed aerobic granules."

"Full 16S rRNA gene sequences were obtained for microbial strains B and F, respectively."

Discussion

"Cells of K. pneumoniae strain B and P. veronii strain F could either form aggregates by themselves or coaggregate with other microbial strains of the activated sludge."

"The cultivation of cells from the dispersed granules on the solid medium showed that $12 \pm 3\%$ of colonies from the granules formed during 10 days of cultivation in the experimental reactor (R2) were colonies of K. pneumonia strain B and $40 \pm 6\%$ colonies were colonies of P. veronii strain F. The concentrations of K. pneumoniae and P. veronii were both 36 mg l^{-1} at the beginning of cultivation."

"P. veronii strain F was selected for further large-scale trials as a starter culture for the facilitated formation of microbial granules in aerobic wastewater treatment."

"Another problem of the bioaugmentation of microbial granules with a selected strain is that not only the duration of granulation but also the specific activities of granules—such as nitrification, spectra of degradation of natural organic compounds and xenobiotics, accumulation of phosphate, cell survivability, and others properties—are important in wastewater treatment."

Achieving High-Rate Partial Nitritation with Aerobic Granular Sludge at Low Temperatures [288]

This is a machine-generated summary of:

Liu, Wenru; Shen, Yaoliang; Yang, Dianhai: Achieving high-rate partial nitritation with aerobic granular sludge at low temperatures [288]

Published in: Biodegradation (2021)

Link to original: https://doi.org/10.1007/s10532-021-09965-8

If you want to cite the papers, please refer to the original.

For technical reasons we could not place the page where the original quote is coming from.

Abstract-Summary

"Partial nitritation is necessary for the implementation of the mainstream anammox (anaerobic ammonium oxidation) process in wastewater treatment plants."

"The difficulty in outcompeting nitrite-oxidizing bacteria (NOB) at mainstream conditions hinders the performance of partial nitritation."

"The present work aimed to develop a high-rate partial nitritation process for low-ammonium wastewater treatment at low temperatures by seeding aerobic granules."

"With the hydraulic retention time progressively shortened from 1.0 to 0.17 h, the influent nitrogen loading rate of the partial nitritation process reached 6.8 ± 0.4 kg N m^{-3} d^{-1} even at 10–15 °C."

"The high concentration (7.5 gVSS L^{-1}) and activity (0.48 g N g^{-1} VSS d^{-1} at 11 °C) of granular sludge made the reactor possess an overcapacity evaluated by the ratio between the actual ammonium oxidation rate of the granules and their maximum potential."

"This work indicated the great potential of applying aerobic granules to achieve high-rate partial nitritation at mainstream conditions."

Introduction

"Partial nitritation can be readily achieved and stably maintained in a segregated reactor with high-rate performance and also without compromising the activity of anammox bacteria (Isanta and others 96; Bian and others 97; Kouba and others 98)."

"Advantages of the DO/ammonium ratio control strategy in fast start-up and long-term maintenance of partial nitritation with biofilm or granules have been demonstrated at mainstream conditions (Isanta and others 96; Poot and others 99; Bian and others 97)."

"The biofilm or granular sludge partial nitritation processes were generally run under high DO concentrations (> 1 mg O_2 L^{-1}) and also with long SRTs, thereby high-rate nitritation could be achieved."

"The deterioration process of the high-rate partial nitritation was investigated in that study, the fast achievement and the stable maintenance of a high-loaded granular partial nitritation process by seeding conventional aerobic granular sludge remains unclear."

"The aim was to achieve a fast initiation of partial nitritation process and maintain its high-rate performance."

Materials and Methods

"The seeding granular sludge was obtained from a continuous aerated reactor, which performed full nitrification (nitrifying the influent ammonium to nitrate) with DO concentration below 1.0 mg O_2 L^{-1} and influent nitrogen loading rate (NLR) of 0.8 kg N m^{-3} d^{-1}."

"Of fluorescence in-situ hybridization (FISH), in-situ batch-tests were performed in the reactor to investigate the extent of stratification of nitrifiers in the granular sludge, as promoted by Soler-Jofra and others (100)."

"During each batch test, DO concentration, temperature, and pH were measured online and controlled according to the normal operating values of the reactor."

"Ex-situ batch-tests were also conducted in a reactor with a working volume of 0.5 L to measure the potential maximum specific ammonium oxidation rate ($SAOR_m$) and maximum specific nitrite oxidation rate ($SNOR_m$) of the granular sludge."

"Representative biomass samples were obtained from the granular sludge reactor on day 130 to assess the microbial community through high-throughput sequencing."

Results and Discussion

"By previous studies (Picioreanu and others 101; Poot and others 99; Soler-Jofra and others 100), a good stratification of nitrifiers in the granules, with AOB dominated the outer layers and NOB in the inner layers, was a precondition for achieving partial nitritation based on the DO/ammonium ratio control."

"As abovementioned, owing to the low biomass concentration and the adverse decrease of AOB activity under some conditions (e.g., aggressive SRTs, low DO), only limited NLRs (below 0.3 kg $N \cdot m^{-3} \cdot d^{-1}$) were obtained in most flocculated sludge partial nitritation processes (Kouba and others 98; Wang and others 102)."

"This also represented an indication that the partial nitritation granular sludge reactor had developed a high degree of functional redundancy or overcapacity during the operation under relatively high temperatures."

"High NLRs (2.4–6.8 $kgN \cdot m^{-3} \cdot d^{-1}$) of partial nitritation were achieved in the granular sludge reactor treating low-strength wastewater (around 55 mg NH_4^+-$N \cdot L^{-1}$) at realistic operating temperatures."

"The growth of anammox bacteria in partial nitritation granules was also reported in previous studies (Wang and others 103; Qian and others 104; Liu and others 105)."

Conclusions

"Stable partial nitritation with high influent nitrogen loading rates (2.4–6.8) was achieved in a granular sludge reactor treating low-strength wastewater at 10–25 °C."

"The HRT of the partial nitritation reactor could be shortened to 0.4–0.17 h. The overcapacity developed in the granular reactor with the high biomass concentration and metabolic activity was supposed to be the reasons helping to face the adverse impacts of the temperature decrease."

"The developed stratification of nitrifiers with AOB predominated the granular surface combined with sufficient ammonium in the bulk liquid played an important role in repressing NOB."

"Future studies are needed to determine the potential of enriching anammox bacteria in partial nitritation granules and the presence and contribution of AOA during the long-time operation at low temperatures."

3.4 Anaerobic

Introduction by Editor

Anaerobic digestion is a biological process that occurs in the absence of oxygen, involving the breakdown of organic matter by anaerobic microorganisms. These microorganisms undergo a series of biochemical reactions to convert complex organic compounds into simpler forms, primarily methane (CH_4) and carbon dioxide (CO_2). Anaerobic digestion takes place within dedicated vessels, such as anaerobic digesters, which provide the necessary conditions for the growth and activity of anaerobic microorganisms.

The successful implementation of anaerobic digestion necessitates careful attention to operational parameters and conditions. Factors such as temperature, pH, hydraulic retention time, and organic loading rate significantly impact the performance of anaerobic digesters. Maintaining optimal operating conditions, such as mesophilic temperatures (around 35–40 °C) or thermophilic temperatures (around 50–60 °C), neutral pH levels, and appropriate hydraulic retention times, is crucial for promoting the growth and activity of anaerobic microorganisms.

Various process configurations are employed for anaerobic digestion in municipal wastewater treatment. The complete-mix digester is the most commonly used configuration, where wastewater and sludge are continuously mixed to ensure uniform conditions throughout the digester. Another configuration is the plug-flow digester, which allows for the sequential movement of wastewater and sludge, creating distinct zones for different stages of anaerobic digestion.

Anaerobic digestion plays a significant role in municipal wastewater treatment, offering several benefits. One key benefit is the recovery of biogas, primarily composed of methane. This biogas can be harnessed as a renewable energy source for the generation of heat and electricity. Utilizing biogas promotes energy self-sufficiency and reduces dependence on fossil fuels. Additionally, anaerobic digestion aids in the stabilization of sludge, reducing its volume and generating a nutrient-rich material that can be used as a valuable soil amendment.

Machine generated keywords: anaerobic, digestion, anaerobic digestion, biogas, reactor, upflow anaerobic, upflow, methane, anaerobic sludge, digestion process, blanket, sludge blanket, ammonia, hrt, bed.

The Challenge of Making Wastewater Treatment Plants Composed by Anaerobic Reactors Capable of Removing Nitrogen [289]

This is a machine-generated summary of:

Magalhães, Taína Martins; Duarte, Natalia Cangussu; de Alencar Neves, Thiago; Bueno, Daniel Augusto Camargo; Delforno, Tiago Palladino; de Oliveira, Valéria

Maia; Tonetti, Adriano Luiz: The Challenge of Making Wastewater Treatment Plants Composed by Anaerobic Reactors Capable of Removing Nitrogen [289]

Published in: Water, Air, & Soil Pollution (2019)

Link to original: https://doi.org/10.1007/s11270-019-4300-0

If you want to cite the papers, please refer to the original.

For technical reasons we could not place the page where the original quote is coming from.

Abstract-Summary

"In this research, an anaerobic filter (AF) and an upflow anaerobic sludge blanket (UASB) reactors were fed with synthetic effluent enriched with nitrate to compare how these reactors would behave if they became denitrifying reactors."

"With the application of 100.0 mg NO_3^--NL^{-1}, the AF presented better efficiency."

"With respect to the biogas production, the composition was significantly altered: from CH_4 and CO_2 concentrations close to 70% and 13% without NO_3N addition to N_2 concentration higher than 85% with addition of 100.0 mg NO_3^--NL^{-1}."

"The UASB hydrodynamic profile was modified due to an increase in the mixing behavior along the denitrification stages by biogas production."

Introduction

"To prevent nitrogen from being disposed into water bodies, techniques that combine the removal of nitrogen and carbonaceous matter in anaerobic reactors can be implemented for wastewater treatment."

"The upflow anaerobic sludge blanket reactor (UASB) and the anaerobic filter (AF) are anaerobic treatment systems commonly used for domestic sewage and industrial wastewater treatment."

"To understand the effect of the denitrification on overall behavior of the reactors, this study aimed at comparing the AF and UASB in terms of organic matter and nitrogen removal efficiencies using synthetic medium to simulate the recirculation of nitrified effluent to an anaerobic reactor in secondary treatment."

"Besides the evaluation of the denitrification efficiency, the research also sought to understand possible interferences of nitrogen removal process in the hydrodynamic behavior of the reactors, the characteristics and diversification of biomass, and the quality of the biogas to be used for energy generation."

Materials and Methods

"The reactors were operated simultaneously at the same theoretical hydraulic retention time (HRT) of 12 h. The superficial velocity was 66.67 mm h^{-1} for both reactors and the feed flow rate of AF and UASB was 1.28 and 1.69 mL min^{-1}, respectively."

"As the same device for gas-liquid-solid separator was introduced in AF, the performance comparison between reactors was accomplished, despite the differences between attached growth in AF and suspended grow in UASB."

"In clean-bed condition (stage 1), the tracer tests were performed only with tap water for both reactors before starting operation with inoculum and synthetic water."

"The tracer test was performed to evaluate dirty-bed condition, in which reactors were in operation with synthetic wastewater (stages 2 to 6)."

"The sludge samples were collected at the bottom of the reactors after steady-state performance was established in stages 2 to 6."

"Sludge samples from both reactors were collected after steady-state performance was established in stage 2 and stage 5."

Results and Discussion

"The presence of the support media in AF also contributed to the maintenance of its similar hydrodynamic behavior throughout the stages, unlike the UASB reactor, which showed an increase in the mixing effect promoted by the biogas production, characterized by the reduction of the parameters HRT_{obs}, T_P, and N, and increase of MDI."

"In the UASB reactor, however, there is no difference between the U2 and U3 heights (83.3 ± 4.5%), which is the only region that differentiates it from the AF reactor in the removal of NO_3^--N. This may indicate that the presence of the biofilm in the region between F2 and F3 heights is fundamental to the improvement of the denitrification process."

"Although the microbiological analysis was not performed in the biofilm, it is possible to assume that it is formed by similar organisms to the UASB, which were fundamental for the denitrification process in the AF, thus complementing the removal promoted by the suspended sludge in this reactor."

Conclusions

"The comparison between the AF and UASB reactors in the simultaneous denitrification and methanogenesis process indicates a better performance of the AF in the removal of NO_3^--N and COD."

"During stage 6, however, when reactors were fed with 100.0 mgNO_3^--NL^{-1}, the predominant compound was N_2, corresponding to more than 85% of the biogas composition in the AF and UASB, thus reducing its energy potential."

"There was no significant difference in the composition of the biogas produced by AF and UASB."

"The greater change in hydrodynamic behavior in the UASB reactor is related to biogas production, which made the reactor behavior closer to a mixed flow."

"The presence of the biofilm in the AF reactor, with potential for NO_3^--N and COD removal, was the most important factor to promote greater denitrification and greater COD removal in this reactor than in UASB, under the same operating conditions."

Coupling Air Stripping Process and Anaerobic Digestion for the Treatment of Landfill Leachate: Organics Degradation and Cytotoxicity Evaluation [290]

This is a machine-generated summary of:

Zayen, Amal; Loukil, Slim; Bouallegui, Zouheir; Sayadi, Sami: Coupling air stripping process and anaerobic digestion for the treatment of landfill leachate: organics degradation and cytotoxicity evaluation [290]

Published in: Euro-Mediterranean Journal for Environmental Integration (2022)

Link to original: https://doi.org/10.1007/s41207-022-00301-x

If you want to cite the papers, please refer to the original.

For technical reasons we could not place the page where the original quote is coming from.

Abstract-Summary

"Optimization of this process, using an orthogonal central composite design, intended to reduce the ammonia concentration and also to fix it in a safe range for the subsequent anaerobic process."

"Optimization showed that, to remove 60% of ammonia, pH, air flow rate, and reaction time should be fixed at 10.8, 6 L min^{-1}, and 18 h, respectively."

"The air stripping process improved the anaerobic digestion (AD) of leachate in an upflow anaerobic fixed bed reactor."

"Cytotoxicity was significantly reduced subsequent to the air stripping and anaerobic process."

Introduction

"Researchers have used a number of anaerobic processes to treat leachate, including conventional digester (Bull et al. 106; Lin 107; Barrantes Leiva et al. 108), anaerobic fixed bed reactor (Ben Dhia et al. 109), sequencing batch reactor (Uygur and Kargi 110; Contrera et al. 111), upflow anaerobic sludge blanket reactor (Kennedy et al. 2000; Sun et al. 112; Abood et al. 113), hybrid bed filter (Nedwell and Reynolds 114), and anaerobic membrane bioreactor (Zayen et al. 115; Xue et al. 116)."

"A wide variety of substances have been reported to be inhibitory to the anaerobic digestion process (Chen et al. 117)."

"Among a wide range of processes, air stripping is one of the most employed process at the commercial scale for the removal of NH_3–N from municipal wastewater, AD supernatant, industrial wastewater, and LFL (Yellezuome et al. 118)."

"It is a relatively easy and low-cost process that permits a certain level of suspended solids and generates no extra sludge or secondary pollutant (Zhang et al. 119)."

"During the air stripping process, NH_3–N is transferred from the waste stream into the air and is then absorbed from the air into a strong acid such as sulfuric acid or directly fluxed into the ambient air (Guo et al. 120; Abood et al. 121; Liu et al. 122)."

Materials and Methods

"The stripping unit for ammonia removal consisted of a Plexiglas column (diameter 20 cm, height 88 cm) filled with plastic carriers (Type HIFLOWE with a specific area of 200 m^2/m^3) offering a large area of contact."

"Leachate was continuously recirculated through the column from top to bottom at a rate of 10 L h^{-1}, while air was blown from the bottom."

"The batch reactors were 500-mL glass vials filled with 15 mL of substrate (raw or air-stripped LFL) and 100 mL of inoculum."

"The cells were harvested by trypsinization, seeded in 96 well-plates at 3000 cells per well in 100 μL of medium, and allowed to attach for 24 h. Once cells were attached, 100 μL of landfill leachate (raw, pretreated, or anaerobically treated), diluted in Roswell Park Memorial Institute fluid (RPMI 1640) medium, was added to obtain final concentrations of 0.1%, 1%, 5%, 10%, 20%, 30%, and 40%."

Results and Discussion

"A relatively high ammonia concentration in LFL has been treated by an air stripping process with an air flow rate of 60 L min^{-1} for the bench test, achieving 98% removal of NH_3^+–N at an HRT of 4–9 days (dos Santos et al. 123)."

"The objective was not to obtain the maximum removal but to reduce the ammonia concentration in the raw leachate and to keep the C/N ratio in a range safe for the subsequent anaerobic process."

"While biogas production was clearly slowed down during the first days in the control test, it rapidly increased with the stripped leachate, showing that methanogenic bacteria were less affected under moderate ammonia concentrations."

"Results obtained during batch anaerobic test showed that the air stripping process is effective to improve leachate digestibility."

"By increasing the OLR to 1.3 g COD L^{-1} day^{-1}, UAFB reactor showed better performances regarding COD removal efficiency and biogas production as microbial consortium became more adapted to the effluent."

Conclusions

"An air stripping process was developed to improve the anaerobic digestion of ammonium-rich leachate by effectively mitigating the ammonia inhibition."

"The increase of OLR above 1.25 g COD L^{-1} day^{-1} would be of a great interest to further evaluate the effect of the air-stripping pretreatment step on the subsequent AD process."

"Cytotoxicity was reduced subsequent to the air stripping and anaerobic digestion."

"These results indicate that the air stripping process combined with anaerobic digestion may be a convenient and efficient method for the treatment of landfill leachate."

Performance Evaluation and Kinetic Modeling of an Upflow Anaerobic Sludge Blanket Septic Tank for Domestic Wastewater Treatment [291]

This is a machine-generated summary of:

Santiago-Díaz, Ángel Luis; Mugica-Álvarez, Violeta; de los Cobos-Vasconcelos, Daniel; Vaca-Mier, Mabel; Salazar-Peláez, Mónica Liliana: Performance evaluation and kinetic modeling of an upflow anaerobic sludge blanket septic tank for domestic wastewater treatment [291]

Published in: Environmental Science and Pollution Research (2021)

Link to original: https://doi.org/10.1007/s11356-021-15141-5

If you want to cite the papers, please refer to the original.

For technical reasons we could not place the page where the original quote is coming from.

Abstract-Summary

"This work evaluated the UASB-septic tank performance using different kinetic models that correlated process efficiency and methane production with hydraulic and organic loading rates through experiments with five different HRT (48 h, 36 h, 24 h, 18 h, and 12 h) using synthetic domestic wastewater."

"Methane yield was 0.3294 L CH_4/g COD removed, being closer to the theoretical value, and the Van der Meer and Heertjes model had the highest R^2 for methane production."

"Pollutant removal markedly decreased when the reactor operated HRT below 24 h; thus, it is recommended to operate the UASB-septic tank at this HRT."

Introduction

"Anaerobic treatments are essential in the DEWATS approach, representing the most cost-effective technologies for domestic wastewater."

"Kinetic modeling is an indispensable tool to evaluate experimental results in terms of simple and convenient mathematical expressions that correlate diverse process variables, providing a rational basis for the design and control of anaerobic digestion, and offering a quantitative description of waste degradation rates and the operating parameters influencing these rates (Maleki and others 124)."

"This research will help design and analyze new systems and improve existing treatment units, positioning the UASB-septic tank as a reliable option for domestic wastewater management within the decentralized approach."

"Although these models differ by adopting different approaches to analyze and forecast the system operation, they can make accurate predictions of anaerobic reactor performance, considering their correlation coefficient (R^2) (Maleki and others 124)."

Materials and Methods

"Some of the important parameters were volatile solids (VS) 65.39 g VS/L, fixed solids (FS) 25.93 g FS/L, total solids (TS) 91.32 g TS/L, sludge volume index (SVI) 18.54 mL sludge/g TS, and settling velocity 5.04 m/h. Synthetic wastewater with domestic characteristics was employed to avoid organic load variations in the reactor influent during the experiment."

"M_{max} is the maximum specific biomass growth rate (g VS/g VS-d), and K_s is the half-velocity constant (g COD/L), which is the substrate concentration at one-half of the maximum specific substrate utilization rate."

"This model relates the substrate utilization rate and the organic loading rate through monomolecular kinetics in biofilm reactors (Stover and Kincannon 125)."

"This model estimates the methane production rate, using the Van der Meer and Heertjes kinetics constant (k_{sg}), the flow rate, and the mass of removed COD, as follows (Van Der Meer and Heertjes 126): where V_{CH4} is methane produced ($mLCH_4$/d)."

Results and Discussion

"This result implies that the UASB-septic tank can operate efficiently at 0.65 kg COD/m^3-d, which is acceptable in a low-rate reactor such as this one."

"The UASB-septic tank performance in the research is similar to other studies regarding modified septic tanks treating domestic or municipal wastewaters (Daija and others 127; Haydar and others 128; Santiago-Díaz and others 129; Singh and others 130), which reported percentage removals varying from 51–86% for COD and 72–95% for SS."

"The maximum substrate utilization rate (U_{max}) obtained in the present experiment falls in the interval from 0.47 to 109.7 g COD/L-d expected for anaerobic reactors (Maleki and others 124)."

"The Y value in this experiment is below the range of 0.05–0.1 reported in Metcalf and Eddy (131) for anaerobic reactors treating domestic wastewater and based on soluble COD, but k_d is over the range (0.02–0.04); while u_{max} and K_s are in the expected range."

Conclusions

"This study showed that the UASB-septic tank reached organic matter and solids removals of 45% for TS, 70% and 68% for total and soluble COD, and 85% for TSS, indicating its good performance and reliability; although, treatment efficiency markedly decreased when the reactor operated under HRT below 24 h. UASB-septic tank kinetic modeling revealed that the modified Stover-Kincannon model fitted better the experimental data, with an R^2 above 98% for linear regression."

"U_{max} obtained with the modified Stover-Kincannon model is higher than the maximum ORL (1.281 g COD/L-d) applied to the reactor, demonstrating its potential to treat domestic wastewater efficiently."

"Methane yield was 0.3294 L CH_4/g COD removed and is close to the theoretical value, and the Van der Meer and Heertjes model had a higher correlation coefficient ($R^2 = 0.979$) between the predicted and the experimental data."

Role of Waste-Based Geopolymer Spheres Addition for pH Control and Efficiency Enhancement of Anaerobic Digestion Process [292]

This is a machine-generated summary of:

Gameiro, Tânia; Novais, Rui M.; Correia, Catarina L.; Carvalheiras, João; Seabra, Maria P.; Tarelho, Luís A. C.; Labrincha, João A.; Capela, Isabel: Role of waste-based geopolymer spheres addition for pH control and efficiency enhancement of anaerobic digestion process [292]

Published in: Bioprocess and Biosystems Engineering (2021)

Link to original: https://doi.org/10.1007/s00449-021-02522-w

If you want to cite the papers, please refer to the original.

For technical reasons we could not place the page where the original quote is coming from.

Abstract-Summary

"An eco-friendly approach has been applied to control pH in anaerobic bioreactors, using waste-containing fly ash geopolymer spheres (GS) instead of powdered chemical compounds, to promote continuous alkalis leaching."

"Results showed that the use of GS with the highest concentration and porosity promoted an increase in methane yield up to 30%, compared to the assay with powdered chemical compounds addition."

"This innovative approach indicates that GS addition can regulate pH in anaerobic digesters treating challenging wastewaters and, simultaneously, improve not only its efficiency but also the sustainability of the entire process."

Introduction

"Due to the direct impacts on microbial activity and community diversity, proper control of the pH is a vital factor for the stability of the AD process."

"To control the acidification phase and keep the pH values in the optimum range for methanogenic microorganisms, alkaline chemicals need to be added to the biological process, such as sodium hydroxide (NaOH) or calcium carbonate ($CaCO_3$), among others."

"To better understand the role of porosity and the concentration of spheres in the pH buffering capacity of AD processes, high organic loads of an easy-acidifying substrate (cheese whey) were used."

"This research contributes to reducing the knowledge gap on the use of geopolymers as pH regulators in AD processes, further demonstrating that this innovative approach not only increases process efficiency, by enhancing CH_4 production but also, due to its simplicity, avoids the need for continuous feeding of commercial alkaline chemicals in sequential AD operating cycles."

Experimental

"Following preliminary studies, in which the concentration of chemical alkalinity or spheres to be added to the bioreactors was evaluated (data not shown for the sake of brevity), two different AnSBR assays were performed: the first assay with two AD cycles and a longer-term assay with four AD cycles under the best conditions obtained in the previous test."

"For the first assay with two AD cycles, four AnSBR reactors were used, one of which was carried out with the addition of a buffer solution with chemical reagents ($NaHCO_3$ and $KHCO_3$), to promote pH regulation, starting with an initial concentration of 4 g/L of alkalinity measured as $CaCO_3$."

"The reactors were then incubated until they reached again a COD concentration below 1 gO_2/L. For the second experiment with four AD cycles, four sequential substrate additions were performed on days 0, 47, 78 and 103 of incubation."

Results and Discussion

"Comparing the pH values of the reactor with chemical alkalinity addition, with those obtained for reactors with 16 g/L of GS (LPGS and HPGS), it is clear that, in reactors with GS addition, the optimum pH range for the methanogenic activity was maintained for longer periods (pH stabilization period between days 30 to 55—1st AD cycle—and between days 65 to 80—2nd AD cycle)."

"To the performance advantages (e.g. ~ twofold increase in methane volume production in a two-cycle AnSBR) in comparison with the use of commercial powdered alkaline chemicals, GS also shows a crucial technical advantage as it can be recovered after the AD process, since the composition of the aluminosilicate gel has not been seriously altered after use for such a long time and adverse environment (more than 120 days and facing relatively low pH values)."

Conclusions

"The feasibility of using geopolymer spheres to control pH in AD systems with multiple AD treatment cycles (AnSBR-type system) was demonstrated for the first time."

"Designing the spheres porosity and tuning their load in the reactor can boost the methane volume generation (up to almost double in a 2-cycle AnSBR system) in comparison with the use of powdered commercial chemicals, promoting the stability of the reactor within a pH range which favours methanogenesis."

"The proposed strategy upcycles a waste material (fly ash from biomass combustion) to produce a bulk-type (not powder) pH regulator, with environmental and performance advantages over commercial powdered alkaline chemicals (produced with virgin resources)."

Performance of Staged and Non-staged Up-Flow Anaerobic Sludge Bed (USSB and UASB) Reactors Treating Low Strength Complex Wastewater [293]

This is a machine-generated summary of:

Sevilla-Espinosa, Susana; Solórzano-Campo, Maricela; Bello-Mendoza, Ricardo: Performance of staged and non-staged up-flow anaerobic sludge bed (USSB and UASB) reactors treating low strength complex wastewater [293]

Published in: Biodegradation (2010)

Link to original: https://doi.org/10.1007/s10532-010-9339-y

If you want to cite the papers, please refer to the original.

For technical reasons we could not place the page where the original quote is coming from.

Abstract-Summary

"This paper reports on the comparative study carried out with two high-rate anaerobic reactor systems with the objective of evaluating their performances when used for the treatment of low-strength, complex wastewater."

"The other is the up-flow staged sludge bed (USSB) system in which the reactor was divided longitudinally into 3, 5 and 7 compartments by the use of baffles."

"Intermediate flow hydraulics, between plug-flow and completely-mixed, in the UASB and 7 stages USSB reactors allowed efficient degradation of substrates with minimum effluent concentrations."

"Low number of compartments in the USSB reactors increased the levels of short-circuiting thus reducing substrate removal efficiencies."

"All reactors showed high COD removal efficiencies (93–98%) and thus can be regarded as suitable for the treatment of low strength, complex wastewater."

"Staged anaerobic reactors can be a good alternative for this kind of application provided they are fitted with a large enough (≥ 7) number of compartments to fully take advantage of their strengths."

Introduction

"Since organic matter concentration is low and influent flow-rates can vary widely, reactor design is based on hydraulic rather than on organic load and the reactor needs high biomass retention (Lettinga and Hul shoff Pol 132)."

"Due to these reasons and, in some cases, because of scale factors, COD removal efficiencies observed in up-flow anaerobic sludge bed (UASB) reactors have not exceeded 60–85% over a wide range of temperatures (10–28°C) and hydraulic retention times (HRT, 2–48 h) (Behling and others 133; Kalogo and Verstraete 134; Lew and others 135; Ruiz and others 136; Singh and Viraraghavan 137)."

"The treatment of a dilute soluble and colloidal waste using anaerobic baffled reactors (ABR) with more than 80% COD removal efficiencies (35°C and 6–80 h HRT) has been reported (Langenhoff and others 138)."

"This paper reports on the comparative study carried out with a UASB, used as reference, and 3, 5 and 7 stages up-flow sludge bed (USSB) reactors with the objective of evaluating their performance when used for the treatment of a low-strength, complex waste."

Materials and Methods

"Several parameters were measured in the feed, the effluent and, eventually, in the compartments of the reactors for which liquid and sludge samples were taken through the installed sampling ports."

"For routinely assessment of reactors performance, total and soluble COD were measured by the closed-reflux colorimetric method (APHA 139)."

"Residence time distribution (RTD) studies were carried out to analyse the hydrodynamics of each reactor by adding a pulse of inert tracer to the inlet (5 g NaCl in 50 ml water)."

"The tests were performed in 120 ml serum bottles filled with 40 ml of a phosphate buffer (0.1 M Na and K) spiked with sodium acetate (3 g COD/l)."

"20 successive data points for each reactor (effluent COD, turbidity and volatile acids) were selected from the whole data set obtained near the end of the run (steady state) and were statistically analysed."

Results and Discussion

"To total COD, all reactors showed significantly different levels of turbidity in the effluents and the values of this parameter decreased with the increase in the number of compartments except for the UASB which showed the lowest level of turbidity."

"Effluent volatile acids too showed the same trend of change with the number of compartments and, as COD, the minimum level of volatile acids was observed in the USSB7 reactor."

"Soluble COD near the inlet port of the UASB reactor was the lowest of all and, similarly to the USSB7, it decreased, initially at a fast rate then slowly, until a value of 12 mg COD/l was attained."

"AMA increased with the number of compartments in the USSB reactors but the sludge from the UASB showed the highest value of all."

Conclusions

"Up-flow staged sludge bed reactors have been shown here to be a good alternative for the anaerobic treatment of complex low-strength wastewater because: (i) plug-flow reactor hydraulics are approached which more effectively remove substrate than a completely-mixed reactor under Monod or first-order biological kinetics, especially at low substrate concentrations, (ii) solids are effectively entrapped and removed in the sludge bed and (iii) phase separation occurs promoting the efficient degradation of substrates."

"A low number of compartments (3) increased the levels of short-circuiting thus reducing substrate removal efficiencies."

"Flow hydraulics in the UASB and 7 stages USSB reactors were similar, intermediate between plug-flow and perfectly mixed with deviation from ideal behaviour due to by-passing and dead spaces, which resulted in comparable reactor performance."

"The UASB reactor did not show good mixing conditions due to low loading rates, low biogas production, reactor geometry and small scale but this might change at an increased scale."

Energy Recovery from Wastewater Treatment Plants Through Sludge Anaerobic Digestion: Effect of Low-Organic-Content Sludge [294]

This is a machine-generated summary of:

Zhang, Yuyao; Li, Huan: Energy recovery from wastewater treatment plants through sludge anaerobic digestion: effect of low-organic-content sludge [294]

Published in: Environmental Science and Pollution Research (2017)

Link to original: https://doi.org/10.1007/s11356-017-0184-y

If you want to cite the papers, please refer to the original.

For technical reasons we could not place the page where the original quote is coming from.

Abstract-Summary

"During anaerobic digestion, low-organic-content sludge sometimes is used as feedstock, resulting in deteriorated digestion performance."

"To investigate the feature of low-organic-content sludge digestion and explain its intrinsic mechanism, batch experiments were conducted using designed feedstock having volatile solids (VS) contents that were 30–64% of total solids (TS)."

"The oligotrophic environment resulting from low-organic-content substrates led to decreased microbial activity and a switch from butyric fermentation to propionic fermentation."

"The results observed here corroborated with data from published literature on anaerobic digestion of low-organic-content sludge and showed that it may not be feasible to recover energy from sludge with an organic content lower than 50% through mono digestion."

Introduction

"Since organic degradation rates have an approximate linear relation with the VS/TS ratios in feed sludge during anaerobic digestion (Liao and Li 140), the corresponding biogas potential of low-organic-content sludge should be much lower than the common values (Appels and others 141)."

"One study reported that the organic removal rate decreased to 30% and the biogas production declined to lower than 0.4 m^3/kg VS added when using this kind of sludge as the feedstock of anaerobic digestion (Duan and others 142)."

"The traditional operational experience is not applicable to those anaerobic digestion processes treating low-organic-content sludge."

"Some researchers have evaluated the biogas potential from sludge anaerobic digestion but they made the assumption that sludge with different organic contents would have the same organic removal rates (Cao and Paw Owski 143)."

"The results were used to characterize anaerobic digestion using low-organic-content sludge as feedstock, investigate the quantitative relation between biogas potential and sludge organic content, and analyze the intrinsic mechanism determining the performance."

Materials and Methods

"The collected sludge was kept at 4 °C, which would be diluted and then used as the raw substrate in the experiments."

"Considering that sludge organic content and composition both determine sludge digestibility, a series of sludge samples (substrate) with identical organic composition but gradient organic contents were designed."

"By excluding the interference of sludge organic composition, only the effect of sludge organic content was exhibited during sludge anaerobic digestion."

"All the bottles were placed in a water bath (DKZ, YiHeng Instrument, Shanghai, China) to keep a constant temperature of 35 ± 2 °C."

"The temperature of the detector was 60 °C, and helium was used as the carrier gas at 220 ml/min and 120–125 kPa."

"To detect the organic composition of the feed sludge, polysaccharide, protein, and lipid contents were measured using a phenol-sulfuric acid colorimetric method (Dubois and others 144), the Lowry method (Lowry and others 145), and the Soxhlet method (Rice and others 146), respectively."

Results and Discussion

"The high concentration of VFAs remaining in the digestate indicated their low conversion rate to biogas, which was consistent with the low biogas yield in the group W1."

"The R^2 (> 0.994) and root-mean-square error (RMSE) values (< 0.1) indicated that the model can describe the biogas production of all the groups very well, and shows that the digestion efficiency increased with the increasing organic content in substrate."

"This digestion time is acceptable for batch or continuous anaerobic digestion of sewage sludge; however, the low VS content in substrate and the corresponding low biogas yield possibly decrease the potential benefits of anaerobic digestion."

"Because no reports were found on anaerobic digestion of sludge with a VS/TS ratio of 40%, it was reasonably assumed that such sludge would have a biogas yield lower than 340 ml/g VS, which is the critical value for sludge of this organic content and implies that this sludge cannot be used for bioenergy recovery."

Conclusions

"Characteristics of low-organic-content sludge digestion and their intrinsic variations were investigated using batch experiments with specially designed feedstock."

"In the batch experiments using low-organic-content substrate, propionic fermentation competed over butyric fermentation."

"A first-order model described the biogas production of low-organic-content sludge very well, and the degradation coefficient k exhibited a positive relation with organic content in the substrate."

Treatment of Domestic Wastewater in an Up-Flow Anaerobic Sludge Blanket Reactor Followed by Moving Bed Biofilm Reactor [295]

This is a machine-generated summary of:

Tawfik, A.; El-Gohary, F.; Temmink, H.: Treatment of domestic wastewater in an up-flow anaerobic sludge blanket reactor followed by moving bed biofilm reactor [295]

Published in: Bioprocess and Biosystems Engineering (2009)

Link to original: https://doi.org/10.1007/s00449-009-0321-1

If you want to cite the papers, please refer to the original.

For technical reasons we could not place the page where the original quote is coming from.

Abstract-Summary

"The entire treatment system was operated at different hydraulic retention times (HRT's) of 13.3, 10 and 5.0 h. An overall reduction of 80–86% for COD_{total}; 51–73% for $COD_{colloidal}$ and 20–55% for $COD_{soluble}$ was found at a total HRT of 5–10 h, respectively."

"By prolonging the HRT to 13.3 h, the removal efficiencies of COD_{total}, $COD_{colloidal}$ and $COD_{soluble}$ increased up to 92, 89 and 80%, respectively."

"The removal efficiency of $COD_{suspended}$ in the combined system remained unaffected when increasing the total HRT from 5 to 10 h and from 10 to 13.3 h. This indicates that, the removal of $COD_{suspended}$ was independent on the imposed HRT."

"Ammonia-nitrogen removal in MBBR treating UASB reactor effluent was significantly influenced by organic loading rate (OLR)."

"The removal efficiency was decreased by a value of 34 and 43% at a higher OLR's of 7.4 and 17.8 g COD m^{-2} day^{-1}, respectively."

"The mean value of the net sludge yield was only 6% in UASB reactor and 7% in the MBBR of the total influent COD at a total HRT of 13.3 h. Accordingly, the use of the combined UASB–MBBR system for sewage treatment is recommended at a total HRT of 13.3 h."

Introduction

"These reactors remain robust high-rate treatment systems, generally without moving mechanical parts, limiting both capital and operating costs [148]."

"The performance of the UASB reactors is affected by operational conditions, one of which is the hydraulic retention time (HRT)."

"Castillo and others [149] investigated the effect of different HRT's on a pilot-scale UASB reactor (750 l), fed with domestic wastewater (COD inf."

"There has been a tendency for this to become constant at a HRT more than 6 h. The reactor achieved 66% for COD removal at an HRT of 8 h. In another study, A'lvarez and others [150] investigated the performance of an UASB reactor treating domestic wastewater at an HRT of 11 h and a temperature of 14 °C."

"The results obtained from lab sale and full sale units suggest useful application of moving bed biofilm reactors (MBBR) systems for aerobic post-treatment [151, 152]."

Materials and Methods

"The reactor had a working volume of 10 l and a height of 1.35 m. Eight ports for obtaining sludge samples are arranged along the reactor height, the first one at 0.1 m above the base of the column."

"The UASB reactor was inoculated with 6 l digested sludge."

"The reactor was filled with 1158 carrier media."

"Representative samples of colonized carrier media were taken from the reactor three times in each run."

"The concentration of biomass is expressed as g VSS l^{-1} media to be able to calculate the sludge residence time (SRT)."

"The sludge residence time (SRT) of the UASB and MBBR was calculated according to the following equation; where: V, reactor volume; X, average biomass concentration of the reactor (mg $VSSl^{-1}$); Q_w, excess sludge (l day^{-1}); X_w, concentration of the excess sludge (mg $VSSl^{-1}$); Q wastewater flow rate (l day^{-1}); X_e effluent concentration (mg $VSSl^{-1}$) and according to Zeeuw [153] $X_e = COD_{suspended}/1.4$."

Results and Discussion

"Approximately, the same result, at a total HRT of 13.3 h, was achieved in the MBBR system treating chemically pretreated sewage [154] at shorter HRT of 8.0 h. Residual COD values at a total HRT of 10 and 5.0 h were 95 and 142 mg l^{-1} for COD_{total}, 65 and 97 mg l^{-1} for $COD_{soluble}$, respectively."

"An almost complete removal of $COD_{colloidal}$ was achieved in the MBBR, i.e. only 6, 12 and 17 mg l^{-1} remained in the final effluent when operated at HRT's of 5.3, 4 and 2 h, respectively."

"At OLR of 4.6, 7.4, and 17.8 g COD m^{-2} day^{-1}, ammonia was removed by a value of 62, 28 and 19%, while at the same time 11, 4.4 and 0.3 mg l^{-1} of nitrate were, respectively produced."

Discussion

"The results obtained in this study indicated that the combined system consisting of UASB–MBBR system treating domestic wastewater at a total HRT of 13.3 h is very effective for removal of COD fractions, ammonia and FC."

"Kim and others [155] reported a similar COD removal value (92%) for aerobic filter in combination with UASB reactor at a total HRT of 13.5 h (8 h for UASB + 5.5 h for aerobic filter)."

"The integrated system achieved a removal efficiency of COD (85–93%) at a total HRT of 7.9 h (4.0 h UASB + 3.9 h aerobic reactor)."

"Similar results has been achieved by Seghezzo and others [156] who found that the excess sludge from the UASB reactor treating pre-settled sewage at an HRT of 6.3 h was 0.18 kg sludge kg COD $removed^{-1}$ day^{-1}."

Conclusions

"COD_{total} removal in the combined UASB–MBBR is significantly influenced by the HRT and OLR."

"The overall removal efficiencies in this study were 92, 86 and 80% at HRT's of 13.3, 10 and 5.0 h, respectively."

"As the overall HRT increased, faecal coliform (FC) removal increased."

"The mean overall log_{10} reduction was 2.3 and 1.4 at total HRT of 13.3 and 10 h, respectively."

"In view of these results, we recommend to use a combined UASB–MBBR system for sewage treatment at an HRT of 8 and 5.3 h, respectively."

Simultaneous Addition of Zero-Valent Iron and Activated Carbon on Enhanced Mesophilic Anaerobic Digestion of Waste-Activated Sludge [296]

This is a machine-generated summary of:

Wang, Tongyu; Qin, Yujie; Cao, Yan; Han, Bin; Ren, Junyi: Simultaneous addition of zero-valent iron and activated carbon on enhanced mesophilic anaerobic digestion of waste-activated sludge [296]

Published in: Environmental Science and Pollution Research (2017)

Link to original: https://doi.org/10.1007/s11356-017-9859-7

If you want to cite the papers, please refer to the original.

For technical reasons we could not place the page where the original quote is coming from.

Abstract-Summary

"The performance of biogas generation and sludge degradation was studied under different zero-valent iron/activated carbon (ZVI/AC) ratios in detail in mesophilic anaerobic digestion of sludge."

"A good enhancement of methane production was obtained at the 10:1 ZVI/AC ratio, and the cumulative methane production was 132.1 mL/g VS, 37.6% higher than the blank."

"The methane content at the 10:1 ZVI/AC ratio reached 68.8%, which was higher than the blank (55.2%) and the sludge-added AC alone (59.6%)."

"The concentrations of available nitrogen (AN) and available potassium (AK) increased after the anaerobic digestion process and showed a gradually decreasing trend with increasing ZVI/AC ratio."

Introduction

"Several researches have been conducted to improve the methane production process, such as adding zero-valent iron (ZVI) or activated carbon (AC) in the anaerobic digestion process."

"Addition of charcoal or AC to anaerobic digestion of different substrates (e.g., cattle dung, paper sludge mixed with wheat husks, swine manure) has been used to cause an improvement in biogas yield (Desai and Madamwar 157; Mumme and others 158)."

"Hydrothermal biochar could increase methane yield by 32.0%, and pyrolysis biochar could prevent mild ammonia inhibition (2.1 g per kg total ammonium nitrogen) in anaerobic digestion of paper sludge mixed with wheat husks (Mumme and others 158)."

"The main objective of this study was to investigate the performance of iron–carbon micro-electrolysis system on methane production and sludge decomposition in the anaerobic digestion process."

"An anaerobic digestion experiment using sludge samples treated with different ZVI/AC ratios was conducted."

Materials and Methods

"Before the digestion, the dewatered sewage sludge was mixed with the inoculum at the 2:1 (volume/volume) ratio, and then diluted to a required solid content with deionized water."

"To assure anaerobic conditions in the anaerobic digestion process, oxygen was removed by ventilating with nitrogen gas for 5 min after feeding mixed sludge."

"A few sludge samples were dried at 105 °C, grounded, and sieved through mesh size < 2 mm for determining Fe, AN, AP, and AK."

"Proteins were measured by Lowry's method using bovine serum albumin as a standard solution (Lowry and others 159)."

"A volume of 1 mL biogas was injected into a gas chromatograph (Agilent GC7890II) to determine the content of CH_4 and CO_2."

"Surface morphology and structure of sludge samples (dried at 105 °C, ground and sieved through a mesh size < 0.25 mm) were analyzed by scanning electron microscopy (SEM, Merlin) with energy-dispersive X-ray (EDX) spectroscopy."

Results and Discussion

"To clarify the influence of ZVI and AC on the degradation of sludge, VS, COD, proteins, and polysaccharides were measured before and after the anaerobic digestion process."

"It suggested that the dosage of ZVI and AC accelerated the degradation of sludge during the anaerobic digestion process, which was accorded with the COD reduction curve."

"To the blank, the lag phase time of groups which added ZVI and AC was observed with values of 4.0 days (B), 4.3 days (C), and 3.7 days (D), which suggested that the addition of ZVI and AC had a good effect on the anaerobic digestion process."

"The changed tendency conformed to the VS reduction curve, suggesting that most of volatile solids were consumed and converted biogas in reactor D. In comparison to adding AC alone in the mesophilic AD system, adding ZVI and AC could be more efficient to promote the organic matter degradation efficiency, which was responsible for enhancement of biogas production and sludge stabilization."

Conclusion

"Great enhancement of the methane production and degradation of sludge were obtained by simultaneous addition of ZVI and AC in mesophilic anaerobic digestion."

"The concentration of NH_4^+-N maintained an appropriate level and did not preclude the anaerobic digestion process, due to the absorption of ZVI and AC."

"After 28 days of anaerobic digestion, the concentration of AP in digested sludge gradually decreased with the increasing dosage of ZVI/AC, which might be caused by chemical precipitation by Fe^{2+}/Fe^{3+} and adsorption by AC."

"These results could suggest that ZVI and AC might be considered in practical application of anaerobic digestion to enhance the methane production and sludge decomposition."

Anaerobic Digestion Process for the Production of Biogas from Cassava and Sewage Treatment Plant Sludge in Brazil [297]

This is a machine-generated summary of:

Peres, Sergio; Monteiro, Marina Rebeca; Ferreira, Micheline Lima; do Nascimento Junior, Adalberto Freire; de Los Angeles Perez Fernandez Palha, Maria: Anaerobic Digestion Process for the Production of Biogas from Cassava and Sewage Treatment Plant Sludge in Brazil [297]

Published in: BioEnergy Research (2018)

Link to original: https://doi.org/10.1007/s12155-018-9942-z

If you want to cite the papers, please refer to the original.

For technical reasons we could not place the page where the original quote is coming from.

Abstract-Summary

"This study aimed to evaluate the process of anaerobic digestion using cassava wastewater, a pollutant and toxic residue of cassava processing, and sludge from a sewage treatment plant, to determine the best conditions for the production of biogas and to identify its chemical composition, specifically the levels of methane and carbon dioxide."

"The best result was achieved using the following media: 80% cassava wastewater and 20% primary sludge, 4:1, which produced the highest methane content in the biogas over the shortest period of time, reaching 81.41% mol/mol of methane in 48 days of fermentation."

"Biogas produced by cassava wastewater, both in large starch-producing factories and in smaller flour factories, can be used as a source of renewable energy, reducing production costs and providing an environmentally correct destination for this waste."

Introduction

"One crop representative of the national and regional agricultural sector that generates considerable waste is cassava, classified as the fourth most cultivated product in Brazil in 2013 [160]."

"During the processing of cassava root, whether in flour or starch manufacturing plants, a large amount of waste is generated."

"Inoue [161] and Santos [162] estimated that processing of 1 ton of cassava produces an average of 300 L of CW in a flour plant and 600 L in a starch plant which is equivalent to the pollution caused by 200–300 inhabitants/day, while that generated from flour is equivalent to the pollution caused by 150–200 inhabitants/day."

"Because of the large volume produced, CW by-product is being considered as a material that can be utilized for other activities, not only because of the environmental damage it can cause, but also because the improper disposal of this waste constitutes a loss of income for the producer, considering the quantities and potentialities of the wastewater generated by this activity."

Materials and Methods

"The primary sludge (PS) and secondary sludge (SS) were supplied by the state of Pernambuco owned Sanitation Company called Companhia Pernambucana de Saneamento (COMPESA), from its Peixinhos Sewage Treatment Plant (STP), located in the Municipality of Olinda, one city of the Metropolitan Area of Recife, Pernambuco."

"The study was divided into the following stages: (1) obtaining and characterizing the composition of the raw materials; (2) implantation of anaerobic digestion systems using CW and sewage sludge (primary or secondary, separately); (3) analysis of the biogas produced in the process, its methane concentration; and (4) evaluation of the results."

"During the characterization of the composition of the raw material, the following were determined for the CW: (1) potential of hydrogen (pH), (2) total solids (TS), (3) fixed solids (FS), (4) volatile solids (VS), (5) COD, and (6) BOD."

Results and Discussion

"As CW presented a COD of 101.28 $g{\cdot}L^{-1}$, that almost 25 times the value Chang and others recommend as suitable for AD."

"It can be seen that, among the reactors, that having a mixture of 80% CW and 20% PS (system M80PS20) produced the greatest concentration of methane-rich biogas in the shortest period of time, with concentrations of CH_4 reaching 81.41% (mol/mol) in 48 days."

"To M80PS20, another system that produced biogas with high methane concentration was M70PS30, composed of 70% CW and 30% primary sludge."

"Regarding the type of inoculum used in the fermentations, it was verified that the reactors with CW and STP sludge, for all the three concentrations studied (80%, 70%, and 60%), presented higher CH_4 concentrations when the systems used primary sludge as inoculum."

Conclusions

"The process of biogas production from the anaerobic biodigestion of CW and STP sludge adapted itself well, producing biogas with a high concentration of methane."

"While all of the systems studied produced biogas with significant methane concentrations, the results indicate that the one composed of 80% CW and 20% primary sludge (M80PS20) was most efficient, generating biogas with 81.41% methane during 48 days of fermentation."

"Based on the observed methane concentration profiles, it was also verified that primary sludge was the inoculum most appropriate for the biodigestion process."

Biogas Generation in Anaerobic Wastewater Treatment Under Tetracycline Antibiotic Pressure [298]

This is a machine-generated summary of:

Lu, Meiqing; Niu, Xiaojun; Liu, Wei; Zhang, Jun; Wang, Jie; Yang, Jia; Wang, Wenqi; Yang, Zhiquan: Biogas generation in anaerobic wastewater treatment under tetracycline antibiotic pressure [298]

Published in: Scientific Reports (2016)

Link to original: https://doi.org/10.1038/srep28336

If you want to cite the papers, please refer to the original.

For technical reasons we could not place the page where the original quote is coming from.

Abstract-Summary

"The effect of tetracycline (TC) antibiotic on biogas generation in anaerobic wastewater treatment was studied."

"The reactor was operated with synthetic wastewater in the absence of TC and in the presence of 250 μg/L TC for 90 days, respectively."

"The removal rate of TC, volatile fatty acids (VFAs), biogas compositions (hydrogen (H_2), methane (CH_4), carbon dioxide (CO_2)), and total biogas production in each compartment were monitored in the two operational conditions."

"The presence of TC had a large negative effect on CH_4 and CO_2 generation, but appeared to have a positive effect on H_2 production and VFAs accumulation."

"This suggested that the presence of TC had less influence on the degradation of organic matter but had a strong influence on biogas generation."

Introduction

"The biogas, such as hydrogen (H_2), methane(CH_4), and carbon dioxide(CO_2), are the end-products of substrates in anaerobic wastewater treatment."

"The emission of biogas (H_2, CH_4, and CO_2) in anaerobic wastewater treatment has a significant environmental impact."

"Evaluation of the possible biogas generation (H_2, CH_4, and CO_2) in anaerobic wastewater treatment is of great significance."

"Most studies of TC in wastewater focused on its removal efficiency [58, 163–165] and its sorption in sewage sludge [166, 167], but there has not been a comprehensive evaluation of biogas generation (H_2, CH_4, and CO_2) in anaerobic wastewater treatment under TC pressure."

"The removal rate of TC, the levels of volatile fatty acids (VFAs), biogas compositions (H_2, CH_4, and CO_2) and total biogas production in each compartment were monitored, and the environmental impact of TC on anaerobic wastewater treatment was evaluated."

Results

"The generation average of H_2 in C1 was 0.0102 L/(g MLSS · d) in the TC-absence wastewater."

"The averages H_2 generation was 0.0413–0.0703 L/(g MLSS · d), which corresponded to 4.05–6.89 times that of the average in the TC-absence wastewater."

"The generation of CH_4 was 0.1407–0.2570 L/(g MLSS d), 0.0398–0.1167 L/(g MLSS d), and 0.0337–0.0737 L/(g MLSS d) in C1, C2, and C3 with TC exposure, which corresponded to 43.90–80.20%, 27.18–79.79%, and 38.24–83.70% of the average in the TC-absence wastewater."

"T-test results showed that TC had a very significant effect on CH_4 generation in all compartments during TC exposure except for days 1–18 in C3 (a significant effect)."

Discussion

"The slight increase of TC removal rate in C3 during days 21–33 resulted because the increase of effluent TC concentration from front compartment forced the sludge to absorb a lot of TC to reach a new sludge-solution equilibrium."

"The accumulation of VFAs resulted from change in the dynamic balance of generation from glucose acidification by acetogens and H_2/CO_2 conversion by homoacetogenic bacteria, and the utilization of acetate by aceticlastic methanogens to produce CH_4 under anaerobic conditions [168]."

"VFAs accumulation can be enhanced by the inhibition of the conversion of VFAs to CH_4 by aceticlastic methanogens, the inhibition generation from glucose acidification by acetogens caused a decrease in VFAs, resulting in a lower concentration of VFAs in TC-presence wastewater treatment."

"Hydrogenotrophic methanogens were inhibited under TC pressure [169], which reduced H_2 consumption and thus resulting in an increase in H_2 accumulation."

Methods

"Biogas compositions (H_2, CH_4, CO_2) and VFAs were analyzed by a gas chromatograph (Agilent Technologies 7820 A, USA) equipped with a thermal conductivity detector (TCD) and a flame ionization detector (FID)."

"Biogas compositions (H_2, CH_4, CO_2) were determined by the thermal conductivity detector (TCD) equipped with stainless steel column (2 m $\times$ 3.2 mm) packed with TDX-01 (80/100 mesh) [170], the operational temperatures of injector, oven and detector were kept at 100, 100, and 200 °C, respectively."

"The sample injection volume was 1 mL. Volatile fatty acids (VFAs) were analyzed in the soluble phase by using the flame ionization detector (FID) equipped with a HP-FFAP capillary column (30 m $\times$ 0.32 mm internal diameter) [171]."

"The sample injection volume was 3 μL. TC were ionized in the positive ionization mode and analyzed with the following parameters: capillary voltage 4000 V; nebulizer pressure 35 psi; drying gas, N_2, 8 L/min; gas temperature 350 °C."

3.5 Sludge Dewatering

Introduction by Editor

The primary objective of sludge dewatering is to reduce the moisture content of the sludge, thereby decreasing its volume and increasing its solid content. This enables easier handling, transportation, and disposal of the sludge while reducing associated costs. Moreover, dewatered sludge can be further processed for beneficial reuse, such as land application as a soil amendment or conversion into biogas for energy generation. Various methods are employed for sludge dewatering in municipal wastewater treatment. Mechanical dewatering processes, including centrifugation, belt filter press, and screw press, utilize mechanical forces to separate water from sludge solids. Advanced techniques such as thermal drying, electro-dewatering, and membrane-based processes offer enhanced dewatering capabilities, albeit with higher energy requirements and specialized equipment.

Successful sludge dewatering necessitates careful attention to operational parameters and conditions. Factors such as sludge characteristics (e.g., solids concentration, particle size), polymer dosage, applied pressure, and filtration rates significantly impact dewatering performance. Monitoring and controlling these parameters are critical for achieving optimal efficiency. Additionally, regular maintenance and cleaning of dewatering equipment are essential to ensure continuous and reliable operation.

Efficient sludge dewatering plays a vital role in sustainable wastewater treatment practices. It reduces the volume of sludge, minimizing the environmental impact associated with transportation and disposal. Furthermore, dewatered sludge can be utilized for beneficial purposes, such as agricultural application as a soil amendment or conversion into biogas for energy generation. Optimizing sludge dewatering processes enhances operational efficiency, reduces costs, and promotes resource recovery.

Machine generated keywords: dewatere, dewaterability, sludge dewaterability, sludge, bind, sludge dewatere, sewage sludge, conditioning, ep, eps, improve sludge, water content, moisture, flocculant, sludge cake.

Simultaneous Enhancement of Sludge Dewaterability and Removal of Sludge-Borne Heavy Metals Through a Novel Oxidative Leaching Induced by Nano-CaO_2 [299]

This is a machine-generated summary of:

Wu, Boran; Dai, Xiaohu; Chai, Xiaoli: Simultaneous enhancement of sludge dewaterability and removal of sludge-borne heavy metals through a novel oxidative leaching induced by nano-CaO_2 [299]

Published in: Environmental Science and Pollution Research (2017)

Link to original: https://doi.org/10.1007/s11356-017-9261-5

If you want to cite the papers, please refer to the original.

For technical reasons we could not place the page where the original quote is coming from.

Abstract-Summary

"To minimize the sludge volume to be handled and limit the potential environmental risk, this study developed a novel oxidative leaching process for enhanced sewage sludge dewatering and simultaneous removal of heavy metals based on nano-CaO_2."

"The speciation partitioning analysis of the heavy metals pre and post nano-CaO_2 peroxidation indicated that the content of organically bound metals decreased and the percentage of soluble fraction increased substantially, which was beneficial for the removal of heavy metals through the dewatering unit."

"Nano-CaO_2 peroxidation could also induce the transformation of extracellular polymeric substances (EPS) from the tightly bound layers to the loosely bound layers of sewage sludge flocs."

"Through the decline of the Ryan–Weber constant of fluorescence titration and the pseudo-first-order kinetic constant of complexation, it was verified that the binding capacity of EPS with metal ions could be damaged by nano-CaO_2 peroxidation, which was the primary mechanism behind the substantial reduction of organically bound metals."

Introduction

"Heavy metals and persistent organic pollutants (POPs) in wastewater from industrial sources can accumulate through sorption to sewage sludge and are the main limitation for the developing countries to utilize and safely dispose sewage sludge (Gianico and others 172)."

"There are few reports focusing on the transformation and migration of heavy metals during conditioning processes of sewage sludge."

"It is widely accepted that the decomposition of EPS certainly influences the solubility of heavy metals in sewage sludge (Hao and others 173, 174; Philippis and Colica 175)."

"As mentioned above, conditioning and dewatering processes widely exist in most wastewater treatment plants and are essential for the subsequent transportation, utilization, and disposal of sewage sludge."

"This study chose nano-CaO_2 as an effective conditioning reagent for improving the dewaterability of sewage sludge from wastewater treatment plants of combined sewer systems."

Experimental Section

"The solution-to-solid ratio is 40 mL/g dry solid (DS), and the extraction was conducted using a rotary shaker apparatus at 200 rpm for 22 h. First, the sewage sludge sample was centrifuged at 10,000×g for 10 min and heavy metals in the supernatant were regarded as the soluble fraction."

"The total solid concentrations of different EPS extraction solutions from raw and pretreated sewage sludge were adjusted to 15 mg/L by dilution."

"This equation has been widely used to characterize the interaction between metals and dissolved organic matters (Chai and others 176; Kumke and others 177; Monteil-rivera and Dumonceau 178):where I_0 and I are the fluorescence peak intensities of EPS with and without metal ions, respectively; the I/I_0 ratio can be obtained from the PARAFAC analysis; K is the conditional stability constant; I_{ML} accounts for the residual intrinsic EPS fluorescence; [L] represents the total ligand concentration, and [M] is the metal concentration."

Results and Discussion

"The speciation of heavy metals in sewage sludge is changeable and affected significantly by oxidation–reduction potential (ORP) and pH. After the treatment by nano-CaO_2 under optimal conditions of the RSM experiment, the leaching of heavy metals was enhanced substantially, and the soluble fractions of Cu, Ni, Pb, and Zn increased by 27.5, 87.9, 30.0, and 250.7%, respectively (The increasing rate was defined as the ratio of the increased content of a specific fraction after nano-CaO_2 treatment and the initial content of the corresponding fraction in raw sewage sludge.)."

"The complexation of metal ions and EPS should occupy the main organic fraction of heavy metals and have significant effects on the heavy metal speciation in sewage sludge."

"EPS from the pretreated sewage sludge would display lower electric affinity for metal ions, and heavy metals in the organic fraction would decrease accordingly."

Conclusion

"An oxidative leaching process based on nano-CaO_2 was first used to enhance sewage sludge dewatering and simultaneous solubilization of heavy metals with the operation under faintly acidic conditions."

"The peroxidation by nano-CaO_2 exerted significant effects on the speciation partitioning of heavy metals and promoted the release of heavy metals from the solid phase of sewage sludge, which was beneficial for the removal of heavy metals through the dewatering unit."

"The substantial decline of the Ryan–Weber constant of fluorescence titration and the pseudo-first-order kinetic constant of the complexation between EPS and metal ions revealed that the treatment by nano-CaO_2 did harm the binding capacity of EPS with metal ions, which was the primary reason behind the substantial reduction of organically bound heavy metals."

"All these results are believed to provide novel insights into embedding the function of toxicity reduction into widely existing dewatering units and improve the management of sewage sludge from wastewater treatment plants of combined sewer systems."

Integrated Treatment of Municipal Sewage Sludge by Deep Dewatering and Anaerobic Fermentation for Biohydrogen Production [300]

This is a machine-generated summary of:

Yu, Li; Yu, Yang; Jiang, Wentian; Wei, Huangzhao; Sun, Chenglin: Integrated treatment of municipal sewage sludge by deep dewatering and anaerobic fermentation for biohydrogen production [300]

Published in: Environmental Science and Pollution Research (2014)

Link to original: https://doi.org/10.1007/s11356-014-3514-3

If you want to cite the papers, please refer to the original.

For technical reasons we could not place the page where the original quote is coming from.

Abstract-Summary

"Based on the traditional processes, sludge dewatered by usual methods was further dewatered by hydraulic compression and the filtrate released was treated by anaerobic fermentation."

"The difficulties in sludge dewatering were associated with the existence of sludge flocs or colloidal materials."

"The filtrate from the dewatering procedure was a potential substrate for biohydrogen production."

"The novel method could greatly reduce the sludge volume and successfully treated filtrate along with the conversion of organics into biohydrogen."

Introduction

"In China, the moisture content of municipal sewage sludge must be reduced to less than 60% (w/w) according to the new stringent sludge treatment and disposal regulations."

"Prior to these sludge treatment processes, sludge volume needs to be reduced and the biodegradable fraction of the organic matter needs to be stabilized."

"The moisture content of sludge is still around 80% after usual mechanical dewatering techniques (centrifugation, belt filter, filter press, etc.) used in conventional wastewater treatment plants."

"Thermo-drying treatment is reported to be efficient in enhancing the sludge dewaterability but is limitedly employed due to high cost and air pollution (Mihoubi 179)."

"Hydraulic dewatering is one simple, efficient, and cost-effective technology for sludge dewatering."

"From the perspective of energy conservation and pollution reduction, anaerobic digestion, a transformation process of organic matter into energy gases, is selected as the treatment of the liquor in our laboratory."

Materials and Methods

"The dewatering liquor was referred to as F1, and the solid phase was marked as S2 for the control group, while the liquor was considered as F2, and dewatered sludge was noted as S3 for the group with the selected dosage of CaO. BESA ($C_2H_4BrO_3SNa$, sodium 2-bromoethanesulfonate), a universal methanogenic inhibitor, has a chemical structure ($BrCH_2CH_2SO_3^-$) that resembles that of the co-enzyme M ($HSCH_2CH_2SO_3^-$) essential for the methanogenic stage (Mohan and others 180)."

"The content of hydrogen in biogas was determined by a gas chromatograph (GC, Techcomp 7890) equipped with a thermal conductivity detector (TCD) and a stainless steel column packed with Carbosieve S (80/100 mesh), 2 m × 3 mm."

"Energy-dispersive X-ray (EDX) spectroscopy was connected with Quanta 200F, which was combined with SEM observation to determine the elemental composition on the surface of sludge."

"Polymerase chain reaction denaturing gradient gel electrophoresis (PCR-DGGE) was used to study the microbial community structure in F2, M1, and M2."

Results and Discussion

"The moisture content of sludge was 58.8 ± 1.2, 55.0 ± 2.0, 50.9 ± 1.0, 49.7 ± 2.0, 50.2 ± 1.5, and 50.0 ± 1.3% with the addition of CaO at 102, 114, 125, 136, 148, and 159 mg/g DS, respectively."

"A lower COD concentration between 1,017 ± 73 and 1,763 ± 43 mg/L was obtained in the steady stage from 7 to 36 days in M2."

"ORP in M2 was further decreased to −356 ± 6 mV on day 11 and kept stabilized around −350 mV during the optimal biohydrogen production period, which was averagely 100 mV lower than that in M1."

"The presence of uncultured Bacteroides sp. (band 9) indicated that hydrogen consumption was effectively inhibited (Shanmugam and others 181), which was in accordance with the result that biohydrogen yield was augmented by dosing ZVI in M2."

Conclusions

"The moisture content of sludge was decreased substantially with CaO addition by compression under high pressure driven by hydraulic power."

"The optimal CaO dosage (125 mg/g DS) was capable of reducing the moisture content of sludge to 50.9%."

"The sludge dewatering liquor was treated effectively by anaerobic fermentation."

Enhancement of Sewage Sludge Dewaterability by Fungal Conditioning with Penicillium simplicissimum NJ12: From Bench- to Pilot-Scale Consecutive Multi-batch Investigations [301]

This is a machine-generated summary of:

Tao, Neng; Wu, Xiu; Zhang, Feng; Pi, Zilei; Wen, Jiaqi; Fang, Di; Zhou, Lixiang: Enhancement of sewage sludge dewaterability by fungal conditioning with Penicillium simplicissimum NJ12: from bench- to pilot-scale consecutive multi-batch investigations [301]

Published in: Environmental Science and Pollution Research (2021)

Link to original: https://doi.org/10.1007/s11356-021-15170-0

If you want to cite the papers, please refer to the original.

For technical reasons we could not place the page where the original quote is coming from.

Abstract-Summary

"Bench- and pilot-scale successive multi-batch trials were conducted to investigate the performance and sustainability of fungal conditioning with Penicillium simplicissimum NJ12 for improving sludge dewatering."

"The dominant factors affecting the sludge dewaterability improvement by P. simplicissimum NJ12 were also identified."

"Fungal treatment with P. simplicissimum NJ12 at a volume fraction of 5% of the inoculum greatly improved the sludge dewaterability."

"These findings demonstrate the possibility of P. simplicissimum NJ12-assisted fungal treatment for enhancing sludge dewatering."

Introduction

"A particularly promising option to improve sludge dewaterability is fungal treatment (Molla and Fakhru'l-Razi 182)."

"improved the dewaterability with a decrease of the capillary suction time (CST) from 83 to 35 s. After characterizing the EPS content and composition and the sludge microstructure, some researchers have concluded that the enhancement of sludge dewaterability with fungal treatment is highly related to degradation of slime EPS and the formation of larger and stronger bioflocs (Fakhru'l-Razi and Molla 183; Liu and others 184; Li and others 185, 186)."

"Because of these excellent performances, fungal treatment is recognized as a cost-effective method for enhancing sludge dewatering and stabilization."

"Although previous studies have examined changes in sludge properties (e.g., rheology, surface charge, particle size, and EPS content and composition) during fungal treatment under different laboratory conditions, the most important factors influencing the sludge dewaterability improvement are still not identified, such as using multivariate statistical analysis."

Materials and Methods

"To investigate the role of the isolate strain P. simplicissimum NJ12 in sludge dewatering and identify the key factors responsible for the dewaterability improvement, four groups of batch experiments of sludge treatment were conducted in 500-mL flasks."

"The best equation is selected according to the highest multiple correlation coefficient (r^2) and takes the following form: where Y is the dependent variable (i.e., the SRF); $X_1, X_2, \ldots, X_n$ are the independent variables (i.e., sludge pH, zeta potential, median particle size (d_{50}), EPS content and composition, organic acid concentration, and fungal cell density); b_0 is a constant given by the point where the regression line intercepts the Y axis, which represents the value of Y when all the explanatory variables are 0; and bi ($1 \leq i \leq n$) is the standard partial regression coefficient, representing the amount that Y changes by when the explanatory variable changes by one unit."

Results and Discussion

"The dominant organic acid produced by P. simplicissimum NJ12 was gluconic acid with a maximum yield of 45 mM. This fungal production of gluconic acid did not result in significant acidification of the treated sludge, possibly because of the strong buffering capacity of the sludge."

"With stepwise MLR, the most important factors determining the sludge dewaterability improvement during fungal treatment with P. simplicissimum NJ12 were the zeta potential and protein content in slime EPS."

"In this trial, six successive batches of fungal treatment were conducted by recirculating the treated sludge rich in P. simplicissimum NJ12 at a recycling rate of 1:2 ($V_{\text{biotreated sludge}}/V_{\text{total sludge}}$) into the next batch treatment as the inoculum."

"These results indicate that P. simplicissimum NJ12 should be replenished periodically at a set batch interval to maintain the activity of Penicillium species in the sludge and ensure the effectiveness of fungal treatment is sustained."

Conclusions

"Both bench- and pilot-scale successive multi-batch investigations indicated that P. simplicissimum NJ12-assisted fungal treatment was technically effective at improving sludge dewatering."

"Consecutive operations of fungal treatment could be realized by recirculating the fungal-treated sludge and periodically re-adding fresh P. simplicissimum NJ12."

"On the basis of statistical analysis by stepwise multiple linear regression, the degradation of protein in slime EPS and the increase in zeta potential were the most important factors affecting the dewaterability improvement of the fungal-treated sludge."

Mechanistic Understanding of Predatory Bacteria-Induced Biolysis for Waste Sludge Dewaterability Improvement [302]

This is a machine-generated summary of:

Wu, Junkang; Chen, Zhoukai; Zhang, Shiwen; Gao, Lei; Yu, Ran; Zhan, Manjun: Mechanistic Understanding of Predatory Bacteria-Induced Biolysis for Waste Sludge Dewaterability Improvement [302]

Published in: Water, Air, & Soil Pollution (2019)

Link to original: https://doi.org/10.1007/s11270-019-4242-6

Copyright of the summarized publication:

If you want to cite the papers, please refer to the original.

For technical reasons we could not place the page where the original quote is coming from.

Abstract-Summary

"Due to the dramatically increasing production of excess sludge during biological wastewater treatment, the development of an economical, efficient, and environmentally friendly sludge dewatering method is highly required."

"Bdellovibrio-and-like organisms (BALOs), a group of predatory bacteria were applied for waste sludge dewaterability enhancement and biomass reduction and the potential biolysis mechanisms were elaborated."

"Within 24-h sludge biolysis with our isolated and enriched BALOs, the sludge specific resistance to filtration value as the dewaterbility index reached the maximal reduction rate of $65.3 \pm 6.4\%$."

"The contents of tightly bound extracellular polymeric substances (EPS) dramatically decreased after sludge biolysis while the loosely bound EPS contents increased, which implied the disruptions of sludge flocs structure for sludge dewaterability improvement."

"High-throughput sequencing results revealed the remarkable shift of sludge's microbial community structure after biolysis treatment."

"BALO predation could disintegrate the waste sludge structure, promote the cell lysis and the intracellular substances release, and cause the variations of microbial community compositions to efficiently improve the sludge dewaterability."

Introduction

"Since sludge generally contains an extremely high water content of more than 95% even after gravity thickening (Christensen and others 189), a sludge dewatering pretreatment process is always required for a wastewater treatment plant (WWTP) before disposal."

"As a potentially economical and environmental-friendly method, the application of BALO predation for sludge biolysis and dewatering pretreatment has seldom been addressed yet."

"EPS could dramatically impact the BALO-cell predation and biolysis efficiencies for sludge dewaterability improvement, and the exact role of EPS in BALO predation biolysis needs to be identified."

"Microbial community structure or compositions could potentially affect BALO predation or sludge biolysis, and the explorations of microbial community dynamic variations in response to exogenous BALO predation are crucial and necessary for mechanistic insight into waste sludge biolysis."

"The objectives of this study were to investigate the impacts of TS content on BALO-induced sludge biolysis efficiency for sludge dewaterability improvement."

Materials and Methods

"The D15 culture at the magnitude of 10^7 PFU/mL was added into the sludge (volume ratio of 1:10) with 0.5, 1.5, 2.5, and 3.0% TS content for biolysis treatment."

"The filter-sterilized D15 culture medium using 0.2 μm Millipore filters (Millipore, USA) was also added in the same series of diluted sludge as the control."

"40 mL sludge suspension was centrifuged at 4000×g for 10 min."

"After centrifugation at 4000×g for another 10 min, the supernatant was collected for LB-EPS quantification."

"The suspension was then centrifuged at 4000×g for 10 min to obtain TB-EPS in the supernatant."

"Sludge samples in triplicate (TS content at 2.5%) with (Bly.) or without (Con.) 24-h BALOs biolysis treatment were chosen for high-throughput sequencing analysis."

Results

"BALO predation activities generally induced positive effects on sludge biolysis efficiencies, and TS content ranging from 1.5 to 2.5% generally were referable for dewaterability improvement."

"The sludge with 2.5 and 3.0% TS content displayed the relatively lower and similar SCOD peak value after biolysis treatment."

"BALO-induced sludge biolysis treatment could not only lyse the cells but also disrupt the sludge polymeric matrix structures with the preference of the moderate TS contained ones."

"The sludge of 2.5% TS content were sampled after 24-h treatment with (Bly.) and without (Con.) BALO addition for high-throughput sequencing to explore the microbial responses to BALO predations."

"Abundant and diverse microbial prey species were expected to be present in the sludge, and BALO biolysis treatment significantly changed the sludge microbial community compositions."

Discussion

"The shift or the abundance reductions of the dominant microbial community members further confirmed the BALO-induced biolysis performance for sludge dewaterability enhancement."

"BALOs should efficiently induce sludge biolysis and cause the microbial community composition changes with the declines of the dominant species' abundance to improve sludge dewaterability."

"The BALO-induced sludge biolysis for dewaterability improvement should be closely associated with the disruptions of sludge EPS structure."

"The BALO-induced disintegration of sludge floc and EPS structure coupled with the encysted cells' lysis would contribute to the interstitial and intracellular water release, sludge biomass reduction, and dewaterability enhancement."

"The re-growth of surviving bacteria could accompany with increasing EPS secretion and re-stabilization potentials of sludge flocs structure to deteriorate sludge dewaterability."

"BALO predators would stimulate the growth of several certain microbes in the sludge and further promote the EPS production for sludge structure re-stabilization with the extension of biolysis time."

Conclusion

"This study investigated the BALO-induced biolysis performance for waste sludge dewaterability enhancement and the potential biolysis mechanisms."

"The results indicated that BALOs could effectively disturb the sludge floc structure and lyse the involved cells to efficiently reduce the SRF by as high as $65.3 \pm 6.4\%$ within 24 h. The moderate TS content (e.g., 1.5–2.5%) generally favored BALO-induced biolysis treatment, probably due to the satisfying cell densities and predator-prey attack/collision likelihoods."

"High-throughput sequencing results indicated that the sludge's microbial community structure remarkably shifted after BALO biolysis treatment despite that the microbial diversity did not significantly change."

"The relative abundances of genus Ferruginibacter and Sphingopyxis related to EPS production remarkably decreased, further indicating the potential disruptions of sludge flocs and EPS structures via BALO predation."

Enhanced Dewatering of Waste-Activated Sludge by Composite Hydrolysis Enzymes [303]

This is a machine-generated summary of:

Wu, Boran; Chai, Xiaoli; Zhao, Youcai: Enhanced dewatering of waste-activated sludge by composite hydrolysis enzymes [303]

Published in: Bioprocess and Biosystems Engineering (2016)

Link to original: https://doi.org/10.1007/s00449-016-1544-6

If you want to cite the papers, please refer to the original.

For technical reasons we could not place the page where the original quote is coming from.

Abstract-Summary

"The results indicated that tightly bound EPS (TB-EPS) was released into the liquid phase consistently during enzymatic hydrolysis to form soluble EPS (S-EPS) and loosely bound EPS and that the TB-EPS content was positively correlated with the capillary suction time of WAS."

"That TB-EPS reduction fit a first-order kinetic model and that mild temperature (25–30 °C) and a slightly acidic condition were favorable for the improvement of enzyme activity."

"Solid phase extraction combined with UV–Vis spectroscopy analysis was used to characterize the processes of migration and transformation of the hydrophobic (HPO), transphilic and hydrophilic (HPI) fractions in EPS during the enzymatic process."

"The results revealed that HPO and HPI were mainly composed of PN and PS, respectively, and that the enzymatic hydrolysis could enhance the transformation of HPI from TB-EPS to S-EPS, which was the dominant mechanism of improving WAS dewaterability."

Introduction

"EPS constitute a significant fraction of sludge mass and play a key role in binding a large volume of water (i.e., bound water) [190, 191]."

"The negatively charged bound EPS network plays a significant role in maintaining a hydrated sludge structure and could effectively prevent the release of water [192]."

"Sludge hydrophobicity increases with the protein content of bound EPS, and an increased hydrophobic generally leads to a worse dewaterability [193]."

"It is recognized that to a certain extent, the degradation of EPS and the lysis of sludge micro-organisms can enhance the release of water from sludge floc [194]."

"A kinetic model of EPS degradation was established to understand the effects of reaction conditions, temperature, pH and enzyme dosage on enzymatic dewatering."

Methods and Materials

"To harvest enzyme under normal conditions for Aspergillus oryzae metabolism, the serum bottles were purged with nitrogen gas and sealed with a rubber cap to maintain the anaerobic conditions used for the preparation of enzyme stock solution and incubation of sludge samples."

"The separation method of different water-extractable EPS fraction by SPE was established based on the modification of a similar procedure to separate the hydrophilic/hydrophobic soluble substances in the effluent of wastewater treatment plants [176, 195], because these dissolved organic matters usually show similar compositions and properties of soluble EPS."

"A Poly-Sery HLB SPE (strong polarity)/neutral Al_2O_3–XAD-2 SPE (non-polarity) (ANPEL, China) was used to fractionate EPS into hydrophobic (HPO), transphilic (TPI) and hydrophilic (HPI) fractions."

"A 3-mL sample of extracted EPS was first passed through the Poly-Sery HLB SPE, which was activated by successive treatments with 6 mL methanol and 6 mL deionized water."

Results and Discussion

"It is excepted that the enzymatic hydrolysis of the dominant macromolecule of EPS, proteins and carbohydrates (75–90% of the total EPS mass) [196] will lead to the cleavage of WAS floc and be beneficial for the water removal from WAS, so the composition variation of different EPS fractions in the presence of enzyme was investigated first at the enzyme dosage of 1 mg/g DS and temperature of 30 °C."

"The CST correlated negatively with PN/PS from S-EPS, which implied that the hydrolysis of protein and transamination of amino acids are beneficial for the removal of hydrophilic function groups (i.e., $-NH_2$) and improve WAS dewaterability [197]."

"The dosage of composite enzymes was controlled at a fixed value, and then the effect of initial EPS concentration on PN and PS removal efficiency was investigated by kinetic analysis."

"The hydrolysis enzyme promoted the release of hydrophilic components (PN) from TB-EPS into S-EPS and resulted in the improvement of WAS dewaterability."

Conclusions

"The content of S-EPS and LB-EPS in WAS increased after the pretreatment with the composite enzymes, but the TB-EPS content decreased greatly during the enzymatic reaction."

"The rate constant for PN and PS reduction increased from 0.0763 to 0.0852 and 0.0892 to 0.1031 h^{-1}, respectively, as the temperature increased from 10 to 30 °C."

"A weakly acidic condition (pH 6–7) is beneficial for improving the rate constant of PN reduction, but the rate constant for PS reduction decreased consistently with increasing pH from 3 to 11."

"The SPE combined with UV–Vis spectroscopy analysis found that HPI and HPO in EPS were mainly composed of PN and PS, respectively, and that enzymatic hydrolysis could enhance the transformation of HPI from TB-EPS to S-EPS, which resulted in the breakage of the WAS micro-floc structure and was the dominant mechanism of improving WAS dewaterability."

Conditioning of Synthetic Sludge and Anaerobically Digested Sludge Using Chitosan, Organic Polyelectrolytes and Inorganic Metal Cations to Enhance Sludge Dewaterability [304]

This is a machine-generated summary of:

Lau, Shiew Wei; Sen, Tushar Kanti; Chua, Han Bing; Ang, Ha Ming: Conditioning of Synthetic Sludge and Anaerobically Digested Sludge Using Chitosan, Organic Polyelectrolytes and Inorganic Metal Cations to Enhance Sludge Dewaterability [304]

Published in: Water, Air, & Soil Pollution (2017)

Link to original: https://doi.org/10.1007/s11270-017-3545-8

If you want to cite the papers, please refer to the original.

For technical reasons we could not place the page where the original quote is coming from.

Abstract-Summary

"This study investigated the dewatering performance of synthetic sludge in comparison with anaerobically digested sludge when conditioned with chitosan, organic polyelectrolytes and inorganic metal cations."

"Capillary suction time (CST), turbidity, electrical conductivity, zeta potential, cake solids content and particle size were used to assess sludge dewatering performance and to determine the optimum conditioner dose."

"Both synthetic sludge and AD sludge exhibited similar trend but little different extent of dewaterability when conditioned with low molecular weight (MW) chitosan."

"The low MW and medium MW chitosans, commercial cationic polyelectrolytes and trivalent metal cations (Al^{3+}, Fe^{3+}) demonstrated as effective conditioning agents with good sludge dewaterability."

"When assessing the dewaterability measurement parameters using synthetic sludge, the optimal dosage was found at the range of 15 to 20 g-chitosan/kg-dry sludge where the values of CST, turbidity and cake solids content were attained between 6.6 and 11.0 s, 35.4–40.6 NTU, and 24.3–25.3%, respectively."

"The application of cationic polyelectrolytes and trivalent metal cations generally improved the sludge dewaterability via charge neutralisation and polymer bridging."

Introduction

"A more recent research by Zemmouri and others (199) revealed the potential use of chitosan with improved filterability and dewatering capacity of conditioned sludge but the study was limited to activated sludge only."

"Our previous limited studies using chitosan in synthetic sludge conditioning have also shown great potential in improving sludge dewaterability (Lau and others 200)."

"Apart from organic conditioning agents, there are much less reported studies on various inorganic monovalent, divalent and trivalent metal cations as effective flocculating/conditioning agents which can enhance the digested sludge dewaterability (Lau and others 201; Lau and others 202; Nguyen and others 203)."

"The main objective of this research is to test the applicability and effectiveness of chitosan, organic polyelectrolytes and various inorganic metal cations in synthetic sludge and real AD sludge under various process conditions."

"Several dewaterability measurement parameters such as CST, supernatant turbidity, zeta potential, floc size and filtered cake solids content (%) were evaluated in assessing sludge dewatering performance when conditioned with chitosan in synthetic sludge."

Materials and Methods

"The desired total solids after dilution were of the average of 15.5 g/L or 1.55%, and the composition of the synthetic sludge can be approximated as follows: 40.2% cellulose, 17.7% yeast, 6.5% alginate, 32.9% KCl and 2.7% $CaCl_2$."

"The characteristics of the sludge sample were determined according to Standard Methods 2540 (APHA 204) for the total solids (TS), total dissolved solids (TDS) and total suspended solids (TSS)."

"The conditioner solution was pipetted into the 100-mL sludge sample at the predetermined dosage, expressed as the weight of conditioner over the weight of total solids in the sludge (g-conditioner/kg-dry sludge; thereafter abbreviated as g/kg for simplicity)."

"V_C is the dosing volume of conditioner (mL), TS the total solids of sludge sample (g/L), V the volume of sludge sample (mL), D_C the desired dosage (g/kg), C_C the conditioner concentration (g/L), and the conversion factor 1000 g/kg."

Results and Discussion

"To determine the optimum conditioning dosage, the CST parameter was correlated to the measurements of supernatant turbidity, EC, zeta potential, filtered cake solids content and floc size using synthetic sludge as controlled experiments."

"Synthetic sludge conditioned with aluminium chloride, ferric chloride and few cationic polyelectrolytes (i.e. the PAM, medium MW and low MW chitosans) demonstrated superior dewaterability where the CST values reduced by 91–95%."

"With addition of chitosan, the findings suggest that both charge neutralisation and polymer bridging mechanisms are important to enhance sludge flocculation and dewatering at low pH while the deficiency of ionised amine groups at pH 7.0 may reduce the capture of fine particles as indicated by the increase in turbidity and decrease in cake solids content at pH 7.0."

Conclusions

"Chitosan conditioning effectively enhanced sludge dewaterability of both synthetic sludge and real AD sludge."

"Low MW and medium MW chitosans, PAM and trivalent metal cations (Al^{3+}, Fe^{3+}) demonstrated good dewaterability in synthetic sludge, whereas for real AD sludge, low MW and medium MW chitosans showed better dewaterability than Al^{3+} and Fe^{3+}."

"Low pH is favourable for sludge conditioning but temperature in the range of 25–55 °C has a minor effect on sludge dewaterability."

"Synthetic sludge and real AD sludge both show more or less similar behaviour in dewaterability by various conditioning agents under various process conditions and hence development of less complex chemically controlled synthetic sludge may also be applicable for studying the final properties of complex unstable digested sludge."

Enhanced Dewaterability of Waste Activated Sludge with Fe(II)-Activated Hypochlorite Treatment [305]

This is a machine-generated summary of:

Zhu, Xiaofei; Yang, Qi; Li, Xiaoming; Zhong, Yu; Wu, You; Hou, Lihua; Wei, Jing; Zhang, Weixuan; Liu, Yu; Chen, Chongyu; Wang, Dongbo: Enhanced dewaterability of waste activated sludge with Fe(II)-activated hypochlorite treatment [305]

Published in: Environmental Science and Pollution Research (2018)

Link to original: https://doi.org/10.1007/s11356-018-2829-x

If you want to cite the papers, please refer to the original.

For technical reasons we could not place the page where the original quote is coming from.

Abstract-Summary

"A novel method was explored to improve the waste-activated sludge (WAS) dewaterability using Fe(II) combined with calcium hypochlorite, and the possible mechanisms were investigated simultaneously."

"Capillary suction time (CST), specific resistance to filtration (SRF) of sludge, and water content (W_C) of dewatered sludge cake were selected as the factors to evaluate the sludge dewaterability."

"Economic analysis suggested Fe(II) combined with calcium hypochlorite treatment had greater potential on enhancing WAS dewaterability compared with the traditional sludge treatment."

Introduction

"Fenton oxidation has been regarded as an effective way to enhance the dewaterability of sludge (Lu and others 205)."

"The efficiency of Fenton and Fenton-like process is closely related to the pH condition and it usually requires the strong acid condition, which is one of the demerits in enhanced sludge dewatering by Fenton process (Behin and others 206)."

"Compared with hydrogen peroxide (H_2O_2) used in traditional Fenton process, hypochlorite salt could be catalyzed by Fe^{2+} in a wider range of pH (Behin and others 206)."

"There are few studies on the improvement of WAS dewaterability by Fe(II)-activated hypochlorite-advanced oxidation."

"The scientific attempt on the application of Fe(II)-catalyzed HOCl oxidation to enhance the sludge dewatering should be useful to enrich the theory of sludge treatment and disposal."

"The aim of this study was to investigate the feasibility of Fe(II)-activated hypochlorite advanced oxidation on improving dewatering performance of WAS."

Materials and Methods

"The $FeSO_4{\cdot}7H_2O$ and $Ca(ClO)_2$ were simultaneously added into the sludge sample to evaluate the influence of Fe(II)/$Ca(ClO)_2$."

"At fixed Fe(II)/$Ca(ClO)_2$ 1.25 and pH 7.3, $Ca(ClO)_2$ at 0.02, 0.03, 0.04, 0.05, and 0.06 g/g TSS was respectively added into the reactor to assess the effect of $Ca(ClO)_2$ dosage on sludge dewaterability."

"Then, the sludge sample was collected to measure the capillary suction time (CST), specific resistance to filtration (SRF), and water content (W_C)."

"Extracellular polymeric substances (EPSs) of sludge samples were extracted by a thermal extraction method according to our previous study (see in supporting information)."

"TSS, VSS, and pH of raw sludge were measured based on the standard methods (APHA, 207)."

"The sludge cakes from the SRF measurement process were trapped, and then dried in an oven at 105 °C for 8 h to determine the W_C of dewatered sludge cake (Luo and others 208)."

Results and Discussion

"After the Fe(II)/$Ca(ClO)_2$ oxidation, the zeta potential of sludge became less negative and the surface charge was markedly neutralized with increasing doses of $Ca(ClO)_2$ and Fe(II)."

"Fe(II)/$Ca(ClO)_2$ oxidation process altered the zeta potential evolution of the flocs through EPS degradation by ·OH radical and enhanced the charged neutralization of sludge surface by Fe^{3+} and Ca^{2+} (He and others 192; Zhen and others 209)."

"After the Fe(II)/$Ca(ClO)_2$ treatment, the concentration of PN and PS declined to 295.7 and 92.5 mg/L, respectively, indicating the powerful ·OH from Fe(II)-activated HClO partly disrupted and degraded EPS fractions trapped in the sludge flocs."

"The powerful oxidation of ·OH produced from the Fe(II)-activated $Ca(ClO)_2$ effectively degrade the particular compounds of fluorescing substances in sludge EPS."

"These advantages make Fe(II)/$Ca(ClO)_2$ treatment be a potential in sludge dewatering."

Conclusion

"The combination of Fe(II) and $Ca(ClO)_2$ was demonstrated to be an effective method to enhance the WAS dewatering in this study."

"After Fe(II)/$Ca(ClO)_2$ treatment, the EPS was destroyed into small fragment and bound water was released."

"Then, the small fragment was re-flocculated by Fe^{3+} and Ca^{2+}, which further improved the WAS dewaterability."

Utilization of Different Plant Species for Dewatering of Sewage Sludge [306]

This is a machine-generated summary of:

El-Gendy, Ahmed Shafik; Ahmed, Amira Galal: Utilization of different plant species for dewatering of sewage sludge [306]

Published in: Environmental Science and Pollution Research (2021)

Link to original: https://doi.org/10.1007/s11356-021-12982-y

If you want to cite the papers, please refer to the original.

For technical reasons we could not place the page where the original quote is coming from.

Abstract-Summary

"Bench-scale and pilot-scale experiments were conducted in an outdoor environment to study the ability of some plant species in dewatering of sewage sludge collected from biological activated sludge treatment."

"Sludge dewatering in the plant reactors was compared with that in the control reactors (no plants)."

"The bench-scale experiments were conducted in reactors with capacities of 17 L. All plants showed a growth in the sewage sludge matrix."

"High dewatering efficiencies of sewage sludge were obtained with the use of each type of plant, as compared with those in the controls."

"Among other plants tested in the current study, water hyacinth proved to have the highest dewatering efficiency and was selected for further testing in a pilot-scale experiment."

"The pilot study showed that the use of water hyacinth in conventional sludge drying beds can triple the sludge dewatering capacity of these beds."

Introduction

"Dewatered sewage sludge from wastewater treatment plants is increasingly facing disposal-related problems."

"In Egypt, the amount of sludge generated from wastewater treatment plants is about 2,900,000 m^3 per year."

"Conventional drying beds are the most commonly used technology in the dewatering of sewage sludge."

"A number of constructed wetland systems have been employed for the treatment of various kinds of wastewaters, including sewage sludge from conventional treatment plants (Heinss and Koottatep 212)."

"A small number of studies have indicated the possibility of using plant drying beds for dewatering of sewage sludge."

"The objective of the current research was to study the potential of four plants to grow in sewage sludge from activated sludge treatment process."

"Another objective of the current research was to define the operating conditions of a plant drying bed that can be used for dewatering of sewage sludge."

Materials and Methods

"The bench-scale experiments, which include Exp.1 and Exp.2, were conducted using liquid sewage sludge collected from El-Katameya Wastewater Treatment plant in New Cairo, (29° 58′ 29.1″ N and 31° 23′ 35.08″ E)."

"The pilot-scale experiment (Exp.3) was conducted using liquid sewage sludge collected from Qalioub Al Kharqaneyah Wastewater Treatment plant in Al Qalyubia, (30° 10′ 54.60″ N and 31° 10′ 37.44″ E)."

"Exp.2 was conducted over a period of 103 days, during which total liquid sewage sludge masses of 90 kg, 60 kg, 50 kg, 55 kg, and 15 kg were added to the water hyacinth, common reed, Samar, El Nesila, and the control reactors, respectively."

"By the end of sludge dewatering in the water hyacinth bed after 26 days from the start of Exp.3, three plants were collected randomly from the bed, and the roots were separated for analysis."

Results and Discussion

"In Figs, 2 and 3, water hyacinth plants were able to dewater 90 kg of liquid sewage sludge at an efficiency of 88.2% (on average) in 103 days, compared with 57.4%, 57.3%, 46.2%, and 15.4% dewatering efficiencies for El Nesila, common reed, Samar and the control reactors, respectively, in the same duration of time."

"The results obtained from the bench-scale experiments revealed that out of the other plants studied in the current work, water hyacinth is the most effective in dewatering liquid sewage sludge."

"The water hyacinth bed received more batches of liquid sewage sludge (1 m^3 each) that were added every 2 days until the bed was filled with semi-dewatered sludge after 22 days from the start of Exp.3."

"The amount of liquid sewage sludge received by the water hyacinth bed during one cycle of dewatering was 67% higher than that received by the control bed."

Conclusion

"It can reduce the time needed for dewatering of liquid sewage sludge as compared with conventional drying beds."

"The bench-scale experiments showed that the ability of the plants used in the current study to dewater liquid sewage sludge followed the following order: water hyacinth >> El Nesila > common reed > Samar."

"The pilot-scale experiment demonstrated that the ability of conventional drying beds in dewatering liquid sewage sludge tripled with the use of water hyacinth plants at an initial density of 24 kg/m^2."

"This makes the dewatered sludge utilizing water hyacinth safer in handling after the removal of plants as compared with sludge dewatered using conventional drying beds."

"The application of water hyacinth drying beds can significantly reduce the area needed for sludge dewatering as compared with conventional drying beds."

Self-recycling of Sewage Sludge as a Coagulant and Mechanism in Sewage Sludge Dewatering [307]

This is a machine-generated summary of:

Wu, Jiahuan; Lu, Tao; Yang, Guang; Meng, Wei; Yuan, Haoran; Chen, Yong: Self-recycling of sewage sludge as a coagulant and mechanism in sewage sludge dewatering [307]

Published in: Journal of Material Cycles and Waste Management (2020)

Link to original: https://doi.org/10.1007/s10163-020-01083-6

If you want to cite the papers, please refer to the original.

For technical reasons we could not place the page where the original quote is coming from.

Abstract-Summary

"When 1 g BC500 was added to 100 mL waste-activated sludge (WAS), the maximum mean settlement rate reached 8.20 mL min^{-1} within 10 min, and the volume of the settlement with settling time decreased to 13 mL. The capillary suction time of

BC500 (12 s) was superior to those of municipal solid waste incineration (MSWI) fly ash (13.40 s), coal fly ash (14.20 s), and activated carbon (12.20 s)."

"Added BC may reduce the moisture content of sludge cake, especially for BC500."

"As results shown, BC was feasible to improve sewage sludge dewatering used as a coagulant."

Introduction

"Chen and others [213] chose coal fly ash modified with sulfuric acid as a skeleton builder, and the results showed that the specific resistance to filtration of the sludge decreased from 1.86×10^{13} to 4.23×10^{11} m kg^{-1}, while the filter cake moisture content decreased from 86.90 to 56.52%."

"Compared to common coagulant aids, such as fly ash, rice husk, lignite, and coal fly ash used in these researches, one obvious advantage of BC is it does not increase the complexity of sewage sludge composition, which will contribute to subsequent sewage sludge disposal and utilization."

"The dewatering effects of BCs, MSWI fly ash, coal fly ash, and activated carbon on waste-activated sludge (WAS) were compared using the settling test, capillary suction time (CST), sludge cake moisture content, and dewatering rate."

"The possibility of improving sewage sludge dewaterability with BC as coagulant was studied."

Experimental

"The heating rate was 10 °C min^{-1}, and the holding time was 4 h. Sewage sludge samples pyrolyzed at 300, 500, 700 and 900 °C were denoted as BC300, BC500, BC700, and BC900, respectively."

"The ash content was determined as the weight loss at 815 °C for 1 h, and the volatile matter content was acquired by the weight loss when dried sample was heated in muffle at 900 °C for 7 min [214, 215]."

"A thermal analyzer (STA PT1600 TG-DSC; TA Instruments, Linseis, Germany) was used to analyze the thermos dynamic properties of sewage sludge under a N_2 atmosphere using the following program: heating rate of 5 °C min^{-1} from room temperature to 50 °C, maintenance for 30 min at 50 °C, and then heating at 10 °C min^{-1} to 100 °C; and a carrier gas (N_2) flow rate of 100 mL min^{-1}."

Results and Discussion

"As was commonly observed for BC derived from both sewage sludge and other feedstock, an increase in temperature led to an augmentation in pH [216]."

"At the temperature range of 155–400 °C, weight loss was result of decomposition of biodegradable organic matter (e.g., proteins and carboxyl groups), which resulted in the losses of mass caused by volatile analysis and by the combined evaporation of moisture [217, 218]."

"BC500 amendment yielded lower V values compared with MSWI fly ash and activated carbon, the settling time was within 10 min, and the efficiency of sludge gravity dewatering was significantly improved."

"For BC500, although this sample did not have the maximum surface area, total pore volume, and average pore size, the zeta potential may have been influenced by the comprehensive effects of various factors, such as abundant pore structures, richer functional groups in BC500, high positive ion content, and surface charge."

Conclusion

"BC500 had the best sludge dewatering effect, which might relate to its rich pores with diameters 2–10 nm."

"Both the CST and sludge cake moisture content of WAS added with BC500 were reduced 12 s and 63.5%."

"Amendment of WAS with sewage sludge-derived BC, in particular that pyrolyzed at 500 °C, may improve sludge dewaterability."

Enhancing the Dewaterability of Anaerobically Digested Sludge Using Fibrous Materials Recovered from Primary Sludge: Demonstration from a Field Study [308]

This is a machine-generated summary of:

Tochioka, Eiji; Yamashita, Manabu; Usui, Jiro; Miyake, Haruo; Terada, Akihiko; Hosomi, Masaaki: Enhancing the dewaterability of anaerobically digested sludge using fibrous materials recovered from primary sludge: demonstration from a field study [308]

Published in: Clean Technologies and Environmental Policy (2019)

Link to original: https://doi.org/10.1007/s10098-019-01698-w

If you want to cite the papers, please refer to the original.

For technical reasons we could not place the page where the original quote is coming from.

Abstract-Summary

"Challenge, we developed a novel process in which fibrous materials initially collected from a primary sedimentation tank, namely recovered fibers, were used as an auxiliary agent to improve the dewaterability of anaerobically digested sludge."

"Three fibrous materials individually recovered from the respective primary sedimentation tanks in different wastewater treatment plants (WWTPs) were used to investigate sludge dewaterability, organic C compositions, and calorific values."

"Six different WWTP sludge samples after anaerobic digestion were continuously supplied to three different dewatering devices, i.e., screw press, centrifuge, and belt press machines, with or without the supply of recovered fibers."

"Irrespective of the type of a dewatering device, the supply of recovered fibers mixed with the tested sludge samples at a feed ratio of 0.18–0.20 g-fibers/g-sludge (dry weight) reduced the amount of polymer flocculant by 13–50%, thereby displaying the superiority of the addition of recovered fibers."

"At fiber feed ratios of 0.20 g-fibers/g-sludge and 0.40 g-fibers/g-sludge when the rates of sludge supply and polymer flocculant were kept constant, the corresponding water contents in the dewatered sludge cakes decreased to about 74–76% and 69–72%, respectively."

"The consistent results regardless of the dewatering device or sludge sample showed the significance of the addition of recovered fibers for improvement in sludge dewaterability."

Introduction

"For the improvement in dewaterability of digested sludge, previous works demonstrated the effectiveness of external additions of polymer flocculant (Thapa and others 219) and cellulose-based fibers (Jaafarzadeh and others 220)."

"The effectiveness of recovered fibers in improvement in sludge dewaterability has been limited to WWTPs employing a screw-type dewatering device."

"To validate the technology for various configurations of WWTPs, the applicability of recovered fibers in improving sludge dewaterability using screw press, belt press, and centrifuge dewatering devices needs to be determined."

"Taking these contexts into consideration, this study investigated the wide applicability of the addition of recovered fibers to anaerobically digested sludge to achieve high dewaterability for the first time."

"Dewaterability tests of digested sludge with recovered fibers using various dewatering devices were performed to confirm the effectiveness of digested sludge as an auxiliary agent to facilitate dewaterability."

Materials and Methods

"The effect of the addition of recovered fibers to the digested sludge mixture on the required amount of polymer flocculant addition and water content of the dewatered sludge cake was investigated using six different sludge samples from full-scale WWTPs in Kyushu Island (WWTPs B, C, E, and F), Shikoku Island (WWTP G), and Honshu Island (WWTP D) in Japan."

"Screw press sludge dewatering machines convey the mixture of recovered fibers, polymer flocculant, and digested sludge to the conic edge to squeeze out the water contents of the mixture by continuous rotation of a cone-shaped screw."

"The retention time of the centrifuge sludge dewatering machine was 10–30 min at approximately 2000 g. Each experiment was performed in intervals of greater than 1 h. The belt press sludge dewatering machine mixed and compressed recovered fibers, anaerobic digestion sludge, and polymer flocculant, conveyed the mixture onto a belt, and sandwiched the top and bottom rolls for sludge dewatering."

Results

"Comparable results of the amount of polymer flocculant and the corresponding water content in the dewatered sludge cake were attained in WWTP G, thereby resulting in a 50% and 1.1% reduction in D_P and water content of the dewatered sludge cake with recovered fibers, respectively."

"The comparative investigation demonstrated that the presence of recovered fibers allowed a decrease in the water content of the dewatered sludge cake with both dewatering machines."

"The addition of recovered fibers improved the dewatering performances irrespective of the tested dewatering machine, thereby contributing to the lower amount of added polymer flocculant and lower water content of the sludge cakes."

"The further addition of polymer flocculant (D_P increase from 0.012 to 0.018 g/g-TS) substantially improved the dewaterability of the digested sludge cake and reduced the water content to 65.8%, thereby demonstrating the synergistic effects of dosages of recovered fibers and polymer flocculant."

Discussion

"Different from the previous researches using the recovered and external cellulose constituents for anaerobic digestion and dewatering, respectively, the central concept advocated in this study lies in recovered fibers that are mixed with anaerobically digested sludge for application in a dewatering machine."

"The addition of recovered fibers improved the dewaterability of digested sludge irrespective of the applied dewatering machine, the proposed concept is of promise to reduce the water content in digested sludge, thereby saving energy and money for sludge incineration."

"Notwithstanding the applied dewatering machine, the addition of recovered fibers can provide an appropriate adjustment of the water content of digested sludge cakes."

"Given the trend that the majority of WWTPs still employ either a belt press or a centrifuge dewatering machine, e.g., in Europe (Kelessidis and Stasinakis 221), our strategy of adding recovered fibers is likely applicable to various WWTPs with different sludge treatment configurations."

Conclusions

"This study demonstrated that, irrespective of the type of dewatering machine, recovered fibers from primary sedimentation tanks could improve the dewaterability of anaerobically digested sludge."

"The addition of recovered fibers improved dewaterability, successfully reducing polymer flocculant dosage to attain an equivalent level of water content by a standard condition."

"We also demonstrated a negative linear correlation between the amount of added recovered fibers and the water content of the digested sludge cake, which showed that the strategy of adding recovered fibers to digested sludge is promising for effectively removing water from the sludge."

Improving Sludge Dewaterability by Combined Conditioning with Fenton's Reagent and Surfactant [309]

This is a machine-generated summary of:

Hong, Chen; Wang, Zhiqiang; Si, Yanxiao; Yang, Qiang; Xing, Yi: Improving sludge dewaterability by combined conditioning with Fenton's reagent and surfactant [309]

Published in: Applied Microbiology and Biotechnology (2016)

Link to original: https://doi.org/10.1007/s00253-016-7939-0

If you want to cite the papers, please refer to the original.

For technical reasons we could not place the page where the original quote is coming from.

Abstract-Summary

"The effect of Fenton's reagent combined with dodecyl dimethyl benzyl ammonium chloride (DDBAC) on sludge dewaterability was studied."

"The capillary suction time (CST) and water content (WC) of sludge cake were used to evaluate sludge dewaterability."

"Bound water content (W_B), extracellular polymeric substance (EPS) concentration, and organic acid concentration in the sludge supernatant were measured to explain the change of dewaterability in the conditioning process."

"Results indicated that Fenton's reagent combined with DDBAC could enhance sludge dewaterability significantly."

Introduction

"Tony and others (222) demonstrated the advantages of using Fenton's reagent for the dewatering efficiency of activated sludge."

"These studies also reported that the addition of Fenton's reagent could increase the extracellular polymeric substance (EPS) concentration in the sludge supernatant, which had a significant impact on sludge dewaterability."

"Our previous studies revealed that the application of surfactant contributed to the conversion of bound water into free water by changing the structure and surface properties of sludge flocs (Hong and others 223; Hong and others 224)."

"The composite application of Fenton's reagent and surfactant in sludge conditioning has not been reported in literatures."

"All the above mentioned studies explained the mechanism of sludge dewatering only through the changes in S-EPS concentration."

"Bound water content (W_B), EPS concentration, and organic acid concentration in the sludge supernatant were measured to clarify the mechanism of Fenton's reagent integrated with surfactant pretreatment in improving sludge dewatering performance."

Materials and Methods

"The sludge before/after conditioning was analyzed in terms of EPS, bound water content (W_B), and organic acid concentration in the sludge supernatant."

"WC was calculated as follows: where W_1 is the weight of the sludge cake, and W_2 is the weight of the sludge cake dried at 105 °C for 24 h. CST was measured with a standard CST apparatus (304 M, Triton Electronics, Ltd.)."

"The dewatered sludge samples were first subjected to a temperature of −20 °C at a rate of 10 °C min^{-1}, at which all free water was assumed to be frozen under this condition, and then warmed to room temperature at the same rate."

"The formula for W_B is as follows: where W_B is the bound water content (g/g, being the matter content in per gram dry sludge, similarly hereinafter); W_T is the total water content (g g^{-1}); ΔH is the DSC endothermic of the sample in joules per gram; and ΔH_0 is the standard fusion heat of ice, 334.7 J g^{-1}."

Results

"The W_B further decreased to 1.02 g g^{-1} after the sludge was conditioned by DDBAC with the dosage of 80 mg g^{-1} (FD-4)."

"The PN and PS concentrations in S-EPS, LB-EPS, and TB-EPS changed obviously after sludge conditioned with Fenton's reagent."

"When the Fe^{2+}/H_2O_2 dosage increased from 0/0 to 10/10 mg g^{-1} and further increased to 40/40 mg g^{-1}, the PN concentration in S−/LB-EPS firstly increased from 363.31/143.16 to 2660.18/1229.82 mg g^{-1} and then decreased to 838.86/639.83 mg g^{-1}, respectively."

"The PS concentration in S-/LB-EPS firstly increased from 72.68/29.06 to 352.28/155.79 mg g^{-1} and then decreased to 267.13/125.89 mg g^{-1}."

"The PN and PS concentrations in TB-EPS decreased monotonously from 3028.17 and 593.59 to 773.18 and 187.62 mg g^{-1}, respectively, when the Fe^{2+}/H_2O_2 dosage increased from 0/0 to 40/40 mg g^{-1}."

Discussion

"The dewatering performance of sludge conditioned by Fenton's reagent reached the limitations, although more Fe^{2+} or H_2O_2 was applied after the dosage 40/40 mg g^{-1}."

"The main reason was that ·OH yield increased with the increase in H_2O_2 dosage, enhancing the efficiency of ·OH oxidizing sludge flocs, releasing the bound water to liquid, and then increasing sludge dewaterability."

"The degree of sludge oxidative decomposition and the content of releasable bound water decreased with increasing H_2O_2 dosage."

"A reasonable explanation was that the water combined with sludge flocs through stronger chemical bonds could not be released into the liquid phase by the Fenton-DDBAC conditioning process."

"When the dewatering performance of sludge reached limitation at the Fe^{2+}/H_2O_2 dosage of 40/40 mg g^{-1}, DDBAC could further promote the releasing of bound water."

"EPS was decomposed into small-molecule organics by the combined action of Fenton's reagent and DDBAC and resulted in substantial release of bound water."

3.6 Composting

Introduction by Editor

Composting plays a vital role in the secondary treatment phase of municipal wastewater treatment or sewage treatment plants. It is a sustainable and environmentally friendly process that involves the controlled decomposition of organic matter found in wastewater sludge, resulting in the production of a stable and nutrient-rich material known as compost. Composting relies on the principles of aerobic decomposition, which occurs when microorganisms break down organic matter in the presence of oxygen. By creating controlled conditions, including appropriate aeration, temperature, and moisture levels, the growth and activity of microorganisms are facilitated, leading to the transformation of organic compounds into more stable forms and the generation of compost.

Composting relies on the fundamental principle of aerobic decomposition, where microorganisms break down organic matter in the presence of oxygen. In the secondary treatment phase, the sludge undergoes controlled decomposition under specific conditions, including adequate aeration, temperature, and moisture levels. These conditions support the growth and activity of microorganisms, enabling the conversion of organic compounds in the sludge into more stable forms, ultimately resulting in the production of compost.

Municipal wastewater treatment plants employ various methods for composting. Common techniques include windrow composting, where sludge is stacked in long rows and periodically turned to ensure proper aeration, and static pile composting, where sludge is piled statically and aerated at regular intervals. In-vessel composting systems, such as enclosed reactors or digesters, provide controlled environments for efficient composting. These methods offer flexibility in terms of scale, process control, and resource utilization.

Successful composting requires careful attention to operational parameters and conditions. Factors such as temperature, moisture content, oxygen availability, carbon-to-nitrogen ratio, and turning frequency significantly influence the composting process. Maintaining optimal operating conditions, including thermophilic temperatures (around 50–60 °C), appropriate moisture levels, and proper aeration, is crucial to promote microbial activity and achieve efficient decomposition. Regular monitoring of key parameters and the implementation of appropriate management practices, such as turning or mixing, are essential for successful composting.

Composting offers numerous benefits in the context of municipal wastewater treatment. Firstly, it allows for the conversion of organic waste into a valuable resource, compost, which can be used as a soil amendment or fertilizer. This contributes to sustainable agriculture and reduces the reliance on chemical fertilizers. Additionally, composting helps to reduce the volume of sludge, facilitating its management and disposal. By diverting organic waste from landfills and harnessing its potential through composting, municipalities can achieve waste reduction targets and promote a circular economy.

Machine generated keywords: compost, sewage sludge, soil, sludge compost, heavy, heavy metal, moisture, final, sludge, waste, ratio, fertilizer, metal, sewage, aeration.

Composting of Sewage Sludge with a Simple Aeration Method and Its Utilization as a Soil Fertilizer [310]

This is a machine-generated summary of:

Nguyen, Thanh-Binh; Shima, Kazuto: Composting of Sewage Sludge with a Simple Aeration Method and its Utilization as a Soil Fertilizer [310]

Published in: Environmental Management (2018)

Link to original: https://doi.org/10.1007/s00267-017-0963-8

If you want to cite the papers, please refer to the original.

For technical reasons we could not place the page where the original quote is coming from.

Abstract-Summary

"Two consecutive composting trials (run A and run B) using Japanese sludge and woodchips (1:1, v/v) were conducted in cubic boxes ($0.45 \times 0.45 \times 0.45$ m^3) made by plywood at Okayama University."

"The changes in physicochemical properties with particular attention to inorganic nitrogen (NH_4–N, NO_3–N) and free amino acid nitrogen (FAA-N) indicated that the biodegradation took place by different mineralization pathways during the composting process."

"The degradation of organic matter into amino acids followed by ammonification was predominant in run B, whereas the nitrification was greater in run A. A pot experiment using the two finished composts and their raw materials was carried out to study their effectiveness as fertilizer to Komatsuna (Brassica rapa var."

"The lowering proportions of FAA-N/T-N, NH_4–N/NO_3–N, and C/N ratios in the composts compared to those in raw materials was found to correlate with the increase in plant biomass."

Introduction

"Composted sludge can be used as a soil amendment to supply nutrients to plants, improve soil physical properties, and increase the percentage of organic matter (Warman and Termeer 225; Pedra and others 226; Alvarenga and others 227), although negative effects associated with heavy metal accumulation and phytotoxicity phenomena in plant–soil systems should be taken into consideration."

"Sludge composting is widely applied in developed countries, which emphasize large-scale and mechanized composting process."

"A simple and cost-effective composting method for recycling of sludge and producing commercial compost is necessary to ensure its applicability in developing countries."

"The main objectives of this study are two-fold: first, to examine the feasibility of sewage sludge composting using a simple aeration method, providing information about the changes in physicochemical properties of sludge during the composting process with particular attention to the temperature and the nitrogen availability; and second, to relate some compost properties to plant biomass as a means to evaluate the quality of the self-making composts."

Materials and Methods

"In Oct 2016, a pot culture experiment using the two finished composts (called compost A and compost B from run A and run B, respectively) and aforementioned sludges was conducted in a greenhouse at Okayama University to study the effectiveness as soil fertilizer to Japanese Komatsuna (Brassica rapa var."

"The sludges and composts were analyzed for the following parameters: moisture content (105 °C for 24 h); pH (1:5 of fresh sample: water, w/v) using pH electrode; total carbon (T-C) and total nitrogen (T-N) were measured by C-N analyzer (CORDER MT-700; Yanaco, Japan); volatile solid was determined by loss on ignition (550 °C for 3 h); loss of organic matter (OM) and loss of T-N were computed using the equations of Paredes and others (228)."

Results

"The total volatile solids substantially decreased from 75.2 to 69.2% in run A and from 83.9 to 69.2% in run B. Most of T-N losses occurred during the first 11 days, while OM loss extended longer in later periods as found in run B. At the end of composting, the percent of OM losses were 25.7% and 50.2% in run A and run B, respectively."

"C/N ratios of the finished composts, which were 6.5 in run A and 7.2 in run B, showed little change during the whole process."

"In run B, the amount of FAA-N was found to increase sharply during the first 4 days, from 2.43 to 3.77 g kg^{-1}, then gradually decreased until the end of the composting process."

Discussion

"In spite of the fact that OM loss and T-N loss in run B were much higher than run A, the results of T-N losses measured during the composting of sewage sludge in the current study were consistent with those reported by Witter and Lopez-Real (229)."

"Our results show that finished composts had smaller C/N ratios (6.5–7.2) than the optimum range, but they were not significantly different from other Japanese commercial composts based sewage sludge (data not shown)."

"No studies regarding to composting of sewage sludge indicated an optimum level of free amino acid-N in the finished compost."

"The inhibition effects on growth of Komatsuna were also investigated by Kubota and others (230), who demonstrated that the presence of low fatty acids and relatively high values of org-C/org-N ratios in the immature composts resulted in lower plant biomass."

Conclusions

"The compost temperature in run A (winter) was not well controlled in comparison to that in run B (summer)."

"The degradation of organic matter into amino acids followed by ammonification was predominant in run B, whereas the nitrification was greater in run A. The finished composts had pH values of 5.7–6.8, C/N ratios of 6.5–7.2, T-N of 46.2–50.6 g kg^{-1}, FAA-N/T-N of 1.23–1.28% and NH_4–N/NO_3–N of 1.0–8.1."

"Although sludge properties varied seasonally, total plant biomass produced by two finished composts was over the control treatment and comparable to commercial chemical fertilizer."

"The lowering of FAA-N/T-N, NH_4–N/NO_3–N, and C/N ratios in composts compared to those in raw materials was found to correlate with the increase in plant biomass."

Characterization of Composted Sewage Sludge During the Maturation Process: A Pilot Scale Study [311]

This is a machine-generated summary of:

Bożym, Marta; Siemiątkowski, Grzegorz: Characterization of composted sewage sludge during the maturation process: a pilot scale study [311]

Published in: Environmental Science and Pollution Research (2018)

Link to original: https://doi.org/10.1007/s11356-018-3335-x

If you want to cite the papers, please refer to the original.

For technical reasons we could not place the page where the original quote is coming from.

Abstract-Summary

"This paper determines the impact of the maturation process of composted sewage sludge on the quality of the final product and assesses the stabilization effect."

"The samples of composted sewage sludge were taken from a wastewater treatment plant located in Pomerania in northern Poland."

"The sewage sludge was composted in an open windrow composting plant with the addition of straw and wood chips in the turning windrow."

"The aeration of the sewage sludge mixture was conducted based on two methods."

"The values of pH, organic matter (OM), total organic carbon (TOC), elementary composition, nutrients, total content, and mobile forms of heavy metals were analyzed in the compost samples."

"The total content of heavy metals in the final product did not exceed the limit values for the agricultural use of sewage sludge, compost from municipal waste, and for organic fertilizers."

"There were no significant changes in the percentage of bioavailable and mobile forms of heavy metals during compost maturation."

"The study has shown that the process of maturation of compost from sewage sludge not affects changes in the content of heavy metal forms."

"The scope of this study has been planned on a wider scale for different variants of sewage sludge composting, in order to evaluate the process."

Introduction

"In Poland, the preferred direction of sewage sludge utilization is associated with its natural use, including using it for agricultural purposes, land reclamation, and the cultivation of plants intended for the production of compost."

"The main biological stabilization process for the agricultural use of sewage sludge is composting."

"The total concentration of heavy metals in sewage sludge cannot provide useful information about the risk of bioavailability, toxicity, and the capacity for immobilization in the environment."

"An increase in the percentage of heavy metal mobile forms in the sewage sludge compost was observed by some authors (Oleszczuk 231; Chen 232; Awasthi and others 233)."

"The AT_4 test was used to assess the stability of composts from sewage sludge during maturation."

"The objective of this pilot study was the assessment of the stability degree, composition, and mobility of heavy metals changes in composted sewage sludge during maturation."

Materials and Methods

"Amount of composted material, a representative laboratory sample was collected by combining and thoroughly mixing together 30 primary samples collected at the same time from different points of the windrow."

"This method of collecting samples allowed access to the entire composted material."

"The samples were delivered to the laboratory within 12 h. The tests for the evaluation of biological stability have to be carried out on samples of pretreated compost of sewage sludge."

"Mobile forms of the metals were analyzed using a 0.5 M hydrochloric acid (0.5 M HCl) (Tracepur®, Merck) according to the Polish Chemical and Agricultural Stations procedure was used based on the Polish Standard (234)."

"The U value for each method is as follows: sampling including sample preparation 5%; OM 12%; pH 5%; C 7%; H 18%; N 16%; S 16%; TOC 15%; AT_4 15%; Cd, Pb, Cu, Zn, Cr, Ni, Mo, Co, and Mn 10%; Fe 12%; Ca 10%; Mg 13%; K 8%; and Na 10%."

Discussion

"Villaseńor and others (235) indicated a higher content of both metals in the final product of composted sewage sludge and straw/sawdust (Zn 877 and Cu 336 mg kg^{-1} DM) compared to substrate mixtures before treatment: Zn 433–438 and Cu 214–242 mg kg^{-1} DM."

"In the studies of Gattullo and others 236, the nickel content in a compost from sewage sludge was 37 mg kg^{-1} DM."

"Limit values for the content of other metals, such as Mo, Co, Mn, and Fe in Polish sewage sludge used for agricultural purposes, composts from municipal wastes, and organic fertilizers have not been determined."

"Heavy metal content increased in composted sewage sludge during maturation."

"The content of heavy metals in the tested samples was lower than the limit values for sewage sludge used in agriculture, solid organic fertilizer, and the first class of quality of compost from municipal waste according to Polish law."

Conclusions

"The results of this study show that the maturation process of composting sewage sludge has an impact on the properties of the final product, such as organic matter and total organic carbon concentrations, total heavy metals, and nutrients."

"An analysis of the raw sewage sludge was not carried out, which would allow an evaluation of the differences within the final product."

"It has been shown that composted sewage sludge after maturation is stable, which may indicate the sufficient stabilization of the organic matter of the tested materials."

"These pilot studies will be continued on a wider scale for different variants of sewage sludge composting, in order to evaluate the process."

Relationships Between Stability, Maturity, Water-Extractable Organic Matter of Municipal Sewage Sludge Composts and Soil Functionality [312]

This is a machine-generated summary of:

Sciubba, Luigi; Cavani, Luciano; Grigatti, Marco; Ciavatta, Claudio; Marzadori, Claudio: Relationships between stability, maturity, water-extractable organic matter of municipal sewage sludge composts and soil functionality [312]

Published in: Environmental Science and Pollution Research (2015)

Link to original: https://doi.org/10.1007/s11356-015-4611-7

If you want to cite the papers, please refer to the original.

For technical reasons we could not place the page where the original quote is coming from.

Abstract-Summary

"WEOMs were characterized through chemical analysis; furthermore, compost stability was evaluated through oxygen uptake rate calculation and maturity was estimated through germination index determination, whereas compost impact on soil fertility was studied, in a lab-scale experiment, through indicators as inorganic nitrogen release, soil microbial biomass carbon, basal respiration rate and fluorescein di-acetate hydrolysis."

"The obtained results indicated that WEOM characterization could be useful to investigate compost stability (which is related to protein and phenol concentrations) and maturity (related to nitrate/ammonium ratio and degree of aromaticity) and then compost impact on soil functionality."

"Compost stability resulted inversely related to soil microbial biomass, basal respiration rate and fluorescein di-acetate hydrolysis when the products were applied to the soil."

Introduction

"Even if stability and maturity were sometimes considered as terms with no real distinction, some differences should be made: stability is an index of compost evolution, expresses the biodegradability of its organic matter and is mainly evaluated by means of respirometric measurements, while maturity is an index of the compost capability of sustaining plant growth, is linked to the elimination of phytotoxic compounds and is evaluated by plant or seed bioassays (Adani and others 237; Bernal and others 238; Said-Pullicino and others 239)."

"Various parameters for evaluating compost stability and maturity, such as anaerobic gas formation potential, volatile solids content or organic carbon (C) content in solids (Shao and others 240) have been proposed."

"The aim of this work was the study of the relationships between the quality of municipal sewage sludge (MSS) composts, as investigated through analysis of their WEOM properties and determination of their stability and maturity, and the quality of a sandy loam soil amended with the same composts, as evaluated through soil quality indicators."

Materials and Methods

"Organic acids were determined in WEOM samples using a HPLC-UV system, constituted by a Waters 1525 Binary Pump (Milford, MA, USA), a Dionex (Sunnyvale, CA, USA) AS50 Autosampler and a Waters 2487 Dual lambda Absorbance Detector (Milford, MA, USA) set at 210 nm wavelength, equipped with a Aminex HPX-87H column (300 × 7.8 mm) by Bio-Rad (Hercules, CA, USA)."

"10 seeds of Lepidium were incubated in a 100-mm diameter Petri dish containing an 80-mm cellulose filter preconditioned with 1.5 mL of WEOM of compost diluted with distilled water (at 50 and 75%)."

"The oxygen uptake rate was calculated according to Grigatti and others (241): 2 g of volatile solids (VS) equivalent of each sample was placed in a 1-L bottle with distilled water (180 mL), nutrient- and micronutrient-rich solution (10 mL), phosphate buffer at pH 7 (10 mL) and N-allylthiourea (50 mg L^{-1}) as nitrification inhibitor."

Results and Discussion

"As this ratio has been proposed as an index of compost stabilization degree, since, during composting N mineralization to NO_3^- occurs (Bernal and others 238), we supposed that the MIX compost should be not stable; nitrate/ammonium ratio was instead 51, 2.5 and 226 for AN, AE and GM, respectively, suggesting a higher stability for AN among MSS composts."

"For this parameter, soil response was significantly affected by composts MIX and AE, which promoted higher microbial growth with respect to the unamended soil; GM and AN amended soils had similar trends for C_{mic}."

"Soil basal respiration (SBR) rate, after a couple of weeks of high microbial activity (graph not shown), became constant, as the increase in CO_2 production was constant in time, stabilizing around a value of 0.2 mg C–CO_2 kg^{-1} h^{-1} (0.149, 0.180, 0.204, 0.180 and 0.170 mg C–CO_2 kg^{-1} h^{-1} for control, AN, AE, MIX and GM, respectively)."

Conclusions

"The study of the quality of composts from MSS and rice husk was carried out through characterization of their water extracts and determination of their stability and maturity, together with the investigation of the quality of a sandy loam soil, amended with the same composts, through soil quality indicators."

"Chemical analyses showed that the composts from MSS and rice husk had the characteristics of common organic amendments, with AE having the highest TOC and TON concentrations and AN the highest degree of aromaticity and nitrate/ammonium ratio."

"All the investigated composts could be considered stable and mature, except MIX, which had a phytotoxic effect on seed germination; AN resulted more stable, as also suggested by its higher degree of aromaticity and nitrate/ammonium ratio."

"Compost stability could be studied on the basis of biodegradable substrate occurrence in WEOMs, while maturity could be predicted on the basis of nitrate/ammonium ratio and degree of aromaticity."

Management of Sewage Sludge by Composting Using Fermented Water Hyacinth [313]

This is a machine-generated summary of:

Tello-Andrade, A. F.; Jiménez-Moleón, M. C.; Sánchez-Galván, G.: Management of sewage sludge by composting using fermented water hyacinth [313]

Published in: Environmental Science and Pollution Research (2015)

Link to original: https://doi.org/10.1007/s11356-015-4683-4

If you want to cite the papers, please refer to the original.

For technical reasons we could not place the page where the original quote is coming from.

Abstract-Summary

"The goal of the present research work was to assess the management of sewage sludge (SS) by composting using fermented water hyacinth (WH_{ferm}) as an amendment."

"The water hyacinth was fermented, and a higher production of volatile fatty acids (VFAs) (782.67 mg L^{-1}) and soluble organic carbon (C_{SOL}) (4788.34 mg L^{-1}) was obtained using a particle size of 7 mm compared to 50 mm."

"For composting, four treatments (10 kg fresh weight each) were evaluated: treatment A (100% SS + 0% WH_{ferm}), treatment B (75% SS + 25% WH_{ferm}), treatment C (50% SS + 50% WH_{ferm}), and treatment D (25% SS + 75% WH_{ferm})."

"The WH_{ferm} added to SS, especially in treatments C (50%) and D (75%), increased the initial contents of organic matter (OM), organic carbon (C_{ORG}), C_{SOL}, the C/N ratio, and the germination index (GI)."

"The treatments with WH_{ferm} (B, C, and D) yielded higher values of electrical conductivity, cation exchange capacity, and GI than SS at day 198."

"No significant differences were observed in GI among the treatments with WH_{ferm}."

"WH_{ferm} as an amendment in the composting of SS improved the characteristics of the final product, especially when it was used in proportions of 25 and 50%."

Introduction

"To improve the initial conditions for the composting of SS, amendments such as sawdust (Ammari and others 242), straw (Muñoz and others 243), or tree bark (Doublet and others 244) have been utilized."

"Materials that contain easily assimilated carbon, such as residues from the food industry such as potatoes (Ammari and others 242) and general food scraps (Komilis and others 245), have also been used as amendments in SS composting."

"Invasive macrophytes such as Eichhornia crassipes (C. Mart) Solms (fresh and dry) have also been proposed as an amendment for SS composting."

"Low or negligible OM biodegradation during SS composting using water hyacinth (WH) as an amendment has been reported due to either its high moisture content (fresh WH) (Tello-Andrade and others 246) or the lack of easily available carbon content (dry WH) (Mahanta and others 247; Peña 248)."

"This finding suggests that WH_{ferm} may be an appropriate amendment for the composting of SS."

"The goal of the present research work was to assess the management of SS by composting using WH_{ferm} as an amendment."

Material and Methods

"The quantification of the VFA was performed using the HACH 8196 esterification method, taking aliquots from the liquid phase of the reactor in which the water hyacinth was fermented."

"For the fresh and fermented water hyacinth, the SS, and all of the mixtures, the moisture content was measured using the gravimetric method at 105 °C (APHA 249)."

"The soluble nutrients (TKN_{SOL}, PT_{SOL}, and K_{SOL}) were quantified in an aqueous extract with a mass concentration ratio of 1:10 (w/v) following El Fels and others (250) and using the aforementioned HACH colorimetric methods."

"The germination index (GI) was determined in an extract with distilled water with a mass concentration ratio 1:10 (w/v) (El Fels and others 250)."

Results and Discussion

"Its increase in treatment D (75% WH_{ferm}) was twice as large as the increase measured in only SS (80.51 and 40.51 g kg^{-1} C_{SOL}, respectively)."

"Treatment A (100% SS) was alkaline, with a final pH above 8, possibly due to the high TKN and the low C/N ratio, which delayed the composting process and caused the pH to increase, allowing for the volatilization of ammonia."

"The consumption of C_{SOL} has also been shown to be higher when WH_{fresh} is added to SS in similar proportions (21.81, 52.64, and 66.53% C_{SOL} for 40, 60, and 80% WH_{fresh}, respectively) after 42 days of composting (Tello-Andrade and others 246)."

"This low ratio could potentially have delayed the composting process and caused the generation and volatilization of ammonia (Huang and others 251; Negre and others 252), especially in the treatments with a higher SS proportion, which may have assisted in the pathogen suppression."

Conclusions

"Its characteristics were appropriate for its use as an amendment in the composting of SS."

"The treatments with 25 to 50% of WH_{ferm} showed the best results for the degradation of OM and the content of organic carbon and nutrients and values of C/N ratio, EC, CEC, GI, and pH. These results guarantee a higher nutrient product quality."

"The final product was improved when WH_{ferm} was added to the SS, especially in terms of the content of OM, organic carbon, nutrients, C/N ratio, EC, CEC, and GI."

"The composting of SS amended with fermented water hyacinth is an adequate alternative for the management of both wastes."

"Further work is necessary to assess larger compost piles and the addition of a second amendment with a high content of C_{SOL}, which could result in thermophile temperatures and pathogen killing in a shorter composting time."

The Impact of Agricultural and Green Waste Treatments on Compost Quality of Dewatered Sludge [314]

This is a machine-generated summary of:

Abbasi, Fariba; Mokhtari, Mehdi; Jalili, Mahrokh: The impact of agricultural and green waste treatments on compost quality of dewatered sludge [314]

Published in: Environmental Science and Pollution Research (2019)

Link to original: https://doi.org/10.1007/s11356-019-06618-5

If you want to cite the papers, please refer to the original.

For technical reasons we could not place the page where the original quote is coming from.

Abstract-Summary

"This study aimed to investigate the effect of two agricultural waste wheat straw (WS) and green waste eucalyptus tree leaves (ETL) treatments on the quality of the compost produced from dewatered sewage sludge."

"That, the variation trend of heavy metals, temperature, moisture, PH, and C/N ratio was considered during the composting process."

"The results showed that the variation of some parameters in WS and ETL such as temperature (24.68 ± 0.26 and 23.41 ± 1.35), moisture (4.5 ± 2.64 and 7.66 ± 2.51), pH (5.66 ± 0.35 and 5.97 ± 0.41), and C/N ratio (250 ± 4.08 and 60 ± 3.21) was suitable in both windrows, respectively."

"Trend of mineralization was suitable in both masses, because TVS (43 to 78 mg/g DW for WS and 45 to 69 mg/g DW for ETL) and TDS (21 to 55 mg/g DW for

WS and 2.6 to 38 mg/g DW for ETL) decreased and increased, respectively, in the composting process."

"Although the quality of both masses was suitable and similar, there are some limitations for using treated compost by WS in agricultural lands due to higher levels of EC, fecal coliforms, and manganese."

"Quality of dewatered sludge treated compared with ETL is higher than WS."

Introduction

"In spite of some benefits of the compost, such as desirable effect on soil qualifications and enhancement of soil nutrient level, there are adverse effects due to high level of coliform, heavy metals (Cherif and others 253), and the leakage of nitrogen and phosphorus in groundwater."

"The occupational health effects are another concern in composting process (Abbasi and others 254, 255)."

"Since produced compost from dewatered sewage sludge (DSS) has nutrients and organic matter, it is a source of nutrients for productivity of lands."

"They should control and monitor the final compost during utilization process."

"The variation of operational parameters should be determined to control the quality of compost."

"For the comparison of compost quality, operational variables are usually determined."

"Some elements and contaminants such as heavy metals and pathogens (fecal coliforms, salmonella, and parasites eggs) decrease the quality of compost."

Materials and Methods

"The final weight of windrow reached to 730.8 kg during composting process."

"The weight of windrow reached to 709.8 kg during composting process."

"According to the above-obtained ratios, two masses of DSS were created with WS treatment and another with ETL treatment with a width of 75 cm and a height of 1.5 m. Then, the weight of mass estimated based on its density."

"At the end of the process, with decreasing of 35% and 26% of windrow volume in ETL and WS treatments, the weight of windrowers in ETL and WS treatments decreased by 39% and 40% and reached to 709.8 and 730.8 kg, respectively."

"Sampling for all parameters was conducted during the composting process from three points of windrows (including surface, intermediate, and near the bottom of windrow)."

"In order to measure the temperature, C/N ratio, heavy metals, nutrients, fecal coliform (FC), parasites eggs, and salmonella, sampling was performed in triplicate."

Results and Discussion

"The variation trend of different parameters such as temperature, moisture, pH, EC, C/N ratio, heavy metals, and microbial contamination (including fecal coliforms, salmonella, and parasites eggs) has been analyzed over 100 days."

"In both compost, the variation trend of fecal coliform, salmonella, and parasite eggs had appropriate correlation with the trend of Zn ($r^2 = 0.97$), Mn ($r^2 = 0.96$), and Cu ($r^2 = 0.99$), respectively."

"Since land use of compost is one of the its most important uses, investigation of the concentration and the variation trend of some metals such as Na, K, Zn, Fe, Mg, and Cu are essential in both masses."

"Decomposition of organic matter containing heavy metals increased the level of pollutants (Manáková and others 256; Song and others 257), so the use of contaminated compost with heavy metals has toxic effects on land."

Conclusion

"The purpose of this study was the investigation of WS and ETL treatment effect on the compost quality of DSS."

"According to the result of the process, in WS and ETL, the variation trend of temperature (24.68 ± 0.26 and 23.41 ± 1.35), moisture (4.5 ± 2.64 and 7.66 ± 2.51), pH (5.66 ± 0.35 and 5.97 ± 0.41), C/N ratio (250 ± 4.08 and 60 ± 3.21), TSS (57.84 ± 14 and 79.83 ± 16), TDS (55.75 ± 9 and 53.77 ± 8.4), TVS (27.7 ± 1.5 and 37.57 ± 4.3), and solid mineral (42.04 ± 13 and 20.12 ± 15) were appropriate for both windrows, respectively."

"The variation of fecal coliform (2.72 MPN/g DW) and EC (2.4-mm hose/cm) in compost generated from WS was higher than standard."

"To, it seems that the quality of compost generated from ETL was higher than WS, especially for land use because its quality was above Iran standard."

Co-vermicomposting of Wastewater Treatment Plant Sludge and Yard Waste: Investigation of Operation Parameters [315]

This is a machine-generated summary of:

Yilmaz Cincin, Roda Gokce; Agdag, Osman Nuri: Co-vermicomposting of Wastewater Treatment Plant Sludge and Yard Waste: Investigation of Operation Parameters [315]

Published in: Waste and Biomass Valorization (2019)

Link to original: https://doi.org/10.1007/s12649-019-00900-w

If you want to cite the papers, please refer to the original.

For technical reasons we could not place the page where the original quote is coming from.

Abstract-Summary

"Vermicomposting of domestic wastewater treatment plant sludge with yard waste was carried out using Eisenia Fetida for 60 days."

"The effects of Bio-One™ (liquid microbial fertilizer), additives (different mature vermicomposts), recirculation and ventilation on decomposition rate of composting were investigated for composting mixtures."

"It was concluded that vermicomposting process was faster than traditional composting process in terms of decomposition of organic matter."

"The result C/N ratio was more suitable to use it as soil remedial material in vermicomposting process."

"It has been shown that Bio-One™ addition has a positive effect on the acceleration of the process."

"It has been found that the process of recirculation some of the leachate formed during the vermicomposting process is an advantageous."

"It has been determined that mature vermicompost addition is an effective method."

"The positive results of the plant growth tests on the final products in all cases may indicate that the products will not cause any toxic effect in the use as soil conditioning material."

Statement of Novelty

"There are many studies on vermicomposting of sewage sludge."

"The effect of operating parameters on vermicomposting of sewage sludge has not been investigated before."

"The effects of Bio-One™ (liquid microbial fertilizer), additives (different mature vermicomposts), recirculation and ventilation on decomposition rate of composting were investigated for composting mixtures in this study."

Introduction

"Different worm species (Eisenia andrei, Eisenia fetida, Eudrilus eugeniae, Perionyx excavatus and Perionyx sansibaricus) are used in vermicomposting process according to the type of waste, the composting method and ambient conditions [258]."

"Because worms are able to easily consume all compostable wastes (kitchen wastes, all kind of animal manure, garden wastes etc.), vermicomposting is a sustainable option for sewage sludge management [259]."

"Garden wastes not only provide these advantages but also cause to increase organic matter and nitrogen content in the composting mixture."

"Composting efficiency can be increased by using compost additives in vermicomposting."

"Animal manure generally is used as a compost additive to provide bulking agent and to set appropriate media for worm life."

"The most important advantages of the compost additives are to increase microbial community in the compost pile and seed effect for vermicomposting process."

Materials and Methods

"E. fetida earthworms were purchased from a commercial supplier and saved in the laboratory for adaptation and acclimation to composting material."

"E. fetida earthworms adopted by these types of compost material were used in the experimental study."

"During the experimental study, nine composting reactors were used."

"Earthworms must have a bedding material to survive when unsuitable conditions (for example high temperature and/or high nitrogen concentrations) occur in the compost reactor [260]."

"After pretreatment, sewage sludge and yard wastes were mixed with the same ratio (w/w) in the reactor and vermicomposting process was started."

"Control 1 and control 2 reactors represent traditional composting and vermicomposting of sludge and yard waste mixtures, respectively."

"Bio-One™ was added to the run 1 reactor as a compost additive."

"One way to measure compost stability is to determine its effect on plants."

Results and Discussion

"The results of this study showed that the last TOC values of the vermicompost reactor is smaller than traditional composting reactor."

"To vermicomposting reactors, the TKN content of our traditional composting reactor (control 1) increased during operation period (except first few days)."

"In a comparative study using industrial sludge as composting material with or without vermiculture, the ultimate C/N ratio of the vermicomposting reactor was lower than traditional composting reactor [261]."

"The final C/N ratio was 14.17 in vermicomposting reactor using Bio-One™ as an additive (run 1)."

"While initial EC values were 755 µs/cm in control runs, this value decreased to 500 µs/cm with the addition of mature vermicompost."

"After composting and vermicomposting process TKN content significantly increased in control 1(45.67%), control 2(53.08%), run 6(62.97%), run 7(72.83%) and the final values were 1.18%, 1.24%, 1.32% and 1.4%, respectively."

Conclusion

"The addition of Bio-One™ as a compost additive accelerated the degradation process of the organic material by increasing the microbial community in the composting reactor and positively effected the composting process."

"The leachate recirculation accelerated the aerobic process and effected the compost quality positively."

"It has been shown that mature compost addition has a positive contribution to organic matter consumption."

"That atmospheric O_2 is sufficient for the composting process especially in small volume reactors in vermicomposting."

"The positive result of plant growth tests on compost products of reactors in all cases may mean that the products will not cause any toxic effect on the plants when used as soil conditioning material."

Pyrolysis of Wastewater Sludge and Composted Organic Fines from Municipal Solid Waste: Laboratory Reactor Characterisation and Product Distribution [316]

This is a machine-generated summary of:

Agar, David A.; Kwapinska, Marzena; Leahy, James J.: Pyrolysis of wastewater sludge and composted organic fines from municipal solid waste: laboratory reactor characterisation and product distribution [316]

Published in: Environmental Science and Pollution Research (2018)

Link to original: https://doi.org/10.1007/s11356-018-1463-y

If you want to cite the papers, please refer to the original.

For technical reasons we could not place the page where the original quote is coming from.

Abstract-Summary

"Pyrolysis yields three output products: solid char, liquid oils and gas."

"The potential of pyrolysis for the conversion of wastewater sludge (SS) and organic fines of MSW (OF) to a combustion gas and a carbon-rich char has been investigated."

"Pyrolysis of SS and OF was done using a laboratory fixed-bed reactor."

"The physical characterisation of the reactor is described, and results on pyrolysis yields are presented."

"Feedstock and chars have been characterised using standard laboratory methods, and the composition of pyrolysis gases was analysed using micro gas chromatography."

"Product distribution (char/liquid/gas) from the pyrolysis of sewage sludge and composted MSW fines at 700°C for 10 min were 45/26/29 and 53/14/33%, respectively."

"The combustible fractions of pyrolysis gases range from 36 to 54% for SS feedstock and 62 to 72% from OF."

Introduction

"Char, however, is only one of the products of pyrolysis."

"The distribution of pyrolysis products depends heavily on several process parameters whose influence follows a general trend for all organic feedstock (Wang and others 262)."

"Temperature, residence time and heating rate are the main process parameters, but particle size of the feedstock and residence time of vapour-phase products are also important as these influence the contact between chars and gases, the extent of which affects char formation and the decomposition (cracking) of long-chain hydrocarbon gases (Mok and Antal Jr 263, 264)."

"Fast pyrolysis uses high heating rates to produce more vapour-phase products and decrease char yields (Bridgwater 265)."

"Minimising liquid yield may be desirable for a decentralised pyrolysis process aiming to utilise syngas on-site through direct combustion and simultaneously producing biochar for subsequent uses."

"Initial investigations focused on the syngas and char potential of waste by determination of yield distribution from pyrolysis."

Materials and Methods

"Pyrolysis was also carried out on SS samples ground to a 1-mm size (20 g 700 °C) to determine the influence of particle size on yields."

"Before each pyrolysis run, the reactor, the connected glassware, the empty sample basket and the rubber stopper were weighed to within ± 0.01 g. A sample of feedstock was poured into the basket, and the basket was weighed again."

"The reactor tube cooled to room temperature after about 2 h. The sample basket and contained char were removed from the reactor and weighed."

"The reactor, the glassware (containing condensed and liquid pyrolysis products), the empty basket and the rubber stopper were weighed again."

"The two most extreme temperature profiles, that of the reactor wall and that within the bulk of the sample, were used to calculate ΔT and the resulting heat transfer coefficient."

Results and Discussion

"Focusing first on the differences between the two feedstock, the char yield for SS ranged from 45 to 48%, while that of OF was 52%."

"The gas yield from OF was the highest observed being 33% while gas fractions from SS had a range of 19 to 29%."

"For SS feedstock, the particle size (pellets vs. ground material) was not observed (within SD) to influence mass yields."

"Five samples, one every 2 min, were taken during SS pyrolysis."

"This is explained by the higher pyrolysis temperature used and by the 10% lower volatile content of OF compared to SS."

"Gas sample collection from OF pyrolysis after the 6-min mark was not possible due to low flow rate."

"The fraction of carbon dioxide was some 6 to 11%, significantly higher than in SS pyrolysis, whereas the oxygen and nitrogen fraction were markedly lower."

Conclusions

"Physical characterisation of a pyrolysis reactor gives information on how process parameters influence heat transfer."

"This is essential knowledge in interpreting experimental results and in identifying differences between laboratory and pilot-scale processes and reactor design."

"Pyrolysis char and gas yields were characterised and show potential for use as fuels in energy recovery from these two abundant feedstocks."

Assessment of the Differences Between Composting and Anaerobic Digestion of Sludge [317]

This is a machine-generated summary of:

Younesian, Maryam; Cheraghi, Mehrdad; Sobhanardakani, Soheil; Lorestani, Bahareh; Merrikhpour, Hajar: Assessment of the differences between composting and anaerobic digestion of sludge [317]

Published in: Arabian Journal of Geosciences (2021)

Link to original: https://doi.org/10.1007/s12517-021-08891-6

If you want to cite the papers, please refer to the original.

For technical reasons we could not place the page where the original quote is coming from.

Abstract-Summary

"The evaluated parameters included temperature, carbon to nitrogen ratio (C:N), total C, total N, moisture content, volatile solids, ash, electrical conductivity (EC), pH, and fecal and total coliforms."

"Properties of the waste sludge and both treatments including windrow composting and anaerobic digestion were assessed at the initial stage, and during and end of the degradation period."

"Based on field observations, it could be concluded that after 14 weeks of degradation, compared with the aerobic degradation, the anaerobic digestion would yield a digestate of a higher quality with the final total nitrogen of 2.8%, final total carbon (TOC) of 44%, and C:N of 15.7."

"A substantial reduction was observed in C:N ratio, total carbon content, and the moisture content in waste sludge following degradation, while an increase in nitrogen content as well as pH was observed in digestate."

Introduction

"Using compost reduces the risks of pollution in the environment compared to the use of sewage sludge because the levels of heavy metals (Hua and others 266), nitrates (Correa and others 267), and pathogens (Khalil and others 268) in the compost production process decrease."

"Differing in terms of the degree of control they exert on the environmental conditions such as aerobic or anaerobic during waste degradation, composting techniques influence the quality of end product, thereby affecting on-farm N cycling (Shah and others 269)."

"Not only do the storage conditions affect the level of N losses, but they also specify the properties of the end product which can be considered decisive for the further decomposition and N mineralization processes (Shah and others 269)."

"This study was performed to assess the adequacy of the composting procedure in promoting the physicochemical and microbiological properties of waste sludge collected from the Serkan treatment plants, Hamedan province, west of Iran, treated aerobically and anaerobically using sawdust and rice bran as bulking agents."

Materials and Methods

"The composting process was screened by measuring the main physical and chemical properties including temperature, pH, moisture content, total carbon (C) content, total nitrogen (N) content, C:N ratio, ash content, electrical conductivity (EC), and microbial population 5 days a week for 100 days."

"Moisture content was determined using the EPA standard method 2540 G by drying the mass at 105 °C for 24 h. pH and EC of the samples were measured using a pH meter (780 pH meter, Metrohm) and conductivity meter (8301 EC meter, Hanna), respectively, after the samples were mixed with distilled water at 1:10 (w:v) ratio."

"The ash content was specified after drying the samples at 105 °C, followed by calcination at 550 °C for 1 h and 750 °C for 2 h. Total carbon (C) content was measured using the following formula (USEPA 270): The modified Kjeldahl method (standard method TCW1:2003-E) was employed for calculating the total N content in three steps including digestion, distillation, and titration."

Results and Discussion

"Sawdust and rice bran, with a low moisture and C:N ratio, are good organic bulking agents, which, when mixed with waste sludge, can not only adjust the water content and decrease the C:N ratio, but also improve the structure and increase the porosity of the compost pile."

"The results of the present study indicated that the entire reduction trend for total C would be as follows: windrow composting process, from 56 to 34% (39.2% reduction rate), and anaerobic digestion process, from 56 to 44% (21.4% reduction rate)."

"By Liu and others (271), in their study, the C:N ratio in the sewage sludge treatments with natural pine wood and modified pine wood as additives decreased from the initial 14.33 and 13.58 to 9.09 and 7.26, respectively, at the end of composting."

Conclusions

"Reducing the carbon content and increasing the nitrogen content per unit weight of the material reduces the C:N ratio during the oxidation process (reaching 17.9 and 15.7 for windrow and anaerobic treatments, respectively)."

"The oxidation process cycle significantly decreased EC and changed the pH of the final compost from 6.28 to 7.2 in windrow treatment and from 7.51 to 7.3 in anaerobic treatment."

"The results also showed that the final values of C:N ratio in the studied treatments were less than 25, which indicates the maturity of the compost."

"It is worth mentioning that, in this study, the produced compost and digestate were added to the soil in 3 levels and their effect on the growth of Lepidium sativum seeds was examined."

Bibliography

1. Wang, H., Wang, T., Zhang, B., Li, F., Toure, B., Omosa, I. B., Chiramba, T., Abdel-Monem, M., & Pradhan, M. (2014). Water and wastewater treatment in Africa–current practices and challenges. Clean-Soil Air Water, 42, 1029–1035.
2. Bedbabis, S., Trigui, D., Ahmed, C. B., Clodoveo, M. L., Camposeo, S., Vivaldi, G. A., & Rouina, B. B. (2015). Long-terms effects of irrigation with treated municipal wastewater on soil, yield and olive oil quality. Agricultural Water Management, 160, 14–21.
3. Khan, S., Cao, Q., Zheng, Y. M., Huang, Y. Z., & Zhu, Y. G. (2008). Health risks of heavy metals in contaminated soils and food crops irrigated with wastewater in Beijing, China. Environmental Pollution, 152(3), 686–692.
4. Sachan, S. A. N. J. A. Y., Singh, S. K., & Srivastava, P. C. (2007). Buildup of heavy metals in soil-water-plant continuum as influenced by irrigation with contaminated effluent. Journal of Environmental Science & Engineering, 49(4), 293–296.
5. Monneveux, P., & Nemmar, M. (1986). Contribution à l'étude de la résistance à la sécheresse chez le blé tendre (Triticum aestivum L.) et chez le blé dur (Triticum durum Desf.): étude de l'accumulation de la proline au cours du cycle de développement. Agronomie, 6(6), 583–590.
6. Albouchi, A., Ghrir, R., & Elaouni, M. H. (1997). "Endurcissement à la sécheresse et accumulation de glucides solubles et d'acides aminés libres dans les phyllodes d'Acacia cyanophylla Lindl." Annales Des Sciences Forestières, 54(2), 155–168.
7. Elias, S. H., Mohamed, M., Nor-Anuar, A., Muda, K., Hassan, M. A. H. M., Othman, M. N., & Chelliapan, S. (2014). Ceramic industry wastewater treatment by rhizofiltration system-application of water hyacinth bioremediation. IIOAB Journal, 5, 6.
8. Kumari, M., & Tripathi, B. (2015). Effect of Phragmitesaustralis and Typhalatifolia on biofiltration of heavy metals from secondary treated effluent. International Journal of Environmental Science and Technology, 12, 1029–1038.
9. Ning, D., Huang, Y., Pan, R., Wang, F., & Wang, H. (2014). Effect of eco-remediation using planted floating bed system on nutrients and heavy metals in urban river water and sediment: A field study in China. Science of the Total Environment, 485, 596–603.
10. Rana, V., & Maiti, S. K. (2018). Municipal wastewater treatment potential and metal accumulation strategies of Colocasia esculenta (L.) Schott and Typha latifolia L. in a constructed wetland. Environmental Monitoring and Assessment, 190(6), 328.
11. Upadhyay, A. R., Mishra, V. K., Pandey, S. K., & Tripathi, B. D. (2007). Biofiltration of secondary treated municipal wastewater in a tropical city. Ecological Engineering, 30(1), 9–15.
12. Baskaran, L., Sundaramoorthy, P., Chidambaram, A. L. A., & Ganesh, K. S. (2009). Growth and physiological activity of greengram (Vignaradiata L.) under effluent stress. Botany Research International, 2(2), 107–114.

13. Esperanza, M., Suidan, M. T., Marfil-Vega, R., Gonzalez, C., Sorial, G. A., McCauley, P., & Brenner, R. (2007). Fate of sex hormones in two pilot-scale municipal wastewater treatment plants: conventional treatment. Chemosphere, 66, 1535–1544.
14. Hamid, H., & Eskicioglu, C. (2012). Fate of estrogenic hormones in wastewater and sludge treatment: a review of properties and analytical detection techniques in sludge matrix. Water Research, 46(18), 5813–5833.
15. Liu, Z. h., Kanjo, Y., & Mizutani, S. (2009). Removal mechanisms for endocrine disrupting compounds (EDCs) in wastewater treatment—physical means, biodegradation, and chemical advanced oxidation: a review. Science of the Total Environment, 407(2), 731–748.
16. Haft DH, Tovchigrechko A (2012) High-speed microbial community profiling. Nat Methods 9:793–794
17. Miescier JJ, Cabelli VJ (1982) Enterococci and other microbial indicators in municipal wastewater effluents. J Water Pollut Control Fed 54:1599–1606
18. Harwood VJ, Levine AD, Scott TM, Chivukula V, Lukasik J, Farrah SR, Rose JB (2005) Validity of the indicator organism paradigm for pathogen reduction in reclaimed water and public health protection. Appl Environ Microbiol 71:3163–3170
19. Knights D, Kuczynski J, Charlson ES, Zaneveld J, Mozer MC, Collman RG, Bushman FD, Knight R, Kelley ST (2011) Bayesian community-wide culture-independent microbial source tracking. Nat Methods 8:761–U107
20. Langille MGI, Zaneveld J, Caporaso JG, McDonald DM, Knights D, Reyes JA, Clemente JC, Burkepile DE, Thurber RLV, Knight R, Beiko RG, Huttenhower C (2013) Predictive functional profiling of microbial communities using 16S rRNA marker gene sequences. Nat Biotechnol 31:814
21. Collado L, Levican A, Perez J, Figueras MJ (2011) Arcobacter defluvii sp. nov., isolated from sewage samples. Int J Syst Evol Microbiol 61:2155–2161
22. Cai L, Ju F, Zhang T (2013) Tracking human sewage microbiome in a municipal wastewater treatment plant. Appl Microbiol Biotechnol 98:3317–3326. https://doi.org/10.1007/s00253-013-5402-z
23. Shchgolkova NM, Krasnov GS, Belova AA, Dmitriev AA, Kharitonov SL, Klimina KM, Melnikova NV, Kudryavtseva AV (2016) Microbial community structure of activated sludge in treatment plants with different wastewater compositions. Front Microbiol 7:90
24. Collet P, Spinelli D, Lardon L, Hélias A, Steyer J, Bernard O (2013) Life-cycle assessment of microalgal-based biofuels. In: Biofuels from Algae, pp. 287–312
25. Lardon L, Hélias A, Sialve B, Steyer J, Bernard O (2009) Life-cycle assessment of biodiesel production from microalgae. Environ Sci Technol 43:6475–6481
26. Rai UN, Tripathi RD, Singh NK, Upadhyay AK, Dwivedi S, Shukla MK, et al. Constructed wetland as an ecotechnological tool for pollution treatment for conservation of Ganga river. Bioresour Technol. 2013;148:535–41.
27. Mishra VK, Otter P, Shukla R, Goldmaier A, Alvarez JA, Khalil N, et al. Application of horizontal flow constructed wetland and solar driven disinfection technologies for wastewater treatment in India. Water Pract Technol. 2018;13:469–80.
28. Gschlößl, T., Steinmann, C., Schleypen, P., & Melzer, A. (1998). Constructed wetlands for effluent polishing of lagoons. Water Research, 32, 2639–2645.
29. Horne, A. J. (1995). Nitrogen removal from waste treatment pond or activated sludge plant effluents with free surface wetlands. Water Science and Technology, 31, 341–351.
30. Masi, F., Martinuzzi, N., Loiselle, S., Peruzzi, P., & Bacci, M. (1999). The tertiary treatment pilot plant of PubliSer SPA (Florence, Tuscany): a multistage experience. Water Science and Technology, 40, 195–202.
31. American Public Health Association (APHA). (1989). Standard methods for the examination of water and wastewater (17th ed.). Washington, DC, USA: APHA–AWWWA–WEF.
32. Giannakis S, Gamarra Vives FA, Grandjean D, Magnet A, De Alencastro LF, Pulgarin C (2015) Effect of advanced oxidation processes on the micropollutants and the effluent organic matter contained in municipal wastewater previously treated by three different secondary methods. Water Res 84:295–306

33. Cai MJ, Lin Y (2016) Effects of effluent organic matter (EfOM) on the removal of emerging contaminants by ozonation. Chemosphere 151:332–338
34. Bing J, Hu C, Nie Y, Yang M, Qu J (2015) Mechanism of catalytic ozonation in Fe2O3/Al2O3@SBA-15 aqueous suspension for destruction of ibuprofen. Environ Sci Technol 49:1690–1697
35. Goncalves AG, Orfao JJ, Pereira MF (2015) Ozonation of bezafibrate over ceria and ceria supported on carbon materials. Environ Technol 36:776–785
36. Huang Y, Cui C, Zhang D, Li L, Pan D (2015) Heterogeneous catalytic ozonation of dibutyl phthalate in aqueous solution in the presence of iron-loaded activated carbon. Chemosphere 119:295–301
37. Ikhlaq A, Brown DR, Kasprzyk-Hordern B (2015) Catalytic ozonation for the removal of organic contaminants in water on alumina. Appl Catal B Environ 165:408–418
38. Bai ZY, Yang Q, Wang JL (2016a) Catalytic ozonation of sulfamethazine using Ce0.1Fe0.9OOH as catalyst: mineralization and catalytic mechanisms. Chem Eng J 300:169–176
39. Park J, Heechul PJ, Choi H, Ahn K, Kang J (2004) Removal mechanism of natural organic matter and organic acid by ozone in the presence of goethite. Ozone Sci Eng 26:141–151
40. Park H, Kim Y, An B, Choi H (2012) Characterization of natural organic matter treated by iron oxide nanoparticle incorporated ceramic membrane-ozonation process. Water Res 46:5861–5870
41. Wu J, Ma L, Chen Y, Cheng Y, Liu Y, Zha X (2016) Catalytic ozonation of organic pollutants from bio-treated dyeing and finishing wastewater using recycled waste iron shavings as a catalyst: removal and pathways. Water Res 92:140–148
42. Chen CM, Chen HS, Guo X, Guo SH, Yan GX (2014) Advanced ozone treatment of heavy oil refining wastewater by activated carbon supported iron oxide. J Ind Eng Chem 20:2782–2791
43. Zhuang H, Han H, Hou B, Jia S, Zhao Q (2014) Heterogeneous catalytic ozonation of biologically pretreated Lurgi coal gasification wastewater using sewage sludge based activated carbon supported manganese and ferric oxides as catalysts. Bioresource Technol 166:178–186
44. Bai ZY, Yang Q, Wang JL (2016b) Fe3O4/multi-walled carbon nanotubes as an efficient catalyst for catalytic ozonation of p-hydroxybenzoic acid. Int J Environ Sci Technol 13:483–492
45. Bai ZY, Yang Q, Wang JL (2016c) Catalytic ozonation of sulfamethazine antibiotics using Ce0.1Fe0.9OOH: catalyst preparation and performance. Chemosphere 161:174–180
46. Wang SZ, Wang JL (2017) Carbamazepine degradation by gamma irradiation coupled to biological treatment. J Hazard Mater 321:639–646
47. Clark, E. A., Sterritt, R. M., & Lester, J. N. (1988). The fate of tributyltin in the aquatic environment: an overview. Environmental Science and Technology, 22, 600–604.
48. Dowson, P. H., Bubb, J. M., & Lester, J. N. (1993a). A study of the partitioning and sorptive behaviour of butyltins in the aquatic environment. Applied Organometallic Chemistry, 7, 623–633.
49. Fent, K., & Muller, M. D. (1991). Occurrence of organotins in municipal wastewater and sewage sludge and behaviour in a treatment plant. Environmental Science and Technology, 25, 489–493.
50. Voulvoulis, N., Scrimshaw, M. D., & Lester, J. N. (2004). Removal of organotins during sewage treatment: a case study. Environmental Technology, 25, 733–740.
51. Vossler, T. L., & Maclas, E. (1986). Contribution of fine particle sulfates to light scattering in St. Louis summer aerosol. Environmental Science and Technology, 20, 1235–1243.
52. Kawai, S., Kurokawa, Y., Harino, H., & Fukushima, M. (1998). Degradation of tributyltin by a bacterial strain isolated from polluted river water. Environmental Pollution, 102, 259–263.
53. Stasinakis, A. S., Thomaidis, N. S., Nikolaou, A., & Kantifes, A. (2005). Aerobic biodegradation of organotin compounds in activated sludge batch reactors. Environmental Pollution, 134, 431–438.
54. Kasprzyk-Hordern B, Dinsdale RM, Guwy AJ (2009) The removal of pharmaceuticals, personal care products, endocrine disruptors and illicit drugs during wastewater treatment and its impact on the quality of receiving waters. Water Res 43:363–380

55. Narumiya M, Nakada N, Yamashita N, Tanaka H (2013) Phase distribution and removal of pharmaceuticals and personal care products during anaerobic sludge digestion. J Hazard Mater 260:305–312
56. Pomiès M, Choubert JM, Wisniewski C, Coquery M (2013) Modelling of micropollutant removal in biological wastewater treatments: a review. Sci Total Environ 443:733–748
57. Ratola N, Cincinelli A, Alvesc A, Katsoyiannisa A (2012) Occurrence of organic microcontaminants in the wastewater treatment process. A mini review. J Hazard Mater 239–240:1–18
58. Verlicchi P, Al Aukidy M, Zambello E (2012) Occurrence of pharmaceutical compounds in urban wastewater: removal, mass load and environmental risk after a secondary treatment—a review. Sci Total Environ 429:123–155
59. Semblante GU, Hai FI, Huang X, Ball AS, Price WE, Nghiem LD (2015) Trace organic contaminants in biosolids: impact of conventional wastewater and sludge processing technologies and emerging alternatives. J Hazard Mater 300:1–17
60. Huang C, Abass OK, Yu C (2016) Triclosan: a review on systematic risk assessment and control from the perspective of substance flow analysis. Sci Total Environ 566–567:771–785
61. Maddela NR, Sheng B, Yuan S, Zhou ZB, Villamar-Torres R, Meng FG (2019) Roles of quorum sensing in biological wastewater treatment: a critical review. Chemosphere 221:616–629. https://doi.org/10.1016/j.chemosphere.2019.01.064
62. Liu H, Fang HHP (2002) Extraction of extracellular polymeric substances (EPS) of sludges. J Biotechnol 95:249–256. https://doi.org/10.1016/S0168-1656(02)00025-1
63. Yan X, Zheng SK, Yang J, Ma JH, Han YP, Feng JL, Su XF (2020b) Effects of hydrodynamic shear stress on sludge properties, N2O generation, and microbial community structure during activated sludge process. J Environ Manag 274:111215. https://doi.org/10.1016/j.jenvman.2020.111215
64. Ye FX, Peng G, Li Y (2011) Influences of influent carbon source on extracellular polymeric substances (EPS) and physicochemical properties of activated sludge. Chemosphere 84:1250–1255. https://doi.org/10.1016/j.chemosphere.2011.05.004
65. RI, Dick: Role of activated sludge final settling tanks. J sanit eng div asce (1970).
66. A, Konopka; T, Zakharova; L, Oliver; E, Paseuth; RF, Turco: Physiological state of a microbial community in a biomass recycle reactor. J ind microbiol biotechnol (1998).
67. TM, LaPara; T, Zakharova; CH, Nakatsu; A, Konopka: Functional and structural adaptations of bacterial communities growing on particulate substrates under stringent nutrient limitation. Microb ecol (2002).
68. A, Konopka; T, Zakharova; C, Nakatsu: Effect of starvation length upon microbial activity in a biomass recycle reactor. J ind microbiol biotechnol (2002).
69. CG, Klatt; TM, LaPara: Aerobic biological treatment of synthetic municipal wastewater in membrane-coupled bioreactors. Biotechnol bioeng (2003).
70. TM, LaPara; A, Konopka; CH, Nakatsu; JE, Alleman: Thermophilic aerobic treatment of a synthetic wastewater in a membrane-coupled bioreactor. J ind microbiol biotechnol (2001).
71. Yakovlev SV, Skirdov IV, Shvetsov VN, Bondarev AA, Andrianov YK (1985) Biological treatment of wastewater. Stroyizdat, Moscow (in Russian)
72. Liu Yu (2008) Wastewater purification: aerobic granulation in sequencing batch reactors. CRC Press, London
73. Liu Y, Tay JH (2004) State of the art of biogranulation technology for wastewater treatment. Biotechnol Adv 22(7):533–563
74. Pronk M, De Kreuk MK, De Bruin B, Kamminga P, Kleerebezem RV, Van Loosdrecht MCM (2015b) Full scale performance of the aerobic granular sludge process for sewage treatment. Water Res 84:207–217
75. Nielsen, P. H. et al. A conceptual ecosystem model of microbial communities in enhanced biological phosphorus removal plants. Water Res. 44, 5070–5088 (2010).
76. Hu M, Wang X, Wen X, Xia Y (2012) Microbial community structures in different wastewater treatment plants as revealed by 454-pyrosequencing analysis. Bioresour Technol 117:72–79. https://doi.org/10.1016/j.biortech.2012.04.061

77. Wang X, Hu M, Xia Y, Wen X, Ding K (2012) Pyrosequencing analysis of bacterial diversity in 14 wastewater treatment systems in China. Bioresour Technol 78(19):7042–7047
78. Noyan K, Allı B, Okutman Taş D, Sözen S, Orhon D (2017) Relationship between COD particle size distribution, COD fractionation and biodegradation characteristics in domestic sewage. Chem Tech Biotech 92:2142–2149
79. Ab Halim MH, Anuar AN, Jamal NSA, Azmi SI, Ujang Z, Bob MM (2016) Influence of high temperature on the performance of aerobic granular sludge in biological treatment of wastewater. J Environ Manag 184:271–280
80. Jiang Y, Shang Y, Gong T, Hu Z, Yang K, Shao S (2020) High concentration of Mn2+ has multiple influences on aerobic granular sludge for aniline wastewater treatment. Chemosphere 253:126664
81. Li C, Liu S, Ma T, Zheng M, Ni J (2019) Simultaneous nitrification, denitrification and phosphorus removal in a sequencing batch reactor (SBR) under low temperature. Chemosphere 229:132–141
82. Świątczak P, Cydzik-Kwiatkowska A (2018) Performance and microbial characteristics of biomass in a full-scale aerobic granular sludge wastewater treatment plant. Environ Sci Pollut R 25:1655–1669
83. Yang X, Xu T, Cao P, Qiao K, Wang L, Zhao T, Zhu J (2019) The viscosity behaviors of bacterial suspensions or extracellular polymeric substances and their effects on aerobic granular sludge. Environ Sci Pollut R 26:30087–30097
84. Gu J, Yang Q, Liu Y (2018) Mainstream anammox in a novel A-2B process for energy-efficient municipal wastewater treatment with minimized sludge production. Water Res 138:1–6
85. Azam HM, Alam ST, Hasan M, Yameogo DDS, Kannan AD, Rahman A, Kwon MJ (2019) Phosphorous in the environment: characteristics with distribution and effects, removal mechanisms, treatment technologies, and factors affecting recovery as minerals in natural and engineered systems. Environ Sci Pollut R 26:20183–20207
86. Pang H, Li L, He J, Yan Z, Ma Y, Nan J, Liu Y (2020) New insight into enhanced production of short-chain fatty acids from waste activated sludge by cation exchange resin-induced hydrolysis. Chem Eng J 388:124235
87. Zhang M, Ji B, Liu Y (2021) Microalgal-bacterial granular sludge process: a game changer of future municipal wastewater treatment? Sci Total Environ 752:141957
88. APHA (2005) Standard methods for the examination of water and wastewater, 21th edn. Washington
89. Ji B, Zhang M, Gu J, Ma Y, Liu Y (2020a) A self-sustaining synergetic microalgal-bacterial granular sludge process towards energy-efficient and environmentally sustainable municipal wastewater treatment. Water Res 179:115884
90. Ji B, Zhang M, Wang L, Wang S, Liu Y (2020b) Removal mechanisms of phosphorus in non-aerated microalgal-bacterial granular sludge process. Bioresour Technol 312:121531
91. Morgenroth E, Sherden T, van Loosdrecht MCM, Heijnen JJ, Wilderer PA (1997) Aerobic granular sludge in a sequencing batch reactor. Water Res 31:3191–3194
92. Beun JJ, van Loosdrecht MCM, Heijnen JJ (2000) Aerobic granulation. Water Sci Technol 41:41–48
93. Tay ST-L, Hemond HF, Polz MF, Cavanaugh CM, Dejesus I, Krumholz LR (1998) Two new Mycobacterium strains and their role in toluene degradation in a contaminated stream. Appl Environ Microbiol 64:1715–1720
94. Malik A, Sakamoto M, Hanazaki S, Osawa M, Suzuki T, Tochigi M, Kakii K (2003) Coaggregation among nonflocculating bacteria isolated from activated sludge. Appl Environ Microbiol 69:6056–6063
95. Beun JJ, Hendriks A, Van Loosdrecht MCM, Morgenroth E, Wilderer PA, Heijnen JJ (1999) Aerobic granulation in a sequencing batch reactor. Water Res 33:2283–2290
96. Isanta E, Reino C, Carrera J, Perez J (2015) Stable partial nitritation for low-strength wastewater at low temperature in an aerobic granular reactor. Water Res 80:149–158. https://doi.org/10.1016/j.watres.2015.04.028

97. Bian W et al (2017) Achieving nitritation in a continuous moving bed biofilm reactor at different temperatures through ratio control. Bioresour Technol 226:73–79. https://doi.org/10.1016/j.biortech.2016.12.014
98. Kouba V, Vejmelkova D, Proksova E, Wiesinger H, Concha M, Dolejs P, Hejnic J, Jenicek P, Bartacek J (2017) High-rate partial nitritation of municipal wastewater after psychrophilic anaerobic pretreatment. Environ Sci Technol 51:11029–11038. https://doi.org/10.1021/acs.est.7b02078
99. Poot V, Hoekstra M, Geleijnse MA, van Loosdrecht MC, Perez J (2016) Effects of the residual ammonium concentration on NOB repression during partial nitritation with granular sludge. Water Res 106:518–530. https://doi.org/10.1016/j.watres.2016.10.028
100. Soler-Jofra A, Wang R, Kleerebezem R, van Loosdrecht MCM, Pérez J (2019) Stratification of nitrifier guilds in granular sludge in relation to nitritation. Water Res 148:479–491. https://doi.org/10.1016/j.watres.2018.10.064
101. Picioreanu C, Perez J, van Loosdrecht MC (2016) Impact of cell cluster size on apparent half-saturation coefficients for oxygen in nitrifying sludge and biofilms. Water Res 106:371–382. https://doi.org/10.1016/j.watres.2016.10.017
102. Wang Q, Duan H, Wei W, Ni BJ, Laloo A, Yuan Z (2017) Achieving stable mainstream nitrogen removal via the nitrite pathway by sludge treatment using free ammonia. Environ Sci Technol. https://doi.org/10.1021/acs.est.7b02776
103. Wang L, Zheng P, Chen TT (2012) Performance of autotrophic nitrogen removal in the granular sludge bed reactor. Bioresour Technol 123:78–85. https://doi.org/10.1016/j.biortech.2012.07.112
104. Qian F, Wang J, Shen Y, Wang Y, Wang S, Chen X (2017) Achieving high performance completely autotrophic nitrogen removal in a continuous granular sludge reactor. Biochem Eng J 118:97–104. https://doi.org/10.1016/j.bej.2016.11.017
105. Liu WR, Ji XM, Wang JF, Yang DH, Shen TL, Chen CJ, Qian FY, Wu P (2018a) Microbial community response to influent shift and lowering temperature in a two-stage mainstream deammonification process. Bioresour Technol 262:132–140. https://doi.org/10.1016/j.biortech.2018.04.082
106. Bull PS, Evans JV, Wechsler RM, Cleland KJ (1983) Biological technology of the treatment of leachate from sanitary landfills. Water Res 17:1473–1481
107. Lin CY (1991) Anaerobic digestion of landfill leachate. Water SA 17:301–306
108. Barrantes Leiva M, Hosseini Koupaie E, Eskicioglu C (2014) Anaerobic co-digestion of wine/fruit-juice production waste with landfill leachate diluted municipal sludge cake under semi-continuous flow operation. Waste Manage 34:1860–1870
109. Ben Dhia TO, Bouallagui H, Cayol JL, Ollivier B, Fardeau ML, Hamdi M (2009) Anaerobic degradation of landfill leachate using an upflow anaerobic fixed-bed reactor with microbial sulfate reduction. J Hazard Mater 167:1133–1140
110. Uygur A, Kargi F (2004) Biological nutrient removal from pre-treated landfill leachate in a sequencing batch reactor. J Environ Manage 71:9–14
111. Contrera RC, da Silva KC, Morita DM, Rodrigues JAD, Zaiat M, Schalch V (2014) First-order kinetics of landfill leachate treatment in a pilot-scale anaerobic sequence batch biofilm reactor. J Environ Manage 145:385–393
112. Sun H, Yang Q, Peng Y, Shi X, Wang S, Zhang S (2010) Advanced landfill leachate treatment using a two-stage UASB-SBR system at low temperature. J Environ Sci 22:481–485
113. Abood AR, Bao J, Abudi ZN (2013) Biological nutrient removal by internal circulation upflow sludge blanket reactor after landfill leachate pretreatment. J Environ Sci 25(10):2130–2137
114. Nedwell DB, Reynolds PJ (1996) Treatment of landfill leachate by methanogenic and sulphate-reducing digestion. Water Res 30:21–28
115. Zayen A, Mnif S, Aloui F, Feki F, Loukil S, Bouaziz M, Sayadi S (2010) Anaerobic membrane bioreactor for the treatment of leachates from Jebel Chakir discharge in Tunisia. J Hazard Mater 177:918–923
116. Xue Y, Zhao H, Ge L, Chen Z, Dang Y, Sun D (2015) Comparison of the performance of waste leachate treatment in submerged and recirculated membrane bioreactors. Int Biodeterior Biodegrad 102:73–80

117. Chen Y, Cheng JJ, Creamer KS (2008) Inhibition of anaerobic digestion process: a review. Bioresource Technol 99:4044–4064
118. Yellezuome D, Zhu X, Wang Z, Liu R (2022) Mitigation of ammonia inhibition in anaerobic digestion of nitrogen-rich substrates for biogas production by ammonia stripping: a review. Renew Sust Energ Rev 157:112043
119. Zhang L, Lee YW, Jahng D (2012) Ammonia stripping for enhanced biomethanization of piggery wastewater. J Hazard Mater 199–200:36–42
120. Guo JS, Abbasa AA, Chena YP, Liu ZP, Fang F, Chen P (2010) Treatment of landfill leachate using a combined stripping, Fenton, SBR, and coagulation process. J Hazard Mater 178:699–705
121. Abood AR, Bao J, Du J, Zheng D, Luo Y (2014) Non-biodegradable landfill leachate treatment by combined process of agitation, coagulation, SBR and filtration. Waste Manage 34:439–447
122. Liu Z, Wu W, Shi P, Guo J, Cheng J (2015) Characterization of dissolved organic matter in landfill leachate during the combined treatment process of air stripping, Fenton, SBR and coagulation. Waste Manage 41:111–118
123. dos Santos HAP, de Castilhos Júnior AB, Nadaleti WC, Lourenço VA (2020) Ammonia recovery from air stripping process applied to landfill leachate treatment. Environ Sci Pollut Res 27:45108–45120
124. Maleki E, Bokhary A, Liao BQ (2018) A review of anaerobic digestion bio-kinetics. Rev Environ Sci Biotechnol 17:691–705
125. Stover EL, Kincannon DF (1982) Rotating bio-logical contactor scale-up and design. In: First International Conference On Fixed-film Biological Processes. Kings-Island, Ohio, pp 1–21
126. Van Der Meer RR, Heertjes PM (1983) Mathematical description of anaerobic treatment of wastewater in upflow reactors. Biotechnol Bioeng 25:2531–2556
127. Daija L, Selberg A, Rikmann E, Zekker I, Tenno T, Tenno T (2016) The influence of lower temperature, influent fluctuations and long retention time on the performance of an upflow mode laboratory-scale septic tank. Desalin Water Treat 57:18679–18687
128. Haydar S, Anis M, Hina GE, Aziz JA, Arbi M (2018) An innovative design of septic tank for wastewater treatment and its performance evaluation: an applicable model for developing countries. Chin J Chem Eng 26:886–891
129. Santiago-Díaz ÁL, García-Albortante J, Salazar-Peláez ML (2018) UASB-septic tank as an alternative for decentralized wastewater treatment in Mexico. Environ Technol 40:1780–1792
130. Singh RP, Kun W, Fu D (2019) Designing process and operational effect of modified septic tank for the pre-treatment of rural domestic sewage. J Environ Manag 251:109552
131. Metcalf, Eddy (2014) Wastewater Engineering: Treatment & Reuse, 5th edn. Mc Graw-Hill, Boston
132. Lettinga G, Hul shoff Pol W (1991) UASB process design for various types of wastewaters. Water Sci Technol 24:87–107
133. Behling E, Díaz A, Colina G, Herrera M, Gutiérrez E, Chacin E, Fernández N, Forster CF (1997) Domestic wastewater treatment using a UASB reactor. Bioresour Technol 61:239–245
134. Kalogo Y, Verstraete W (2000) Technical feasibility of the treatment of domestic wastewater by a CEPS-UASB system. Environmental Technology 21(1):55–65. https://doi.org/10.1080/09593332108618138
135. Lew B, Belavski M, Admon S, Tarre S, Gree M (2003) Temperature effect on UASB reactor operation for domestic wastewater treatment in temperate climate regions. Water Sci Technol 48:25–30
136. Ruiz I, Soto M, Veiga MC, Ligero P, Vega A, Blázquez R (1998) Performance of and biomass characterisation in a UASB reactor treating domestic waste water at ambient temperature. Water SA 24:215–222
137. Singh KS, Viraraghavan T (1998) Start-up and operation of UASB reactors at 20°C for municipal wastewater treatment. J Ferm Bioeng 85:609–614
138. Langenhoff AAM, Intrachandra N, Stuckey DC (2000) Treatment of dilute soluble and colloidal wastewater using an anaerobic baffled reactor: influence of hydraulic retention time. Water Res 34:1307–1317

139. APHA (1995) Standard methods for the examination of water and wastewater, 19th edn. American Public Health Association, Washington, DC
140. Liao X, Li H (2015) Biogas production from low-organic-content sludge using a high-solids anaerobic digester with improved agitation. Appl Energ 148:252–259
141. Appels L, Baeyens J, Degrève J, Dewil R (2008) Principles and potential of the anaerobic digestion of waste-activated sludge. Prog Energ Combust 34:755–781
142. Duan N, Dong B, Wu B, Dai X (2012) High-solid anaerobic digestion of sewage sludge under mesophilic conditions: feasibility study. Bioresour Technol 104:150–156
143. Cao Y, Paw Owski A (2012) Sewage sludge-to-energy approaches based on anaerobic digestion and pyrolysis: brief overview and energy efficiency assessment. Renew Sust Energ Rev 16:1657–1665
144. Dubois M, Gilles KA, Hamilton JK, Rebers PA, Smith F (1956) Colorimetric method for determination of sugars and related substances. Anal Chem 28:350–356 https://pubs.acs.org/sharingguidelines
145. Lowry OH, Rosebrough NJ, Farr AL, Randall RJ (1951) Protein measurement with the folin phenol reagent. J Biol Chem 193:265–275
146. Rice EW, Baird RB, Eaton AD, Clesceri LS (2012) Standard methods for examination of water and wastewater. APHA AWWA WEF, Washington, DC
147. Tawfik A, Ohashi A, Harada H (2006) Sewage treatment in a combined up-flow anaerobic sludge blanket (UASB)—down-flow hanging sponge (DHS) system. Biochem Eng J 29(3):210–219
148. Bodík I, Kratochvíl K, Gapariková E, Hutňan M (2003) Nitrogen removal in an anaerobic baffled filter reactor with aerobic post-treatment. Bioresour Technol 86(1):79–84
149. Castillo A, Cecchi F, Mata-Alvarez J (1997) A combined anaerobic–aerobic system to treat domestic sewage in coastal areas. Water Res 31(12):3057–3063
150. A′lvarez JA, Armstrong E, Go′mez M, Soto M (2008) Anaerobic treatment of low-strength municipal wastewater by a two-stage pilot plant under psychrophilic conditions. Bioresour Technol 99:7051–7062
151. Luostarinen S, Rintala J (2006) Anaerobic on-site black water and kitchen waste treatment using UASB-septic tanks at low temperatures. Water Sci Technol 54(2):143–149
152. Ji M, Yu J, Chen H, Yue PL (2001) Removal of slowly biodegradable COD in combined thermophilic UASB and MBBR Systems. Environ Technol 22(9):1069–1079
153. Zeeuw WJ de (1984) Acclimatization of anaerobic sludge for UASB reactor start-up. Ph-D thesis, Department of Environmental Technology, Wageningen University, The Netherlands
154. Wang XJ, Xia SQ, Chen L, Zhao JF, Renault NJ, Chovelon JM (2006) Nutrients removal from municipal wastewater by chemical precipitation in a moving bed biofilm reactor. Process Biochem 41:824–828
155. Kim Y, Mikawa K, Saito T, Tanaka K, Emori H (1997) Development of novel anaerobic/aerobic filter process for nitrogen removal using immobilized nitrifiers pellets. Water Sci Tech 36(12):151–158
156. Seghezzo L, Gutierrez MA, Trupiano AP, Figueroa ME, Cuevas CM, Zeeman G, Lettinga G (2002) The effect of sludge discharges and up-flow velocity on the removal of suspended solids in a UASB reactor treating settled sewage at moderate temperatures. In: Proceedings of the VII Latin-American work-shop and seminar on anaerobic digestion, Merida, Mexico, 22–25 October
157. Desai M, Madamwar D (1994) Anaerobic digestion of a mixture of cheese whey, poultry waste and cattle dung: a study of the use of adsorbents to improve digester performance. Environ Pollut 86:337–340. https://doi.org/10.1016/0269-7491(94)90174-0
158. Mumme J, Srocke F, Heeg K, Werner M (2014) Use of biochars in anaerobic digestion. Bioresour Technol 164:189–197. https://doi.org/10.1016/j.biortech.2014.05.008
159. Lowry OH, Rosebrough NJ, Farr AL, Randall RJ (1951) Protein measurement with the folin phenol reagent. J Biol Chem 193:265–275
160. Instituto Brasileiro De Geografia E Estatística. Levantamento Sistemático da Produção Agrícola (2014). Fundação IBGE. http://www.ibge.gov.br/home/estatistica/indicadores/agropecuaria/ispa/ispa_201412_5.shttm. Accessed 9 Dec 2017

161. Inoue KRA (2008) Produção de biogás, caracterização e aproveitamento agrícola do biofertilizante obtido na digestão da manipueira. UFV Master's Dissertation. Viçosa, MG-Brazil., 2008
162. Santos A (2009) Usos e impactos ambientais causados pela manipueira na microrregião sudoeste da Bahia-Brazil. http://www.publicacions.ub.edu/pubDigital/desarrolloLocalRegional/articulos.asp. Accessed 9 Dec 2017
163. Michael, I. et al. Urban wastewater treatment plants as hotspots for the release of antibiotics in the environment: a review. Water Res. 47, 957–995 (2013).
164. Li, B. & Zhang T. Mass flows and removal of antibiotics in two municipal wastewater treatment plants. Chemosphere 83, 1284–1289 (2011).
165. Le-Minh, N., Khan, S. J., Drewes, J. E. & Stuetz, R. M. Fate of antibiotics during municipal water recycling treatment processes. Water Res. 44, 4295–4323 (2010).
166. Kim, S. et al. Removal of antibiotics in wastewater: effect of hydraulic and solid retention times on the fate of tetracycline in the activated sludge process. Environ Sci Technol. 39, 5816–5823 (2005).
167. Prado, N., Ochoa, J. & Amrane, A. Biodegradation and biosorption of tetracycline and tylosin antibiotics in activated sludge system. Process Biochem. 44, 1302–1306 (2009).
168. Chynoweth D. P., Owens J. M. & Legrand R. Renewable methane from anaerobic digestion of biomass. Renew Energy 22, 1–8 (2001).
169. Aydin, S., Ince, B. & Ince, O. Application of Real-Time PCR to determination of combined effect of antibiotics on Bacteria, Methanogenic Archaea, Archaea in anaerobic sequencing batch reactors. Water Res. 76, 88–98 (2015).
170. Liu, Q. et al. Phosphate enhancing fermentative hydrogen production from substrate with municipal solid waste composting leachate as a nutrient. Bioresour. Technol. 190, 431–437 (2015).
171. Zhang, F. et al. Fatty acids production from hydrogen and carbon dioxide by mixed culture in the membrane biofilm reactor. Water Res. 47, 6122–6129 (2013).
172. Gianico A, Braguglia C, Mascolo G, Mininni G (2013) Partitioning of nutrients and micropollutants along the sludge treatment line: a case study. Environ Sci Pollut Res 20:6256–6265
173. Hao L, Kappler A, Li J, Obst M (2013) Mapping of heavy metal ion sorption to cell-mineral aggregates using metal-selective fluorescent probes and CLSM. Appl Environ Microbiol 79:6524–6534
174. Hao L, Guo Y, Byrne JM, Zeitvogel F, Schmid G, Ingino P, Li J, Neu TR, Swanner ED, Kappler A (2016) Binding of heavy metal ions in aggregates of microbial cells, EPS and biogenic iron minerals measured in-situ using metal- and glycoconjugates-specific fluorophores. Geochim Cosmochim Acta 180:66–96
175. Philippis RD, Colica G (2011) Exopolysaccharide-producing cyanobacteria in heavy metal removal from water: molecular basis and practical applicability of the biosorption process. Appl Microbiol Biotechnol 92:697–708
176. Chai X, Liu G, Xin Z, Hao Y, Zhao Y (2012) Complexion between mercury and humic substances from different landfill stabilization processes and its implication for the environment. J Hazard Mater 209–210:59–66
177. Kumke MU, Eidner S, Krüger T (2006) Fluorescence quenching and luminescence sensitization in complexes of Tb3+ and Eu3+ with humic substances. Environ Sci Technol 39:9528–9533
178. Monteilrivera F, Dumonceau J (2002) Fluorescence spectrometry for quantitative characterization of cobalt(II) complexation by Leonardite humic acid. Anal Bioanal Chem 374:1105–1112
179. Mihoubi D (2004) Mechanical and thermal dewatering of residual sludge. Desalination 167:135–139
180. Mohan SV, Babu VL, Sarma PN (2008) Effect of various pretreatment methods on anaerobic mixed microflora to enhance biohydrogen production utilizing dairy wastewater as substrate. Bioresour Technol 99:59–67

181. Shanmugam SR, Chaganti SR, Lalman JA, Heath DD (2014) Effect of inhibitors on hydrogen consumption and microbial population dynamics in mixed anaerobic cultures. Int J Hydrog Energy 39:249–257
182. Molla AH, Fakhru'l-Razi A (2012) Mycoremediation-a prospective environmental friendly technique of bioseparation and dewatering of domestic wastewater sludge. Environ Sci Pollut Res 19:1612–1619
183. Fakhru'l-Razi A, Molla AH (2007) Enhancement of bioseparation and dewaterability of domestic wastewater sludge by fungal treated dewatered sludge. J Hazard Mater 147:350–356
184. Liu H, Shi JS, Xu XY, Zhan XM, Fu B, Li YF (2017) Enhancement of sludge dewaterability with filamentous fungi Talaromyces flavus S1 by depletion of extracellular polymeric substances or mycelium entrapment. Bioresour Technol 245:977–983
185. Li YY, Dong LM, Li YY, Zheng XY, Zhang C, Ding N (2019a) Enhancement of biological fermented sludge dewaterability by inoculation of filamentous fungi Mucor circinelloides XY-Z and Penicillium oxalicum LY-1. Dry Technol 37:1678–1687
186. Li SM, Li RW, Tang YN, Chen G (2019b) Microwave-induced heavy metal removal from dewatered biosolids for cost-effective composting. J Clean Prod 241:118342–118350
187. Abram, D., & Davis, B. K. (1970). Structural properties and features of parasitic Bdellovibrio bacteriovorus. Journal of Bacteriology, 104(2), 948–965.
188. Ristuccia, P. A., & Cunha, B. A. (1984). Klebsiella. Infection Control, 5(7), 343–348.
189. Christensen ML, Keiding K, Nielsen PH, Jørgensen MK (2015) Dewatering in biological wastewater treatment: a review. Water Res 82:85–98
190. Mikkelsen LH, Keiding K (2002) Physico-chemical characteristics of full scale sewage sludges with implications to dewatering. Water Res 36(10):2451–2462
191. Sheng GP, Yu HQ, Li XY (2010) Extracellular polymeric substances (EPS) of microbial aggregates in biological wastewater treatment systems: A review. Biotechnol Adv 28:882–894. https://doi.org/10.1016/j.biotechadv.2010.08.001
192. He DQ, Wang LF, Jiang H, Yu HQ (2015) A Fenton-like process for the enhanced activated sludge dewatering. Chem Eng J 272:128–134
193. Liu Y, Fang HHP (2003) Influences of extracellular polymeric substances (EPS) on flocculation, settling, and dewatering of activated sludge. Crit Rev Environ Sci Technol 33:237–273. https://doi.org/10.1080/10643380390814479
194. Neyens E, Baeyens J, Dewil R (2004) Advanced sludge treatment affects extracellular polymeric substances to improve activated sludge dewatering. J Hazard Mater 106:83–92
195. Malcolm RL, Maccarthy P (1992) Quantitative evaluation of XAD-8 and XAD-4 resins used in tandem for removing organic solutes from water. Environ Int 18(6):597–607
196. More TT, Yadav JSS, Yan S, Tyagi RD, Surampalli RY (2014) Extracellular polymeric substances of bacteria and their potential environmental applications. J Environ Manage 144(21):1–25
197. Mowla, D., Tran, H. N., & Allen, D. G. (2013). A review of the properties of biosludge and its relevance to enhanced dewatering processes. Biomass and Bioenergy, 58, 365–378. https://doi.org/10.1016/j.biombioe.2013.09.002.
198. Liu, J., Wei, Y., Li, K., Tong, J., Wang, Y., & Jia, R. (2016). Microwave-acid pretreatment: a potential process for enhancing sludge dewaterability. Water Research, 90, 225–234. https://doi.org/10.1016/j.watres.2015.12.012.
199. Zemmouri, H., Mameri, N., & Lounici, H. (2015). Chitosan use in chemical conditioning for dewatering municipal-activated sludge. [article]. Water Science and Technology, 71(6), 810–816. https://doi.org/10.2166/wst.2014.532.
200. Lau, S. W., Ang, H. M., Sen, T. K., & Chua, H. B. (2014) Assessment of chitosan conditioning in a synthetic anaerobic digested sludge system. In Chemeca 2014, Perth, 28 Sep-1 Oct 2014.
201. Lau, S. W., Chong, S. H., Ang, H. M., Sen, T. K., & Chua, H. B. (2013). Dewaterability of anaerobic digested sludge with cations and chitosan as dual-conditioners. In R. Pogaku, A. Bono, & C. Chu (Eds.), Developments in sustainable chemical and bioprocess technology (pp. 11–17). US: Springer.

202. Lau, S. W., Sen, T. K., Chua, H. B., & Ang, H. M. (2015b). Conditioning aids and their applications for sludge dewatering. In T. K. Sen (Ed.), Physical chemical and biological treatment processes for water and wastewater. USA: Nova Publisher.
203. Nguyen, T. P., Hilal, N., Hankins, N. P., & Novak, J. T. (2008). Characterization of synthetic and activated sludge and conditioning with cationic polyelectrolytes. Desalination, 227(1–3), 103–110.
204. APHA. (2012). Standard methods for the examination of water and wastewater (22nd ed.). Washington, DC: American Public Health Association, American Water Works Association, and Water Environment Federation.
205. Lu MC, Lin CJ, Liao CH, Huang RY, Ting WP (2003) Dewatering of activated sludge by Fenton's reagent. Adv Environ Res 7:667–670
206. Behin J, Akbari A, Mahmoudi M, Khajeh M (2017) Sodium hypochlorite as an alternative to hydrogen peroxide in Fenton process for industrial scale. Water Res 121:120–128
207. APHA (1998) Standard methods for examination of water and wastewater, 20th edn. American Public Health Association, Washington, DC, USA
208. Luo K, Yang Q, Li XM, Zhang SY, Pang Y, Li X, Liao XS (2015) Effect of calcium ions on dewaterability of enzymatic-enhanced anaerobic digestion sludge. Appl Biochem Biotechnol 176:2346–2357
209. Zhen G, Lu X, Zhao Y, Chai X, Niu D (2012) Enhanced dewaterability of sewage sludge in the presence of Fe(II)-activated persulfate oxidation. Bioresour Technol 116:259–265. https://doi.org/10.1016/j.biortech.2012.01.170
210. Gagnon V, Chazarenc F, Comeau Y, Brisson J (2013) Effect of plant species on sludge dewatering and fate of pollutants in sludge treatment wetlands. Ecol Eng 61P:593–600. https://doi.org/10.1016/j.ecoleng.2013.02.017
211. Kołecka K, Gajewska M, Obarska-Pempkowiak H, Rohde D (2017) Integrated dewatering and stabilization system as an environmentally friendly technology in sewage sludge management in Poland. Ecol Eng 98:346–353. https://doi.org/10.1016/j.ecoleng.2016.08.011
212. Heinss U, Koottatep T (1998) Use of reed beds for faecal sludge dewatering: a synopsis of reviewed literature and interim results of pilot investigations with septage treatment in Bangkok. Thailand, EAWAG/AIT
213. Chen C, Zhang P, Zeng G et al (2010) Sewage sludge conditioning with coal fly ash modified by sulfuric acid. Chem Eng J 158:616–622. https://doi.org/10.1016/j.cej.2010.02.021
214. Yuan H, Lu T, Wang Y et al (2014) Influence of pyrolysis temperature and holding time on properties of biochar derived from medicinal herb (radix isatidis) residue and its effect on soil CO2 emission. J Anal Appl Pyrolysis 110:277–284. https://doi.org/10.1016/j.jaap.2014.09.016
215. Méndez A, Terradillos M, Gascó G (2013) Physicochemical and agronomic properties of biochar from sewage sludge pyrolysed at different temperatures. J Anal Appl Pyrolysis 102:124–130. https://doi.org/10.1016/j.jaap.2013.03.006
216. Novak JM, Lima I, Xing B et al (2009) Characterization of designer biochar produced at different temperatures and their effects on a loamy sand. Ann Environ Sci 3:195–206
217. Zhai Y, Peng W, Zeng G et al (2012) Pyrolysis characteristics and kinetics of sewage sludge for different sizes and heating rates. J Therm Anal Calorim 107:1015–1022. https://doi.org/10.1007/s10973-011-1644-0
218. De Oliveira Silva J, Filho GR, Da Silva Meireles C et al (2012) Thermal analysis and FTIR studies of sewage sludge produced in treatment plants. the case of sludge in the city of Uberlândia-MG, Brazil. Thermochim Acta 528:72–75
219. Thapa KB, Qi Y, Clayton SA, Hoadley AFA (2009) Lignite aided dewatering of digested sewage sludge. Water Res 43:623–634. https://doi.org/10.1016/j.watres.2008.11.005
220. Jaafarzadeh N, Hashempour Y, Takdastan A, Moghadam MA, Goodarzi G (2016) Evaluation of bagasse pith as a skeleton builder for improvement of sludge dewatering. Environ Eng Manag J 15:725–732
221. Kelessidis A, Stasinakis AS (2012) Comparative study of the methods used for treatment and final disposal of sewage sludge in European countries. Waste Manag 32:1186–1195

222. Tony M, Zhao YQ, Aghareed T (2009) Exploitation of Fenton and Fenton-like reagents as alternative conditioners for alum sludge conditioning. J Environ Sci 21:101–105
223. Hong C, Si Y, Xing Y, Wang Z, Qiao Q, Liu M (2015a) Effect of surfactant on bound water content and extracellular polymers substances distribution in sludge. RSC Adv 5:23383–23390
224. Hong C, Xing Y, Hua X, Si Y, Qiao G, Wang Z (2015b) Dewaterability of sludge conditioned with surfactant DDBAC pretreatment by acid/alkali. Appl Microbiol Biotechnol 99:6103–6111
225. Warman PR, Termeer WC (2005) Evaluation of sewage sludge, septic waste and sludge compost applications to corn and forage: yields and N, P and K content of crops and soils. Bioresour Technol 96:955–961. https://doi.org/10.1016/j.biortech.2004.08.003
226. Pedra F, Polo A, Ribeiro A, Domingues H (2007) Effects of municipal solid waste compost and sewage sludge on mineralization of soil organic matter. Soil Biol Biochem 39:1375–1382. https://doi.org/10.1016/j.soilbio.2006.12.014
227. Alvarenga P et al (2015) Sewage sludge, compost and other representative organic wastes as agricultural soil amendments: benefits versus limiting factors. Waste Manag 40:44–52. https://doi.org/10.1016/j.wasman.2015.01.027
228. Paredes C, Roig A, Bernal MP et al. (2000) Evolution of organic matter and nitrogen duing co-compostiing of olive mill wastwater with solid organic wastes. Biol Fertil Soils 32:222–227
229. Witter E, Lopez-Real JM (1987) The potential of sewage sludge and composting in a nitrogen recycling strategy for agriculture. Biol Agric Hortic 5:1–23. https://doi.org/10.1080/01448765.1987.9755122
230. Kubota H, Hirai I, Satoshi MF1983) Effects of compost maturity on growth of komatsuna (Brassica Rapa var. pervidis) in neubauer's pot. Soil Sci Plant Nutr 29:251–259. https://doi.org/10.1080/00380768.1983.10434626
231. Oleszczuk P (2008) Phytotoxicity of municipal sewage sludge composts related to physico–chemical properties PAHs and heavy metals. Ecotox Environ Safe 69:496–505
232. Chen Y (2012) Sewage sludge aerobic composting technology research progress. AASRI Procedia 1:339–343
233. Awasthi MK, Wanga Q, Huang H, Ren X, Lahori AH, Mahar A, Ali A, Shen F, Li R, Zhang Z (2016) Influence of zeolite and lime as additives on greenhouse gas emissions and maturity evolution during sewage sludge composting. Bioresour Technol 216:172–181
234. Polish Standard PN–R–04024:1997 Chemical and agricultural analysis of soil –determination of a content of available phosphorus potassium magnesium and manganese in organic soils (in Polish)
235. Villaseńor J, Pérez MA, Fernández FJ, Puchalski CM (2011) Monitoring respiration and biological stability during sludge composting with a modified dynamic respirometer. Bioresour Technol 102:6562–6568
236. Gattullo CE, Mininni C, Parente A, Montesano FF, Allegretta I, Terzano R (2017) Effects of municipal solid waste and sewage sludge compost based growing media on the yield and heavy metal content of four lettuce cultivars. Environ Sci Pollut Res 24:25406–25415
237. Adani F, Ubbiali C, Genevini P (2006) The determination of biological stability of composts using the dynamic respiration index: the results of experience after two years. Waste Manag 26:41–48
238. Bernal MP, Navarro AF, Sanchez-Monedero MA, Roig A, Cegarra J (1998) Influence of sewage sludge compost stability and maturity on carbon and nitrogen mineralization in soil. Soil Biol Biochem 30:305–313
239. Said-Pullicino D, Erriquens FG, Gigliotti G (2007) Changes in the chemical characteristics of water-extractable organic matter during composting and their influence on compost stability and maturity. Bioresour Technol 98:1822–1831. https://doi.org/10.1016/j.biortech.2006.06.018
240. Shao ZH, He PJ, Zhang DQ, Shao LM (2009) Characterization of water-extractable organic matter during the biostabilization of municipal solid waste. J Hazard Mater 164:1191–1997

241. Grigatti M, Dios Perez M, Blok WJ, Ciavatta C, Veeken A (2007) A standardized method for the determination of the intrinsic carbon and nitrogen mineralization capacity of natural organic matter sources. Soil Biol Biochem 39:1493–1503
242. Ammari TG, Al-Omari Q, Abbassi BE (2012) Composting sewage sludge amended with different sawdust proportions and textures and organic waste of food industry – assessment of quality. Environmental Technol 1:1–9
243. Muñoz M, Gómez-Rico MF, Font R (2013) Use of thermogravimetry for single characterisation of samples of the composting process from sewage sludge. J Anal Appl Pyrol 103:261–267
244. Doublet J, Francou C, Poitrenaud M, Houot S (2010) Sewage sludge composting: Influence of initial mixtures on organic matter evolution and N availability in the final composts. Waste Manag 30:1922–1930
245. Komilis D, Evangelou A, Voudrias E (2011) Monitoring and optimizing the co-composting of dewatered sludge: A mixture experimental design approach. J Environ Manag 92:2241–2249
246. Tello-Andrade AF, Jiménez-Moleón MC, Sánchez-Galván G (2013) Compostaje de lodo residual y lirio acuático: efecto de la presentación de la planta (fresca y parcialmente digerida). III Congreso Latinoamericano de Biotecnología Ambiental y Algal. Abril 2013 en David, Chiriquí, República de Panamá
247. Mahanta K, Jha DK, Rajkhowa DJ, Kumar M (2014) Isolation end evaluation of native cellulose degrading microorganisms for efficient bioconversion of weed biomass and rice straw. J Environ Biol 5:721–725
248. Peña PP (2014) Efecto de la dosis adicionada y la presentación de lirio acuático (fresco y libre de ácidos grasos volátiles) sobre la composta de lodo residual municipal. Master of Science Thesis, Universidad Autónoma del Estado de México, Toluca, México
249. APHA (2005) Standard methods for the examination of water and wastewater, 21st edn. American Public Health Association, Washington
250. El Fels L, Zamama M, El Asli A, Hafidi M (2014) Assessment of biotransformation of organic matter during co-composting of sewage sludge-lignocelullosic waste by chemical, FTIR analyzes, and phytotoxicity tests. Int Biodeterior Biodegrad 87:128–137. https://doi.org/10.1016/j.ibiod.2013.09.024
251. Huang GF, Wong JWC, Wu QT, Nagar BB (2004) Effect of C/N on composting of pig manure with sawdust. Waste Manag 24:805–813
252. Negre M, Mozzetti CH, Vindrola D, Piccone G (2011) Changes in chemical and biological parameters during co-composting of anaerobically digested sewage sludges with lignocellulosic material. J Environ Sci Health 46:509–517
253. Cherif H, Fathia A, Hadda O, Massimo M, Lorenzo B, Naceur J, Abdennaceur H, Daniele D (2009) Effects of municipal solid waste compost, farmyard manure and chemical fertilizers on wheat growth, soil composition and soil bacterial characteristics under Tunisian arid climate. Eur J Soil Biol 45:138–145
254. Abbasi F, Mohammadreza S, Hossein K, Ali K, Hoshang M (2016a) Effects of materials recovery facility construction on the release of fungal bioaerosols: a case study in southern of Iran. Fresenius Environ Bull 25(5):1513–1519
255. Abbasi F, Samaei M, KhodadadiH KA, Maleknia H (2016b) Effects of materials recovery facility construction on the release of fungal bioaerosols: a case study in southern Iran. Fresenius Environ Bull 5:1512–1518
256. Manáková B, Jan K, Markéta S, Jakub H (2014) Effects of combined composting and vermicomposting of waste sludge on arsenic fate and bioavailability. J Hazard Mater 280:544–551
257. Song X, Manqiang L, Wu D, Lin Q, Chenglong Y, Jiaguo J, Hu F (2014) Heavy metal and nutrient changes during vermicomposting animal manure spiked with mushroom residues. Waste Manag 34:1977–1983
258. Suthar, S.: Vermicomposting potential of Perionyx sansibaricus (Perrier) in different waste materials. Bioresour. Technol. 98, 1231–1237 (2007)

259. Khwairakpam, M., Bhargava, R.: Vermitechnology for sewage sludge recycling. J. Hazard. Mater. 161, 948–954 (2009)
260. Lalander, C.H., Komakech, A.J., Vinnerås, B.: Vermicomposting as manure management strategy for urban small-holder animal farms—Kampala case study. Waste Manag. 39, 96–103 (2015)
261. Sangwan, P., Kaushik, C.P., Garg, V.K.: Vermiconversion of industrial sludge for recycling the nutrients. Bioresour. Technol. 99, 8699–8704 (2008)
262. Wang L, Trninic M, Skreiberg Ø, Gronli M, Considine R, Antal M (2011) Is elevated pressure required to achieve a high fixed-carbon yield of charcoal from biomass? Part 1: round-Robin results for three different corncob materials. Energy Fuels 25(4):3251–3265. https://doi.org/10.1021/ef400041h
263. Mok WSL, Antal MJ Jr (1983a) Effects of pressure on biomass pyrolysis. I. Cellulose pyrolysis products. Thermochim Acta 68:155–164
264. Mok WSL, Antal MJ Jr (1983b) Effects of pressure on biomass pyrolysis. II. Heats of reaction of cellulose pyrolysis. Thermochim Acta 68:165–186
265. Bridgwater AV (2012) Review of fast pyrolysis of biomass and product upgrading. Biomass Bioenergy 38:68–94. https://doi.org/10.1016/j.biombioe.2011.01.048
266. Hua L, Wu W, Liu Y, McBride MB, Chen Y (2009) Reduction of nitrogen loss and Cu and Zn mobility during sludge composting with bamboo charcoal amendment. Environ Sci Pollut Res 16:1–9
267. Correa RS, White RE, Weatherley AJ (2006) Effect of compost treatment of sewage sludge on nitrogen behavior in two soils. Waste Manage 26:614–619
268. Khalil AI, Hassouna MS, El-Ashqar HMA, Fawzi M (2011) Changes in physical, chemical, and microbial parameters during the composting of municipal sewage sludge. World J Microbiol Biotechnol 27(10):2359–2369
269. Shah GM, Shah GA, Groot JCJ, Oenema O, Raza AS, Lantinga EA (2016a) Effect of storage conditions on losses and crop utilization of nitrogen from solid cattle manure. J Agric Sci 154(1):1–14
270. USEPA (1983) Methods for chemical analysis of water and wastes. United States Environmental Protection Agency, Office of Research and Development, Washington, DC, 20450
271. Liu Y, Ding L, Wang B, He Q, Wan D (2020) Using the modified pine wood as a novel recyclable bulking agent for sewage sludge composting: Effect on nitrogen conversion and microbial community structures. Bioresour Technol 309:123357
272. Slaimi, Ridha; Abassi, Mejda; Béjaoui, Zoubeir Assessment of Casuarina glauca as biofiltration model of secondary treated urban wastewater: effect on growth performances and heavy metals tolerance. Environmental Monitoring and Assessment (2021). https://doi.org/10.1007/s10661-021-09438-8
273. Maurício, R.; Dias, R.; Ribeiro, V.; Fernandes, S.; Vicente, A. C.; Pinto, M. I.; Noronha, J. P.; Amaral, L.; Coelho, P.; Mano, A. P. 17α-Ethinylestradiol and 17β-estradiol removal from a secondary urban wastewater using an RBC treatment system. Environmental Monitoring and Assessment (2018). https://doi.org/10.1007/s10661-018-6701-8
274. Ahmed, Warish; Staley, Christopher; Sidhu, Jatinder; Sadowsky, Michael; Toze, Simon Amplicon-based profiling of bacteria in raw and secondary treated wastewater from treatment plants across Australia. Applied Microbiology and Biotechnology (2016). https://doi.org/10.1007/s00253-016-7959-9
275. Gómez-Serrano, C.; Morales-Amaral, M. M.; Acién, F. G.; Escudero, R.; Fernández-Sevilla, J. M.; Molina-Grima, E. Utilization of secondary-treated wastewater for the production of freshwater microalgae. Applied Microbiology and Biotechnology (2015). https://doi.org/10.1007/s00253-015-6694-y
276. Shukla, Reetika; Gupta, Deepak; Singh, Gurudatta; Mishra, Virendra Kumar Performance of horizontal flow constructed wetland for secondary treatment of domestic wastewater in a remote tribal area of Central India. Sustainable Environment Research (2021). https://doi.org/10.1186/s42834-021-00087-7

277. Kim, Dong-Gill; Park, Joohyun; Lee, Dowon; Kang, Hojeong Removal of Nitrogen and Phosphorus from Effluent of a Secondary Wastewater Treatment Plant Using a Pond–Marsh Wetland System. Water, Air, & Soil Pollution (2010). https://doi.org/10.1007/s11270-010-0399-8
278. Bai, Zhiyong; Wang, Jianlong; Yang, Qi Advanced treatment of municipal secondary effluent by catalytic ozonation using Fe_3O_4-CeO_2/MWCNTs as efficient catalyst. Environmental Science and Pollution Research (2017). https://doi.org/10.1007/s11356-017-8619-z
279. Scrimshaw, M. D.; Hala, D.; Okiemute, A. E.; Cartmell, E.; Lester, J. N. Removal Processes for Tributyltin During Municipal Wastewater Treatment. Water, Air, & Soil Pollution (2013). https://doi.org/10.1007/s11270-012-1400-5
280. Kramer, Rafael D.; Filippe, Tais C.; Prado, Marcelo R.; de Azevedo, Júlio César R. The influence of solid-liquid coefficient in the fate of pharmaceuticals and personal care products in aerobic wastewater treatment. Environmental Science and Pollution Research (2018). https://doi.org/10.1007/s11356-018-2609-7
281. Yan, Xu; Yang, Jie; Guo, Dongli; Ma, Jiahui; Su, Xianfa; Sun, Jianhui Effect of carbon source on nitrous oxide emission characteristics and sludge properties during anoxic/aerobic wastewater treatment process. Environmental Science and Pollution Research (2021). https://doi.org/10.1007/s11356-021-14713-9
282. Chen, Ruoyu; LaPara, Timothy M. Aerobic Biological Treatment of Low-Strength Synthetic Wastewater in Membrane-Coupled Bioreactors: The Structure and Function of Bacterial Enrichment Cultures as the Net Growth Rate Approaches Zero. Microbial Ecology (2006). https://doi.org/10.1007/s00248-005-0081-7
283. Khokhlachev, Nikolay S.; Kalenov, Sergei V.; Zanina, Olga S.; Tyupa, Dmitry V.; Baurina, Marina M.; Kuznetsov, Alexander Ye. The role of stress agents as operating factors in formation and functioning of granular aerobic activated sludge at model domestic wastewater treatment. Bioprocess and Biosystems Engineering (2014). https://doi.org/10.1007/s00449-014-1149-x
284. Świątczak, Piotr; Cydzik-Kwiatkowska, Agnieszka Performance and microbial characteristics of biomass in a full-scale aerobic granular sludge wastewater treatment plant. Environmental Science and Pollution Research (2017). https://doi.org/10.1007/s11356-017-0615-9
285. Yılmaz, F.; Otuzaltı, M. M.; Perendeci, N. A.; Karatay, M.; Ünşar, E. Kökdemir; Ateş, M.; Akin, R.; Yıldız, O.; Orhon, D. Potential of aerobic membrane bioreactor for recycling and reuse of domestic wastewater for irrigation. Environmental Earth Sciences (2020). https://doi.org/10.1007/s12665-020-09006-2
286. Guo, Dabin; Zhang, Xuechun; Shi, Yuting; Cui, Baihui; Fan, Jie; Ji, Bin; Yuan, Julin Microalgal-bacterial granular sludge process outperformed aerobic granular sludge process in municipal wastewater treatment with less carbon dioxide emissions. Environmental Science and Pollution Research (2020). https://doi.org/10.1007/s11356-020-11565-7
287. Ivanov, Volodymyr; Wang, Xiao-Hui; Tay, Stephen Tiong-Lee; Tay, Joo-Hwa Bioaugmentation and enhanced formation of microbial granules used in aerobic wastewater treatment. Applied Microbiology and Biotechnology (2005). https://doi.org/10.1007/s00253-005-0088-5
288. Liu, Wenru; Shen, Yaoliang; Yang, Dianhai Achieving high-rate partial nitritation with aerobic granular sludge at low temperatures. Biodegradation (2021). https://doi.org/10.1007/s10532-021-09965-8
289. Magalhães, Taína Martins; Duarte, Natalia Cangussu; de Alencar Neves, Thiago; Bueno, Daniel Augusto Camargo; Delforno, Tiago Palladino; de Oliveira, Valéria Maia; Tonetti, Adriano Luiz The Challenge of Making Wastewater Treatment Plants Composed by Anaerobic Reactors Capable of Removing Nitrogen. Water, Air, & Soil Pollution (2019). https://doi.org/10.1007/s11270-019-4300-0
290. Zayen, Amal; Loukil, Slim; Bouallegui, Zouheir; Sayadi, Sami Coupling air stripping process and anaerobic digestion for the treatment of landfill leachate: organics degradation and cytotoxicity evaluation. Euro-Mediterranean Journal for Environmental Integration (2022). https://doi.org/10.1007/s41207-022-00301-x

291. Santiago-Díaz, Ángel Luis; Mugica-Álvarez, Violeta; de los Cobos-Vasconcelos, Daniel; Vaca-Mier, Mabel; Salazar-Peláez, Mónica Liliana Performance evaluation and kinetic modeling of an upflow anaerobic sludge blanket septic tank for domestic wastewater treatment. Environmental Science and Pollution Research (2021). https://doi.org/10.1007/s11356-021-15141-5
292. Gameiro, Tânia; Novais, Rui M.; Correia, Catarina L.; Carvalheiras, João; Seabra, Maria P.; Tarelho, Luís A. C.; Labrincha, João A.; Capela, Isabel Role of waste-based geopolymer spheres addition for pH control and efficiency enhancement of anaerobic digestion process. Bioprocess and Biosystems Engineering (2021). https://doi.org/10.1007/s00449-021-02522-w
293. Sevilla-Espinosa, Susana; Solórzano-Campo, Maricela; Bello-Mendoza, Ricardo Performance of staged and non-staged up-flow anaerobic sludge bed (USSB and UASB) reactors treating low strength complex wastewater. Biodegradation (2010). https://doi.org/10.1007/s10532-010-9339-y
294. Zhang, Yuyao; Li, Huan Energy recovery from wastewater treatment plants through sludge anaerobic digestion: effect of low-organic-content sludge. Environmental Science and Pollution Research (2017). https://doi.org/10.1007/s11356-017-0184-y
295. Tawfik, A.; El-Gohary, F.; Temmink, H. Treatment of domestic wastewater in an up-flow anaerobic sludge blanket reactor followed by moving bed biofilm reactor. Bioprocess and Biosystems Engineering (2009). https://doi.org/10.1007/s00449-009-0321-1
296. Wang, Tongyu; Qin, Yujie; Cao, Yan; Han, Bin; Ren, Junyi Simultaneous addition of zero-valent iron and activated carbon on enhanced mesophilic anaerobic digestion of waste-activated sludge. Environmental Science and Pollution Research (2017). https://doi.org/10.1007/s11356-017-9859-7
297. Peres, Sergio; Monteiro, Marina Rebeca; Ferreira, Micheline Lima; do Nascimento Junior, Adalberto Freire; de Los Angeles Perez Fernandez Palha, Maria Anaerobic Digestion Process for the Production of Biogas from Cassava and Sewage Treatment Plant Sludge in Brazil. BioEnergy Research (2018). https://doi.org/10.1007/s12155-018-9942-z
298. Lu, Meiqing; Niu, Xiaojun; Liu, Wei; Zhang, Jun; Wang, Jie; Yang, Jia; Wang, Wenqi; Yang, Zhiquan Biogas generation in anaerobic wastewater treatment under tetracycline antibiotic pressure. Scientific Reports (2016). https://doi.org/10.1038/srep28336
299. Wu, Boran; Dai, Xiaohu; Chai, Xiaoli Simultaneous enhancement of sludge dewaterability and removal of sludge-borne heavy metals through a novel oxidative leaching induced by nano-CaO_2. Environmental Science and Pollution Research (2017). https://doi.org/10.1007/s11356-017-9261-5
300. Yu, Li; Yu, Yang; Jiang, Wentian; Wei, Huangzhao; Sun, Chenglin Integrated treatment of municipal sewage sludge by deep dewatering and anaerobic fermentation for biohydrogen production. Environmental Science and Pollution Research (2014). https://doi.org/10.1007/s11356-014-3514-3
301. Tao, Neng; Wu, Xiu; Zhang, Feng; Pi, Zilei; Wen, Jiaqi; Fang, Di; Zhou, Lixiang Enhancement of sewage sludge dewaterability by fungal conditioning with Penicillium simplicissimum NJ12: from bench- to pilot-scale consecutive multi-batch investigations. Environmental Science and Pollution Research (2021). https://doi.org/10.1007/s11356-021-15170-0
302. Wu, Junkang; Chen, Zhoukai; Zhang, Shiwen; Gao, Lei; Yu, Ran; Zhan, Manjun Mechanistic Understanding of Predatory Bacteria-Induced Biolysis for Waste Sludge Dewaterability Improvement. Water, Air, & Soil Pollution (2019). https://doi.org/10.1007/s11270-019-4242-6
303. Wu, Boran; Chai, Xiaoli; Zhao, Youcai Enhanced dewatering of waste-activated sludge by composite hydrolysis enzymes. Bioprocess and Biosystems Engineering (2016). https://doi.org/10.1007/s00449-016-1544-6
304. Lau, Shiew Wei; Sen, Tushar Kanti; Chua, Han Bing; Ang, Ha Ming Conditioning of Synthetic Sludge and Anaerobically Digested Sludge Using Chitosan, Organic Polyelectrolytes and Inorganic Metal Cations to Enhance Sludge Dewaterability. Water, Air, & Soil Pollution (2017). https://doi.org/10.1007/s11270-017-3545-8

305. Zhu, Xiaofei; Yang, Qi; Li, Xiaoming; Zhong, Yu; Wu, You; Hou, Lihua; Wei, Jing; Zhang, Weixuan; Liu, Yu; Chen, Chongyu; Wang, Dongbo Enhanced dewaterability of waste activated sludge with Fe(II)-activated hypochlorite treatment. Environmental Science and Pollution Research (2018). https://doi.org/10.1007/s11356-018-2829-x
306. El-Gendy, Ahmed Shafik; Ahmed, Amira Galal Utilization of different plant species for dewatering of sewage sludge. Environmental Science and Pollution Research (2021). https://doi.org/10.1007/s11356-021-12982-y
307. Wu, Jiahuan; Lu, Tao; Yang, Guang; Meng, Wei; Yuan, Haoran; Chen, Yong Self-recycling of sewage sludge as a coagulant and mechanism in sewage sludge dewatering. Journal of Material Cycles and Waste Management (2020). https://doi.org/10.1007/s10163-020-01083-6
308. Tochioka, Eiji; Yamashita, Manabu; Usui, Jiro; Miyake, Haruo; Terada, Akihiko; Hosomi, Masaaki Enhancing the dewaterability of anaerobically digested sludge using fibrous materials recovered from primary sludge: demonstration from a field study. Clean Technologies and Environmental Policy (2019). https://doi.org/10.1007/s10098-019-01698-w
309. Hong, Chen; Wang, Zhiqiang; Si, Yanxiao; Yang, Qiang; Xing, Yi Improving sludge dewaterability by combined conditioning with Fenton's reagent and surfactant. Applied Microbiology and Biotechnology (2016). https://doi.org/10.1007/s00253-016-7939-0
310. Nguyen, Thanh-Binh; Shima, Kazuto Composting of Sewage Sludge with a Simple Aeration Method and its Utilization as a Soil Fertilizer. Environmental Management (2018). https://doi.org/10.1007/s00267-017-0963-8
311. Bożym, Marta; Siemiątkowski, Grzegorz Characterization of composted sewage sludge during the maturation process: a pilot scale study. Environmental Science and Pollution Research (2018). https://doi.org/10.1007/s11356-018-3335-x
312. Sciubba, Luigi; Cavani, Luciano; Grigatti, Marco; Ciavatta, Claudio; Marzadori, Claudio Relationships between stability, maturity, water-extractable organic matter of municipal sewage sludge composts and soil functionality. Environmental Science and Pollution Research (2015). https://doi.org/10.1007/s11356-015-4611-7
313. Tello-Andrade, A. F.; Jiménez-Moleón, M. C.; Sánchez-Galván, G. Management of sewage sludge by composting using fermented water hyacinth. Environmental Science and Pollution Research (2015). https://doi.org/10.1007/s11356-015-4683-4
314. Abbasi, Fariba; Mokhtari, Mehdi; Jalili, Mahrokh The impact of agricultural and green waste treatments on compost quality of dewatered sludge. Environmental Science and Pollution Research (2019). https://doi.org/10.1007/s11356-019-06618-5
315. Yilmaz Cincin, Roda Gokce; Agdag, Osman Nuri Co-vermicomposting of Wastewater Treatment Plant Sludge and Yard Waste: Investigation of Operation Parameters. Waste and Biomass Valorization (2019). https://doi.org/10.1007/s12649-019-00900-w
316. Agar, David A.; Kwapinska, Marzena; Leahy, James J. Pyrolysis of wastewater sludge and composted organic fines from municipal solid waste: laboratory reactor characterisation and product distribution. Environmental Science and Pollution Research (2018). https://doi.org/10.1007/s11356-018-1463-y
317. Younesian, Maryam; Cheraghi, Mehrdad; Sobhanardakani, Soheil; Lorestani, Bahareh; Merrikhpour, Hajar Assessment of the differences between composting and anaerobic digestion of sludge. Arabian Journal of Geosciences (2021). https://doi.org/10.1007/s12517-021-08891-6

Chapter 4
Tertiary Treatment

Nitin Vitthalrao Thombre

4.1 Introduction by Editor

Tertiary treatment is a crucial phase in the overall process of municipal wastewater treatment or sewage treatment plants. It aims to remove residual contaminants that persist after primary and secondary treatment processes. This chapter provides a comprehensive analysis of tertiary treatment, including its objectives, methods, and operational considerations. By emphasizing the significance of tertiary treatment in meeting stringent water quality standards, this chapter aims to enhance understanding and facilitate the effective implementation of tertiary treatment strategies in municipal wastewater treatment systems.

The primary objectives of tertiary treatment in municipal wastewater treatment plants encompass the removal of suspended solids, dissolved constituents, nutrients, and pathogens from the treated effluent. Suspended solids consist of fine particles, organic matter, and microorganisms that remain in the water after primary and secondary treatment stages. Dissolved constituents may include residual pharmaceuticals, pesticides, heavy metals, and other trace contaminants. Nutrients, particularly nitrogen and phosphorus, require targeting to prevent eutrophication and maintain the ecological balance of receiving water bodies. Lastly, reducing pathogens is essential for safeguarding public health and ensuring the safety of downstream water resource.

Various methods and technologies are employed for tertiary treatment in municipal wastewater treatment plants. These include filtration techniques such as rapid sand filtration, granular media filtration, and membrane filtration (including microfiltration, ultrafiltration, and reverse osmosis). Additionally, advanced oxidation processes, such as ozonation and ultraviolet (UV) disinfection, are commonly utilized to remove dissolved contaminants and pathogens. Chemical processes,

N. V. Thombre (✉)
VA Tech Wabag Ltd., Pune, Maharashtra, India
e-mail: nitin01thombre@gmail.com

N. V. Thombre (ed.), *Municipal Wastewater Reclamation and Reuse*,
https://doi.org/10.1007/978-981-96-9112-8_4

including coagulation, flocculation, and advanced precipitation methods (e.g., phosphorus removal through chemical precipitation), enhance the removal of specific constituents.

Efficient operation and design considerations play a pivotal role in optimizing the performance of tertiary treatment processes. Monitoring and controlling operational parameters such as hydraulic loading rates, contact time, chemical dosing, and process configuration are essential to ensure effective contaminant removal. Proper selection and dosage of chemicals, along with adequate mixing and settling times, are critical for achieving desired treatment objectives. Furthermore, regular equipment maintenance, including filters and membranes, is necessary to prevent fouling and maintain optimal treatment efficiency.

Tertiary treatment acts as the final barrier in municipal wastewater treatment, ensuring that the treated effluent meets stringent water quality standards. By removing residual contaminants, nutrients, and pathogens, it safeguards the environment and protects public health. Tertiary treatment is particularly vital in areas where the treated effluent is discharged into sensitive water bodies, used for irrigation, or reclaimed for beneficial purposes. Moreover, tertiary treatment plays a crucial role in meeting regulatory requirements and complying with water quality regulations, thereby promoting sustainable and responsible wastewater management practices.

4.2 Machine Generated Summaries

Disclaimer: The summaries in this chapter were generated from Springer Nature publications using extractive AI auto-summarization: An extraction-based summarizer aims to identify the most important sentences of a text using an algorithm and uses those original sentences to create the auto-summary (unlike generative AI). As the constituted sentences are machine selected, they may not fully reflect the body of the work, so we strongly advise that the original content is read and cited. The auto generated summaries were curated by the editor to meet Springer Nature publication standards. To cite this content, please refer to the original papers.

Machine generated keywords: filter, ozonation, wetland, dose, antibiotic, construct wetland, reuse, gene, hybrid, medium, horizontal, bacteria, construct, subsurface, ozone

Full-Scale Hybrid Constructed Wetlands Monitoring for Decentralized Tertiary Treatment of Municipal Wastewater [125]

This is a machine-generated summary of:

Ergaieg, Karim; Ben Miled, Taieb: Full-scale hybrid constructed wetlands monitoring for decentralized tertiary treatment of municipal wastewater [125]

Published in: Arabian Journal of Geosciences (2021)

Link to original: https://doi.org/https://doi.org/10.1007/s12517-021-07776-y

If you want to cite the papers, please refer to the original.

For technical reasons we could not place the page where the original quote is coming from.

Abstract-Summary

"This paper explores the feasibility and efficiency of a full-scale hybrid constructed wetland as a decentralized tertiary treatment facility to improve the quality of municipal secondary wastewater for reuse."

"The treatment wetland consists of a horizontal subsurface flow system (HSFCW) combined with a maturation pond (MP), with a treatment capability of 520 m^3/d and operational process since November 2017."

"The removal efficiencies achieved by the multi-stage system (HSFCW-MP) were TSS, 47.31± 30.90%; BOD_5, 45.24± 36.33%; COD, 37.52± 23.01%; TKN, 56.53± 15.70%; and TP, 69.09± 32.78%."

"HSFCW-MP achieved satisfactory removal rates and provided high-quality water from HSFCW and (HSFCW-) MP effluents that fit well with agricultural needs."

Introduction

"Due to the nature of secondary treatment processes applied to wastewater, the reuse of treated wastewater has remained very modest for municipal uses and has been extensively developed for irrigation purposes."

"The purpose of this study is to evaluate the performances of a hybrid system (full-scale pilot plant) established recently in northern Tunisia (Sidi Amor, Borj Touil) using a horizontal subsurface flow constructed wetlands combined with a maturation pond for post-treatment of domestic secondary effluents toward further agricultural reuse."

"The focus of this case study is to provide information and factual data about the potential of post-treatment by the subsurface flow system (alone) or combined with the maturation pond and reuse of the treated wastewater."

"Water quality monitoring was carried out to assess the performance of each part of the wetlands treatment system during the first 3 years of operation, with special regard to compare the final quality of treated water (to be disposed) and reclaimed water (to be reused) to the Tunisian required standards (NT106-02 and NT106-03)."

Materials and Methods

"MP was designed to meet in priority the discharge standard limits for fecal coliforms (FC: 2000 MPN per 100 ml), based on the completely mixed flow reactor model defined as follows: where B_0 is the effluent concentration of coliforms per 100 ml, B_i is the influent concentration of coliforms per 100 ml, θ_m (d) is the hydraulic retention time, and K_T is the first order rate constants for fecal coliform die-off (d^{-1}) at a given T temperature defined as follows: Moreover, MP sizing was performed taking into account both the reclaimed water quality criteria for agricultural reuse and the frequency of intended downstream irrigation, considering also the WHO guidelines setting values $\leq$ 1000 CF/100 ml and $\leq$ 1 intestinal nematode egg/l for unrestricted irrigation (Xanthoulis and others 1)."

"The HSFCW-MP post-treatment system was designed to upgrade the quality of water reclaimed from the transfer network of effluents generated by domestic wastewater plants (secondary level treatment) for agricultural irrigation purposes."

Results and Discussion

"The removal of organics and solids by the horizontal subsurface flow system was the main target at this treatment stage according to the initial design specifications and intended irrigation applications (drip and subsurface irrigation systems)."

"A comparable average removal rates to our study, i.e., 51%, was observed, by Puigagut and others (2), dealing with Kjeldahl nitrogen loading rates applied to horizontal subsurface systems, ranging from 0.4 to 13.1 g (m^2/d)."

"The monitoring of constructed wetlands, focused on horizontal subsurface systems or rather extended in this study to the combined system HSFCW-MP, was carried out over the first 3 years of operation, from December 2017 to December 2020 as described above while microbiological data set up to 2018."

"The main performances for most pollutants abatement of the HSFCW-MP hybrid system were provided by the horizontal subsurface flow system (1st treatment stage), except a further removal of total phosphorous performed by the maturation pond (2nd treatment stage)."

Conclusion

"This study investigated the performance of a two-stage hybrid CW (HSFCW-MP) including a horizontal subsurface flow system (HSFCW) and a maturation pond (MP) as tertiary treatment for municipal wastewater reuse."

"Pollutants removal efficiency by HSFCW alone and/or combined with MP was assessed during seasonal changes."

"The obtained results revealed that HSFCW-MP has the potential to provide high-quality effluents that meet the standards for irrigation uses and could be a potential alternative for tertiary wastewater treatment and reuse in developing countries and water-scarce areas."

"No significant difference was observed between HSFCW and HSFCW-MP dealing with the combined effect of season or constructed wetlands performance."

Effect of Sludge Characteristics on Membrane Fouling During Start-Up of a Tertiary Submerged Membrane Bioreactor [126]

This is a machine-generated summary of:

Díaz, Oliver; Vera, Luisa; González, Enrique; García, Elisa; Rodríguez-Sevilla, Juan: Effect of sludge characteristics on membrane fouling during start-up of a tertiary submerged membrane bioreactor [126]

Published in: Environmental Science and Pollution Research (2016)

Link to original: https://doi.org/https://doi.org/10.1007/s11356-016-6138-y

If you want to cite the papers, please refer to the original.

For technical reasons we could not place the page where the original quote is coming from.

Abstract-Summary

"In membrane bioreactors applied to wastewater treatment, fouling is typically a complex function of sludge characteristics."

"A pilot-scale tertiary submerged membrane bioreactor (tMBR) was continuously operated for over 200 days to assess the effect of biomass physiological state and environmental stress on process performance."

"During the initial period of the tMBR start-up, when MLSS was below 3000 mg/L, the biomass was found to be very sensitive to environmental stress by sudden oxygen increase or organic shock loading, resulting in temporary biomass deflocculation and BPC release, and consequently, severe induced membrane fouling."

"This period was also characterised by low bioactivity, BPC content and membrane fouling."

Introduction

"The rate and extent of fouling is typically a complex function of membrane properties, hydrodynamic and operating conditions, feedwater nature and microbial suspension characteristics (Le-Clech and others 3)."

"Several authors have tried to identify the characteristics of the suspensions that have the greatest influence on fouling, among which are concentration of suspended solids in the mixed liquor (MLSS), viscosity, particle size distribution (PSD), biomass activity and biopolymer content (Le-Clech and others 3; Meng and others 4)."

"To MLSS, other factors such as sludge morphology (i.e. particle-size distribution) and BPCs are expected to cause membrane fouling (Wang and Li 5)."

"Despite extensive research efforts to understand the effect of operating conditions on biomass characteristics and membrane fouling propensity, the variety of installations and feedwater characteristics sometimes lead to contradictory results, as in the case of sludge retention time (SRT) (Sabia and others 6; Villain and Marrot 7)."

Materials and Methods

"This effluent exhibited a high variable chemical oxygen demand (COD = 559 mg/L, on average), mainly as particulate matter (CODs/COD = 0.14 and COD/SST = 1.35), with a low readily biodegradable fraction (5-day biochemical oxygen demand BOD_5 to COD ratio was approximately 0.22)."

"Throughout the experimental time, ultrafiltration was carried out at constant permeate flux ($J = 35$ L/h m^2), registering transmembrane pressure as a function of time."

"Backwashing was applied each 7 min at constant conditions ($J_B = 60$ L/h m^2 and $t_B = 30$ s) for fouling removal."

"Membrane chemical cleaning was carried out by soaking with NaOCl (1000 ppm) for 12 h. Total and soluble chemical oxygen demand (COD and COD_s, respectively), 5-day biochemical oxygen demand (BOD_5), total suspended solids (TSS), mixed liquor suspended solids (MLSS), mixed liquor volatile suspended solids (MLVSS), time to filter (TTF) and turbidity were determined in conformity with the standard methods (8)."

Results and Discussion

"This range may be attributable to influent COD fluctuations (110–960 mg/L) since permeate COD values decreased from 81 mg/L (on average) in the initial phase to less than 35 mg/L during phases III–V. Nitrifying biomass followed the typical trend, always achieving ammonium removal rates higher than 98 % after the first 20 days, with stable behaviour after increasing airflow rate (17th day)."

"During the stable biomass growth period (phases III–V), low average SOUR values were observed (from 4.9 to 2.8 mg O_2/g MLVSS h), similar to those usually reported for MBRs operated in endogenous conditions (Pollice and others 9)."

"Since the membrane was operated at high permeate flux and low shear rates (i.e. without air scouring at the membrane surface), a major contribution from MLSS was expected in contrast with previous studies (Bae and Tak 10) that remark the importance of colloidal/soluble biopolymer rate."

Conclusions

"AOB, ammonium-oxidising bacteria; BOD_5, 5-day biochemical oxygen demand, mg/L; BPC, biopolymer clusters, mg DOC/L; CIP, chemical cleaning in place; COD, total chemical oxygen demand, mg/L; COD_s, soluble chemical oxygen demand, mg/L; d_{10}, measured particle size at 10 % of volume distribution, µm; d_{50}, measured particle size at 50 % of volume distribution, µm; d_{90}, measured particle size at 90 % of volume distribution, µm; DFC_m, Delft filtration characterisation method; DOC, dissolved organic carbon, mg/L; EfOM, organic matter present in secondary effluents, mg DOC/L; EPS, extracellular polymeric substances; $(F/M)_C$, carbon feed to microorganism ratio, kg COD_s/kg MLVSS day; $(F/M)_N$, nitrogen feed to microorganism ratio, kg N/kg MLVSS day; HRT, hydraulic retention time, h; J, filtrate flux, L/h m^2; J_B, backwashing flux, L/h m^2; MBR, membrane bioreactor; MLSS, mixed liquor total suspended solids, mg/L; MLVSS, mixed liquor volatile suspended solids, mg/L; $N–NH_3$, ammonium–nitrogen, mg/L; $N–NO_2^-$, nitrite–nitrogen, mg/L; $N–NO_3^-$, nitrate–nitrogen, mg/L; NOB, nitrite-oxidising bacteria; OPEX, operating expenditure; PSD, particle size distribution; PVDF, polyvinylidene fluoride; r, Pearson's coefficient; r_f, reversible fouling rate, Pa/s; SMP, soluble microbial products; $SOUR_{AOB}$, specific AOB oxygen uptake rate, mg O_2/g MLVSS h; $SOUR_{endogenous}$, specific endogenous oxygen uptake rate, mg O_2/g MLVSS h; $SOUR_{NOB}$, specific NOB oxygen uptake rate, mg O_2/g MLVSS h; $SOUR_{OHO}$, specific heterotrophic oxygen uptake rate, mg O_2/g MLVSS h; SRT, sludge retention time; t_B, backwashing phase duration, s; tMBR, tertiary submerged membrane bioreactor; TMP, transmembrane pressure, kPa; TMP_f, transmembrane pressure final filtration phase, kPa; TMP_i, transmembrane pressure initial filtration phase, kPa; TSS total suspended solids, mg/L; TTF time-to-filter, s; Y_{obs}, low sludge-yield coefficient, kg MLVSS/kg COD"

Monitoring of Horizontal Subsurface Flow Constructed Wetlands for Tertiary Treatment of Municipal Wastewater [127]

This is a machine-generated summary of:

Ergaieg, Karim; Msaddek, Mohamed Haythem; Kallel, Amjad; Trabelsi, Ismail: Monitoring of horizontal subsurface flow constructed wetlands for tertiary treatment of municipal wastewater [127]

Published in: Arabian Journal of Geosciences (2021)

Link to original: https://doi.org/https://doi.org/10.1007/s12517-021-08419-y

If you want to cite the papers, please refer to the original.

For technical reasons we could not place the page where the original quote is coming from.

Abstract-Summary

"This paper explores the feasibility of the electrical resistivity (ER) method as non-invasive technique to detect dysfunctions of full-scale tertiary wastewater treatment wetlands, such as clogging, that ultimately limit the system lifetime and performance."

"These results suggest that the ER method is a promising and feasible technique, as simplified then adapted/adopted to local context, for better monitoring and assessment of treatment wetlands with early process failure."

Introduction

"Clogging occurs when the pore spaces in the filter media of a HSF-CW bed are filled with either organic or inorganic solids, thus limiting the contact area and time between the biofilm (microbial) associated with plants' root systems and the water to be treated (Knowles and others 11)."

"The available clogging detection methods tested in the last 10 years (e.g., hydraulic conductivity measurement, tracer testing, clogging matter characterization) are usually complex and expensive or time-consuming to enable actual and wide large-scale application of these techniques, or to ensure sustainable monitoring of operational wetlands (Knowles and others 12; Nivala and others 13; Aiello and others 14; De Matos and others 15)."

"The distinguishing feature of our work is that it relates both to the first full-scale decentralized treatment wetlands set up in Tunisia for wastewater treatment and reuse (design features, performance, operation, and maintenance gaps), and to the feasibility of the electrical resistivity method for in situ monitoring of HSF-CW units and the detection of possible premature clogging."

Materials and Methods

"Each HSF-CW unit was fed separately by a perforated inlet pipe (Ø 120 mm) located upstream along the width of a gabion zone to allow a uniform distribution of the wastewater across the constructed wetland beds."

"The treatment efficiency achieved by HSF-CW units in the removal of suspended particles and organic matter and the applied surface loading rate (SLR, $g/m^2/day$) to each unit was respectively evaluated as follows (Herrera-Melián and others 16): where C_i is the concentration (mg/l) in the influent; C_o is the concentration (mg/l) in the effluent; Q is the inflow (m^3/day); S is the surface area of the wetland bed (m^2)."

"The vertical electrical survey was performed in this study to gain knowledge about the internal structure of the treatment wetlands (HSF-CW1, HSF-CW2), with the aim of detecting any operational anomalies commonly reported for reed beds such as clogging, lateral losses or short-circuited flow, and dead zones that ineluctably influence the wetlands' performance."

Results and Discussion

"While still achieving satisfactory performance, it is interesting to note here that during the first years of operation, probable early signs of clogging were noticed for both HSF-CW units, mainly dealing with both a partial ponding in the inflow zone and a subsequent partial overland flow with deposits on the surface of reed beds."

"According to field observations (visual monitoring) carried out during the last years of reed bed operation, partial ponding in the inflow zone of HSF-CW1 and surface runoff with subsequent deposits on the wetlands surface often first occurred in compartment A towards the outlet zone by crossing compartment C. Both compartments show anomalies dealing with the lowest ER values, while compartments B, D, and E have the lowest density of vegetation."

Conclusions

"This case study examined both the performance of long-term operated reed beds (two parallel HSF-CW units) as tertiary wastewater treatment, and the feasibility of the electrical resistivity (ER) method for in situ monitoring and detection of common operating anomalies such as clogging."

"Caution should be paid to these preliminary, although promising, results as further investigation in the future is needed both to upgrade and to explore the use of the resistivity method for the detection (quantification) of clogging in the field and the characterization of clogging materials in subsurface reed beds."

"The current resistivity results would be foremost useful to the managers of HSF-CW systems, both to explore and then correct any suspected operational failures (detected) and to guide or rather prioritize the improvement actions to be undertaken in the short range."

Reuse of Tertiary Municipal Wastewater Effluent for Irrigation of Cucumis melo L. [128]

This is a machine-generated summary of:

Martínez, S.; Suay, R.; Moreno, J.; Segura, M. L.: Reuse of tertiary municipal wastewater effluent for irrigation of Cucumis melo L. [128]

Published in: Irrigation Science (2012)

Link to original: https://doi.org/https://doi.org/10.1007/s00271-012-0342-4

If you want to cite the papers, please refer to the original.

For technical reasons we could not place the page where the original quote is coming from.

Abstract-Summary

"This study focuses on the effect of irrigation with ozonized wastewater on the melon crop."

"Melon crops were irrigated with either ozonized wastewater or ground water commonly used in the agricultural area (control treatment)."

"The results obtained show that irrigation with ozonized wastewater, in these conditions of crop growth, produces similar effects as ground water on the soil properties, on the concentration of macroelements in the leaf and fruit and on the yield of melon fruit."

"The reuse of this effluent leads to a greater saving in nitrogenous and potassium fertilizers compared to fertigation of the crop with ground water."

Introduction

"It is essential to bear in mind that the use of wastewater in the irrigation of crops is one of the main sources of contamination of pathogenic microorganisms (Shuval and others 17; Assadian and others 18)."

"The potential risk deriving from the use of this type of water in the irrigation of crops depends on a large number of factors such as the level of contamination, the species of pathogenic microorganisms and the chemical substances present in it."

"In order to do so, two melon crops were monitored and results compared: a melon crop irrigated with ozonized wastewater (OW) and a control treatment consisting of a melon crop irrigated with ground water (GW) commonly used in the area."

"The use of regenerated water could cause serious problems for human health, so that the study includes the objective of possible microbiological contamination of fruit irrigated with ozonized wastewater."

Materials and Methods

"We specify the chemical, physicochemical and microbiological analyses carried out on the irrigation waters, on the soil and on the melon crop (leaf and fruit)."

"Soil samples were analysed in the pre-planting period and at the end of the crop cycle to show the consequences of using both irrigation waters (OW and GW) on soil chemical properties."

"Saturated soil extract measurements were taken during pre-planting (0 DAS), flowering (58 DAS) and harvesting (122 DAS) and comprised pH, EC and common soluble salts in soils like anions (carbonate, bicarbonate, chloride, nitrate and sulphate) and cations (sodium, calcium, magnesium and potassium) following methods from MAPA (19), and finally available boron by the azomethine-H method (Gárate and others 20)."

"Analysis of variance (ANOVA) was used to discriminate the main treatment effects on measured properties of soil and saturated soil extract during the crop cycle, measured properties of plant leaf and fruit, and yield."

Results and Discussion

"For these parameters, both irrigation waters (OW and GW) were within the ranges considered as normal for irrigation water (Ayers and Westcot 21), so they did not result in any problem to the melon crop EC and SAR enable us to assess the risk of occurrence of infiltration problems in the soil."

"According to this, there were significant differences between treatments at the end of the crop cycle, with a higher concentration of boron in the soil irrigated with OW, surely because OW contained a higher concentration of B than did GW."

"There were no significant differences between treatments in the fruit due to the fact that the concentration of boron in both irrigation waters (OW and GW) was not too high for the melon crop, which is moderately tolerant to boron (Ayers and Westcot 21)."

Conclusions

"The study was carried out on a melon crop cultivated in a greenhouse, using sand mulch soil and fertigation by means of drip irrigation."

"The high concentration of boron in the ozonized wastewater did not limit its use for the irrigation of tolerant crops like melon."

"According to the analyses of soil, fruit and leaf carried out during the crop cycle, we can conclude that, using conventional agricultural practices (sand mulch soil, drip irrigation, plastic covered greenhouse), the use of disinfected wastewater for irrigation is apt for growing melon crops in these conditions."

"Obtained, the ozonized wastewater exceeded; on certain occasions, the level of faecal coliforms established for the irrigation of crops for consumption as fresh food."

Evaluation of Horizontal–Vertical Subsurface Hybrid Constructed Wetlands for Tertiary Treatment of Conventional Treatment Facilities Effluents in Developing Countries [129]

This is a machine-generated summary of:

Haghshenas-Adarmanabadi, Amir; Heidarpour, Manouchehr; Tarkesh-Esfahani, Saleh: Evaluation of Horizontal–Vertical Subsurface Hybrid Constructed Wetlands for Tertiary Treatment of Conventional Treatment Facilities Effluents in Developing Countries [129]

Published in: Water, Air, & Soil Pollution (2015)

Link to original: https://doi.org/https://doi.org/10.1007/s11270-015-2718-6

If you want to cite the papers, please refer to the original.

For technical reasons we could not place the page where the original quote is coming from.

Abstract-Summary

"Four large pilot-scale horizontal–vertical hybrid constructed wetlands (CWs) were constructed for the tertiary treatment of effluent of North Wastewater Treatment Plant in Isfahan, Iran."

"The plant effluent did not meet the regulation limits of wastewater reuse for various applications (e.g., irrigation of food and non-food crops, forests and green areas, groundwater recharge, and urban reuse applications)."

"The k-C* first-order model constant was computed for seven physical, chemical, and microbiological parameters—BOD_5, chemical oxygen demand (COD), total suspended solids (TSS), NO_3-N, NH_4-N, total phosphorus (TP), and fecal coliforms—based on the influent/effluent concentration data for estimation of required surface area of full-scale CWs in the future."

"The results of 12-month sampling showed that the hybrid HF-VF CWs are highly efficient in removing of the BOD_5 (85 % medium), COD (80 % medium), TSS (79 % medium), NH_4-N (78 % medium), TP (74 % medium), and fecal coliforms (99 % medium)."

"There were no significant differences between various planted hybrid CWs in removal efficiency and first-order model constant for BOD_5, COD, TSS, and coliforms, nor were there significant differences between planted and unplanted CWs for these parameters."

"For nutrients, the removal efficiencies of planted CWs were higher than those of control CW during the operation time."

"Among the CWs, the Phragmites showed the best efficiency for removal of nutrients followed by Arundo."

"The results of this pilot-scale research study showed that the performance of a single HFCW was probably not sufficient to achieve a suitable water quality for reuse of effluents and the hybrid CWs are more efficient and feasible systems for this purpose."

Introduction

"The use of a single HFCW or VFCW could not be enough to achieve the recommended standards for wastewater reuse (Toscano and others 22), so various types of constructed wetlands can be combined to complete each other as a hybrid CW and produce a low polluted effluent."

"In Slovenia, a large-scale HF-VF-HF hybrid CW was constructed for the treatment of a mixed industrial wastewater (Justin and others 23) and a VF-3HF hybrid CW was used for winery wastewater treatment by Serrano and others (24) in Spain."

"There are relatively few studies concerning model parameter calculation for hybrid subsurface CWs with different plant species (especially Arundo) and for different pollutants."

"The aim of this study was to evaluate the HF-VF hybrid CW performance for tertiary treatment of domestic wastewater through a large-scale pilot system with different plant species."

Materials and Methods

"The samples were collected every month from nine fixed sampling points at the influent and effluents of HF and VF CWs periodically for analysis."

"Because of the continuous discharge to the hybrid CWs and also low fluctuation in the quality of the wastewater treatment plant effluent during the CW retention time, the individual samples were collected from the inlet and outlets in the same days."

"The analysis of variance (ANOVA) was applied to determine the differences between the treatment efficiencies and effluent pollutant concentrations in all CWs."

"The relationship between the influent and effluent concentrations and loading rate can be defined by k-C* first-order model according to Kadlec (25):in which Q is the hydraulic loading rate (m^3/day), A is the CW area (m^2), C^* is the background concentration (mg/L), and k is the first-order removal rate constant (m/year)."

Results and Discussion

"The BOD_5 removal efficiency in Arundo, Typha, and unplanted control hybrid CWs was 87 $\pm$ 9, 86 $\pm$ 8, and 85 $\pm$ 11 %, respectively, and the Phragmites CW got the lowest removal treatment equal to 84 $\pm$ 10 %."

"The best mean COD removal efficiency was related to Arundo hybrid CWs with 83 $\pm$ 11 % followed by Phragmites with 79 $\pm$ 14 %, Typha with 77 $\pm$ 17 %, and the unplanted CW with 77 $\pm$ 15 %."

"The mean nitrate removal efficiency was 52 $\pm$ 17.8, 50 $\pm$ 14.4, 47 $\pm$ 24.4, and 48 $\pm$ 17 % for Phragmites, Arundo, Typha, and unplanted hybrid CW, respectively."

"During passing of wastewater throughout the VFCWs, more reduction of TP was observed and the mean TP removal efficiency reached to 74 $\pm$ 19 % for Phragmites, 55 $\pm$ 27 % for Typha, 54 $\pm$ 33 % for Arundo, and 37 $\pm$ 33 % for unplanted hybrid CWs."

Conclusions

"The results of analysis showed that hybrid systems successfully treated the main physico-chemical and microbiological pollutants (BOD_5, COD, TSS, NO_3-N, NH_4-N, TP, TC, and FC) in the municipal wastewater."

"The concentration of pollutants in the effluents showed that use of a single HFCW was not enough to satisfy the regulation limits for reuse of effluents, and supplementary treatment of HFCW effluents with VFCWs is essential for this purpose."

"The locating of VFCWs after HFCWs had a significant effect on the pollutant concentration reduction in the effluents except for nitrates."

"The longer hydraulic retention time in the HFCWs caused higher removal efficiencies than VFCWs for most of the pollutants, but for microbial parameters, the VFCWs showed higher efficiencies."

"There were no significant differences between Phragmites, Typha, and Arundo hybrid CWs in removal efficiency for BOD_5, COD, TSS, and coliforms, nor were there significant differences between planted and unplanted CWs for these pollutants."

Performance Evaluation of Horizontal and Vertical Flow Constructed Wetlands as Tertiary Treatment Option for Secondary Effluents [130]

This is a machine-generated summary of:

Thalla, Arun Kumar; Devatha, C. P.; Anagh, K.; Sony, Elsa: Performance evaluation of horizontal and vertical flow constructed wetlands as tertiary treatment option for secondary effluents [130]

Published in: Applied Water Science (2019)

Link to original: https://doi.org/https://doi.org/10.1007/s13201-019-1014-9

If you want to cite the papers, please refer to the original.

For technical reasons we could not place the page where the original quote is coming from.

Abstract-Summary

"Constructed wetlands (CWs) are simple low-cost wastewater treatment units that use natural process to improve the effluent water quality and make it possible for its reuse."

"A comparison is made between horizontal subsurface flow (HSSF-CW) and vertical flow (VFCW) constructed wetland in effectively post-treating the effluents from the secondary biological treatment system."

"VFCW showed a better overall removal efficiency than HSSF-CW by 7.14%."

"Constructed wetland can be considered as a sustainable alternative to the tertiary conventional treatment of domestic wastewater, thus making it possible for reuse."

Introduction

"There has been a growing interest in using constructed wetlands (CWs) for the treatment of polluted waters because of its low greenhouse effects, low operational and maintenance costs and energy efficiency."

"Constructed wetland utilization can provide sustainable wastewater treatment because they rely on natural processes and are less expensive to build, operate and maintain when compared with conventional sewage treatment systems."

"A pre-treatment with conventional processes is usually desirable before discharge into a wetland because of the potential solids or oxygen demand overload that create nuisance conditions within a wetland receiving raw or inadequately treated wastewaters (Kadlec and Wallace 26)."

"The pilot-scale model was developed to apply the approach of constructed wetland system under the local circumstances of National Institute of Technology, Karnataka (NITK), and similar regions as a post-treatment option, where we can reclaim the wastewater and use it for alternate purposes."

Materials and Methods

"The pilot-scale study is done for the effluent taken out from the existing treatment plant."

"The wetland is operated as a batch flow reactor at two HRTs of 24 h and 12 h from the effluent of existing NITK treatment plant."

"The water depth was maintained at 0.4 m within the gravel bed and with the aid of fixed outlet pipes (collecting effluent from the floor level of the wetland cell)."

"The pilot-scale VFCW is designed for an influent flow of 0.25 m^3/day considering an average BOD loading of 100 mg/l. The areal loading rate (ALR) is provided of 35 kg/ha day for attaining a discharge limit of BOD less than 30 mg/l (Wallace and Knight 27)."

"The pilot plants used in the vegetation were Cyperus rotundus and Pennisetum pedicellatum."

"The plant individuals reaching an average height between 30 and 150 cm might show more growth rate on the influent water characteristics, commonly used as fodder locally."

Results and Discussion

"The BOD_5 removal efficiency was found to be stable from the fourth week of operation of the pilot model VFCW and from the fifth week in HSSF-CW."

"In 12-h retention time, the average removal efficiencies were 68% and 64% in VFCW and HSSF-CW, respectively."

"The average removal efficiency for 12-h HRT is 50% and 48% in VFCW and HSSF-CW, respectively."

"Nitrate nitrogen average removal efficiency was 47.54% and 66.75% for 12-h and 24-h HRTs in VFCW and 69% and 51% in HSSF-CW."

"To the results obtained in the reduction of BOD and COD for 24-h pilot-scale VFCW and HSSF-CW, TSS average removal efficiency was significant and found to be 90–95%."

"VFCW showed significantly higher plant growth than compared to HSSF-CW, which is in line with the variations in the removal efficiencies of various parameters for the treated wastewater."

Conclusion

"The treatment process in the wetland is generally difficult to understand due to the complex physical, chemical and biological processes involved in it along with variations in the real-time wastewater."

"Phosphates removal efficiency was promising and stable in the treatment process and hence can be adapted in small-scale purposes."

"Constructed wetland process being complex there should be regional-specific design and vegetation selection studies to be carried out for proper guidelines for upcoming CWs."

"Constructed wetland can be considered as a sustainable alternative to the tertiary conventional treatment of domestic wastewater, thus making it possible for reuse."

Performance Linked to Residence Time Distribution by a Novel Wool-Based Bioreactor for Tertiary Sewage Treatment [131]

This is a machine-generated summary of:

Hu, Bibo; Wheatley, Andrew; Ishtchenko, Vera; Huddersman, Katherine: Performance linked to residence time distribution by a novel wool-based bioreactor for tertiary sewage treatment [131]

Published in: Applied Microbiology and Biotechnology (2011)

Link to original: https://doi.org/https://doi.org/10.1007/s00253-011-3659-7

If you want to cite the papers, please refer to the original.

For technical reasons we could not place the page where the original quote is coming from.

Abstract-Summary

"Laboratory-scale experiments were carried out using up-flow 7 L Submerged Aerated Filter reactors packed with wool fibre or commercial plastic pall rings, Kaldnes, (70% by volume) support media for the tertiary treatment of sewage."

"The performance of the wool bioreactor was more consistent than that with Kaldnes medium, for both TOC removal (93%) and SS removal (90%)."

"Both plastic and wool-packed bioreactors achieved complete nitrification at the load of about 0.4 kgCOD/m^3/day."

"The sludge yield of the wool bioreactor was almost half that of the bioreactor with Kaldnes suggesting that wool could retain residual organics and particulates."

"Biomass growth increased the retention of the tracer in the wool reactor by, it was suggested, exposing a greater surface area."

Introduction

"The application of Submerged Aerated Filters (SAFs) for wastewater treatment has been more successful for the removal of recalcitrant organics, suspended particles, phosphorus and nitrification/denitrification than conventional trickling filters (Chen and others 29; Masuda and others 30; Watanabe and others 31)."

"The filter medium in SAFs is as important as in trickling filters to support and retain the growth of the biofilm and maximise organic and nutrient removal."

"We suggested that wool might be better than plastic, such as polyethylene (PE) polyvinyl chloride pall rings (Lessel 32) or plastic and stainless meshes (Watanabe and others 33; Cannon and others 34) for adsorption of recalcitrant organics especially if the media surface characteristics were more important in tertiary than secondary treatment."

"This paper reports on a comparison of the performance of the SAF bioreactors with conventional PE media (Kaldnes) and wool for the aerobic treatment of synthetic domestic wastewater."

Materials and Methods

"The total height of each reactor was 610 mm and the empty bed liquid volume of the filter was 6.93 L. The SAF reactors were packed either with 35 g wool fleece or with 35 g Kaldnes."

"Tracer studies were carried out using a pulse input of lithium chloride at a theoretical uniform concentration of 2 mg Li^+/L of reactor volume."

"Tracer studies were carried out before inoculation and used tap water flowing through the reactor, the subsequent tests with established biofilm used the synthetic sewage as the feed."

"The analysis was carried out in accordance with standard methods (APHA and others 35) and the precision is reported to be ±5% at these volumes and concentrations."

"Ten milliliters of sample were diluted from experience to bring concentration of ammonia nitrogen between 0 and 10 mg/L. The analysis was carried out according to the instructions and standard methods."

Results

"Both reactors exhibited stable performance the standard deviation in the wool effluent SS was 3.66, for Kaldnes 14.11."

"Turbidity was also monitored (more sensitive to smaller particle sizes) for evidence of the impact of the breakdown of the wool (Hu 36) but with similar results to the SS (wool effluent 1.0 SD 0.6 Kaldnes 2.1 SD 1.5 NTU)."

"The sludge yield from Kaldnes rings 0.31 kg SS/kg TOC removed, was similar to results achieved by Broch-Due and others (28), where the sludge yield of two SAF reactors in series was 0.39 kg TSS/kg TCOD."

"The peaks in the ammonia effluent concentration for the wool SAF on day 42, and with Kaldnes on days 20 and 65 was due to sludge removal."

Discussion

"The characteristics of the wools in SAFs with different wool media are similar to the previous wool experiments (prior to inoculation) suggesting that laboratory scouring or chemical modification to the surface of the fibres did not increase tracer retention."

"The wool started as a packed and stationary media which became more retentive after biofilm growth and with unravelling of the fibres, whereas Kaldnes started as a floating media which then moved more with biofilm to reduce dead space."

"There was no difference in tracer retention between the surface treated wools and all the aerated SAFs (Kaldnes and wool) had similar overall well-mixed tracer patterns."

"Moving bed SAFs are reported to achieve a lower SS removal compared to fixed-bed backwashed BAF reactors because of the mixing, but potentially, they can achieve a lower sludge yield because of the better degradation of the solids."

A Combined Electrocoagulation-Electroperoxidation Process for the Tertiary Treatment of Domestic Wastewaters [132]

This is a machine-generated summary of:

Senghor, Fatou; Drogui, Patrick; Seyhi, Brahima: A Combined Electrocoagulation-Electroperoxidation Process for the Tertiary Treatment of Domestic Wastewaters [132]

Published in: Water, Air, & Soil Pollution (2015)

Link to original: https://doi.org/https://doi.org/10.1007/s11270-015-2637-6

If you want to cite the papers, please refer to the original.

For technical reasons we could not place the page where the original quote is coming from.

Abstract-Summary

"Electrolysis was performed using a parallelepipedic electrolytic cell (0.5 L) containing one sacrificial anode (mild steel or aluminum) and one cathode (vitreous carbon)."

"The effects of the treatment time, current density, and type of anode electrode on the process performance were examined."

"There was a decrease in the chemical oxygen demand (COD), suspended solid (SS), turbidity, color, and total phosphorus (P_{tot}) by 67 ± 9, 98 ± 2, 55 ± 10, 61 ± 9, and 97 ± 0 %, respectively, for a treatment time of 60 min in the electrolysis cell in the presence of mild steel (anode) and vitreous carbon (cathode) electrodes."

Introduction

"Advanced electro-oxidation (AEO) processes can be advantageously used for domestic wastewater treatment."

"The AEO techniques are processes based on the destruction of pollutants either at the anode electrode or by using Fenton's reagent (Fe^{2+} and H_2O_2) that is partially or completely generated by electrode reactions."

"The anodic oxidation (AO) pertains to AEO processes, and it can also be performed by H_2O_2 generation (Brillas and others 37)."

"AEO processes based on the cathodic electrogeneration of H_2O_2 have been successfully applied to the treatment of wastewaters containing persistent organic pollutants (Brillas and others 38; Daghrir and others 39; Drogui and others 40; Garrido and others 41)."

"Among AEO processes, the electrocoagulation-electroperoxidation (EC-EP) process is particularly interesting for wastewater treatment (Brillas and Casado 42; Flox and others 43; Guivarch and others 44; Panizza and Cerisola 45; Zarei and others 46; Zhou and others 47)."

"The objective of this work is to evaluate the effectiveness of a method of treatment that combines EC and EP processes in the same reactor to remove different types of pollutants (organic, inorganic, and microbial pollutants) from DWW simultaneously."

Materials and Methods

"The sample had an initial content of fecal coliform of $(4 \pm 5) \times 10^6$ colony-forming units (CFU) 100 mL^{-1}, COD of 190 $\pm$ 133 mg L^{-1}, turbidity of 86 $\pm$ 20 NTU, pH of 7.2 $\pm$ 0.2, and conductivity of 800 $\pm$ 133 μS cm^{-1}."

"This cell (0.5 L capacity) was made of Plexiglas (17.7 × 2 × 15.2 cm)."

"The first set of experiments consisted of an evaluation of the electrolytic cell's capacity to produce H_2O_2 and Fe^{2+}/Fe^{3+} (or Al^{3+}) ions."

"The subsequent set of experiments was performed to evaluate the effectiveness of the electrolytic cell as a tertiary DWW treatment."

"The effects of several parameters, such as the current density (2–9 mA cm^{-2}), anode material (Fe and Al), and treatment time (0–60 min) were tested, whereas the initial pH (7 $\pm$ 0.3) and temperature (20 $\pm$ 2 °C) were maintained at constant levels."

Results and Discussion

"The concentrations of inorganic coagulant (Fe^{2+}/Fe^{3+} and Al^{3+}) were measured in solution and in metallic sludge residues at current densities between 2.0 and 9.0 mA cm^{-2}."

"FC removal was due to the synergetic effect of inorganic coagulant (Fe^{2+}/Fe^{3+}) and oxidant agent (H_2O_2) on bacteria while applying EC-EP system, compared to conventional EC method where only Fe^{2+}/Fe^{3+} was generated."

"The anode material influence on EC-EP process efficiency was evaluated by operating in Fe/VC and Al/VC configurations at a current density of 4 mA cm^{-2} for a treatment time of 60 min."

"Relatively high removal rates for color (84.7 %) and turbidity (88.6 %) were measured by using an Al/VC configuration and comparing the results with the values recorded (25.1–27.6 % of color removal and 39.1–53.0 % of turbidity removal) in the Fe/VC configuration."

Conclusions

"This study provided a feasibility evaluation of the EC-EP process as a tertiary treatment for domestic wastewaters containing different types of pollutants (organic, inorganic, and microbial pollutants)."

"The EC-EP process was operated under the best operating conditions and came to a total cost of US$0.211 $\pm$ 0.02 m^{-3} of treated DWW."

“The next study will provide an evaluation of the mineralization degree from the EC-EP process.”

“The EC-EP process could form the basis of a process (as a tertiary treatment) that is capable of simultaneously removing inorganic, organic, and microbial pollutants from many wastewaters.”

Feasibility of Reuse Filter Backwash Water as Primary/Aid Coagulant in Coagulation–Sedimentation Process for Tertiary Wastewater Treatment [133]

This is a machine-generated summary of:

Mazari, Lilia; Abdessemed, Djamal: Feasibility of Reuse Filter Backwash Water as Primary/Aid Coagulant in Coagulation–Sedimentation Process for Tertiary Wastewater Treatment [133]

Published in: Arabian Journal for Science and Engineering (2020)

Link to original: https://doi.org/https://doi.org/10.1007/s13369-020-04597-1

If you want to cite the papers, please refer to the original.

For technical reasons we could not place the page where the original quote is coming from.

Abstract-Summary

“The purpose of this study is to evaluate the feasibility of reusing filter backwash water (FBWW) from the water treatment plant as a substitute for conventional coagulants (primary coagulant) and coagulant aid with aluminum sulfate ($Al_2(SO_4)_3$) in the coagulation–sedimentation process as tertiary treatment urban effluents for reuse.”

“The impact of reuse filter backwash water has been studied on the water quality in coagulation process.”

“For a dose of 22.5 mg Al/L of FBWW as primary coagulant, the removal of turbidity and UV_{254} absorbance was significant, with percentages of 65.8% and 31.7%, respectively, during the treatment of the secondary effluent.”

"The best results were obtained for a neutral pH. With FBWW as coagulant aid, the optimal dose of $Al_2(SO_4)_3$ decreased from 9.5 to 6.3 mg Al/L with better percentages of turbidity reduction and UV_{254} of 92.9 and 71.4%."

Introduction

"After multiple studies in six different water treatment plants, it was shown that FBWW recycling does not require additional coagulant compared to raw water recycling, the study focused on the impact of recycling FBWW and MBWW on the quality of drinking water in coagulation-sedimentation processes."

"It was shown that there was no statistically significant difference between conventional treatment and recycling processes and the toxicity of water samples in the recycling process did not increase during the 15-day continuous recycling trial."

"Although promising results have been reported on the FBWW recycling in WTP; at present, there are no available studies regarding the use of FBWW as the primary coagulant in wastewater treatment."

"This study aims to experimentally evaluate the feasibility of reusing FBWW as the primary coagulant and coagulant aid in the removal of both the turbidity and water organic matter from secondary municipal wastewater by the coagulation–sedimentation process."

Materials and Methods

"To facilitate the understanding of turbidity and dissolved organic matter removal mechanism, synthetic water composed of humic acid (HA) and kaolin was also used."

"A stock HA solution was prepared by dissolving 1 g of HA in 300 mL 0.01 M NaOH and mixed by a magnetic stirrer for 2 h. The solution was then diluted with deionized water to 1 L as the stock HA solution and stored in the dark."

"The kaolin stock solution was prepared by dissolving 1 g of kaolin in 1 L of deionized water under stirring for 24 h to give better dispersion."

"Coagulation experiments were performed with synthetic water containing 50 mg/L of HA prepared in tap water."

"Jar tests were used in order to determine the optimum conditions of pH and concentration giving the best efficiency on the turbidity."

Results and Discussion

"The turbidity decreases down to 3.3 and 3 NTU for FBWW doses of 16.9 and 22.5 mg Al/L and a removal percentage of 80.5 and 81.5%, respectively, for the synthetic effluent."

"Several researchers have found that the presence of particles with FBWW recycling from rivers has increased the particle aggregation during flocculation and improved removal performance of the organic matter and turbidity [48–50]."

"At the optimal coagulant dose (6.3 mg Al/L), the turbidity removal of the synthetic effluent reaches 84.7%, without addition of FBWW, which corresponds to 2.6 NTU, and a UV_{254} reduction of 56."

"A similar observation has already been reported by Sagar and others, who suggested to recycle FBWW containing destabilized particles that instead of increasing the coagulant doses to eliminate turbidity and organic matter in the surface water treatment could be used to achieve the same results [51]."

Conclusion

"In the course of this work, the feasibility of the reuse process of filter backwash water as primary coagulant and coagulant aid was demonstrated."

"The best yields were obtained for a neutral pH. The reuse of FBWW as coagulant aid minimized the use of $Al_2(SO_4)_3$ by half with a turbidity and UV_{254} removals of 94.6 and 40%, respectively."

"The filter backwash water has a significant potential to be used as substitute to the high cost conventional coagulants when treating a wastewater."

4.3 Filtration

Introduction by Editor

Filtration in tertiary treatment is based on the principle of physical separation, where water is passed through a medium that selectively retains particles, microorganisms, and dissolved constituents. This process efficiently removes residual suspended solids, fine organic matter, and trace contaminants that may persist after primary and secondary treatment stages. By enhancing water clarity and reducing turbidity, filtration significantly contributes to overall water quality improvement.

Several filtration techniques are utilized in the tertiary treatment of municipal wastewater. Rapid sand filtration is a widely employed method where water is forced through a bed of fine sand acting as the filtration medium. This process effectively removes suspended solids and microorganisms, resulting in clarified effluent. Another technique, granular media filtration, employs a layer of granular material such as anthracite, activated carbon, or garnet as the filtration media. This method enhances the removal of fine particles, organic matter, and trace contaminants, ensuring the production of high-quality effluent. Membrane filtration, including microfiltration, ultrafiltration, and reverse osmosis, utilizes semipermeable membranes to separate particles, microorganisms, and dissolved constituents from water, resulting in excellent water quality.

Efficient operation of filtration processes in tertiary treatment necessitates careful consideration of operational parameters. Key factors include selecting the appropriate filter media, managing hydraulic loading rates, determining optimal filter bed depth, scheduling regular backwashing, and implementing chemical dosing for improved performance. Proper maintenance of filter media, routine backwashing, and periodic media replacement or cleaning are essential to prevent clogging, maintain optimal filtration rates, and ensure consistent treatment efficiency. Monitoring and controlling operational parameters play a crucial role in achieving desired filtration outcomes and prolonging the lifespan of filtration systems.

Filtration holds significant importance in the tertiary treatment of municipal wastewater, as it plays a vital role in achieving stringent water quality standards. By removing suspended solids, microorganisms, and trace contaminants, filtration ensures the production of clarified and safe effluent. The treated water can be discharged into receiving water bodies, used for irrigation purposes, or reclaimed for various beneficial uses. Effective filtration not only protects the environment but also safeguards public health, contributing to sustainable and responsible wastewater management practices.

Machine generated keywords: filter, sand, trickling, trickling filter, human animal, loading, cod, coli, hybrid, animal, sand filtration, consume, eutrophication, filtration, coliform

A Shallow Constructed Wetland Combining Porous Filter Material and Rotala rotundifolia for Advanced Treatment of Municipal Sewage at Low HRT [134]

This is a machine-generated summary of:

Xiao, Jibo; Lin, Guo; Cao, Zhuangzhuang; Chu, Shuyi; Cui, Lingzhou; Yang, Yunlong; Wu, Xiangting: A shallow constructed wetland combining porous filter material and Rotala rotundifolia for advanced treatment of municipal sewage at low HRT [134]

Published in: Environmental Science and Pollution Research (2022)

Link to original: https://doi.org/https://doi.org/10.1007/s11356-022-24111-4

If you want to cite the papers, please refer to the original.

For technical reasons we could not place the page where the original quote is coming from.

Abstract-Summary

"A shallow constructed wetland (SCW) with porous filter material and Rotala rotundifolia was used for advanced municipal sewage treatment."

"The wetland without plant was set as the control (SCW-C)."

"The pollutant removal performance of the system at different hydraulic retention times (HRTs) was investigated."

"Results showed that SCW was efficient in pollutant removal."

"The richness and evenness of the bacterial community were significantly higher in the plant-rooted SCW."

"These findings suggested that the SCW could remove pollutants from the municipal sewage effluent and meet the standard requirement at low HRT."

Introduction

"Many municipal sewage treatment plants are required to upgrade the effluent quality from the A level of the discharge standard of pollutants for municipal waste (GB18918-2002) to the fourth class of environmental quality standards for surface water (GB3838-2002)."

"It is a great challenge for municipal wastewater treatment plants because concentrations of chemical oxygen demand (COD), ammonium nitrogen (NH_4^+-N), and total phosphorus (TP) have to be reduced by 40%, 70% (81.2% for water temperature ≦12°C), and 40%, separately."

"The objective of the paper was to (1) investigate the effect of hydraulic retention times (HRTs) on pollutant removal performance in the shallow wetland system; (2) identify microbial community and its variation in the system; and (3) explore the fate of N and P in the system."

"This research may provide a novel wetland structure and a feasible approach to reaching and maintaining high pollutant removal efficiency at low HRT."

Materials and Methods

"The sturdy plants (approximately 10 cm in length) were selected and transplanted into the tank, and the density was 80 strains per square meter."

"The sewage was fed into the Eco-tank continuously using a diaphragm metering pump at a hydraulic retention time (HRT) of 2 days at the initial test period for system adaptation."

"The system operated for 1 month and seven water samples were collected at each HRT."

"The mass of plants and substrate and N and P contents in the tissues and substrate were analyzed before and after the test."

"The ground plants and substrates were digested with $HClO_4$-H_2SO_4 and H_2SO_4-H_2O_2, separately, and determined for the phosphorus content using Mo-Sb colorimetric method."

"Substrate samples were collected from the front (WF), middle (WM), and back (WB) of SCW and SCW-C (CK) system, separately."

Results and Discussion

"TN removals in SCW at HRTs 12, 18, 24, and 30 h were 5.3%, 14.6%, 17.6%, and 18.7% higher than those in SCW-C. The difference between SCW and SCW-C at HRT 12 h was significantly lower than that at other HRTs ($p < 0.05$)."

"The relative abundance of Massilia in the SCW (34.8%) was 8.1 times of that in the SCW-C. The decline in the P difference between SCW and SCW-C at HRT from 24 to 30 h may be correlated with the decrease in activity of Massilia due to the insufficient carbon at longer HRT since it preferred to a copiotrophic environment (Ofek and others 52)."

"According to the aforementioned results, an increase in HRT from 12 to 24 h led to higher difference in TP removal between SCW and SCW-C. It indicated that HRT might influence the distribution of P in the substrate, plant, and effluent."

Conclusions

"Influence of HRT on removals of organic matter, N, and P was obvious at lower HRT 12 to 24 h. Plant rooted systems greatly improved the richness and evenness of the bacterial community."

"The P in the influent mainly flowed to the substrate and plant."

"Most N was removed by nitrification and denitrification."

Slow Sand Filtration of Raw Wastewater Using Biochar as an Alternative Filtration Media [135]

This is a machine-generated summary of:

Kaetzl, Korbinian; Lübken, Manfred; Nettmann, Edith; Krimmler, Stefan; Wichern, Marc: Slow sand filtration of raw wastewater using biochar as an alternative filtration media [135]

Published in: Scientific Reports (2020)

Link to original: https://doi.org/https://doi.org/10.1038/s41598-020-57981-0

If you want to cite the papers, please refer to the original.

For technical reasons we could not place the page where the original quote is coming from.

Abstract-Summary

"Miscanthus-biochar was used as filtration media and compared with sand as a common reference material."

"Mean removal of chemical oxygen demand (COD) of biochar filters was with 74 ± 18% significantly higher than of sand filters (61 ± 12%)."

"To sand filters with a mean reduction of 1.18 ± 0.31 log-units, E. coli removal through biochar was with 1.35 ± 0.27 log-units significantly higher and increased with experimental time."

"Since the E. coli contamination of both filter materials was equal, the higher removal efficiency of biochar filters is probably a result of an improved biodegradation within deeper zones of the filter bed."

"Performance of biochar filters was better or equal compared to sand and have thus demonstrated the suitability of Miscanthus-biochar as filter media for wastewater treatment."

Introduction

"Dalahmeh and others [54] used biochar filters as an on-site sewage treatment for the removal of pharmaceuticals from wastewater and found, particularly due to adsorption of persistent substances, a significant higher elimination of the investigated substances in biochar columns in comparison to the tested sand filters."

"Perez-Mercado and others [55] investigated the suitability of biochar filters for wastewater treatment and reported higher total nitrogen removal for biochar filters compared to sand filters."

"There is still a significant knowledge gap regarding the suitability of SSF as anaerobic treatment of wastewater and the reduction of fecal indicator bacteria (FIB) [56]."

"Most of the experiments with biochar for wastewater treatment have so far been carried out at very low organic loading rates and with relatively long hydraulic residence times."

Results and Discussion

"The increase in materials' removal efficiency towards the end of the experiment indicates an increased biological activity and thus a more efficient biological degradation of organic matter in deeper filter zones as well as an improved particle retention in the developed schmutzdecke on the filter surface."

"While NH_4-N was not significantly affected by the filtration, probably due to the reduction of organic nitrogen under anaerobic conditions, phosphorous (P_{tot}) removal was with roughly 35% significantly higher in biochar columns than in sand filters (around 25%)."

"The higher removal rates of phosphorous in biochar filters can be explained by the high adsorption potential of biochar as reported elsewhere [57, 58]."

"Physical surface filtration and subsequently biodegradation within the schmutzdecke can be considered as the main removal mechanisms in AnBFs, as confirmed by other studies [59–61] and explains the relatively high removal rate shortly after filter activation with still clean filter material."

"Higher removal rates of E. coli, COD and TOC through biochar filters indicates enhanced microbial decomposition during the filtration."

Conclusion

"Our results could show that AnBFs constitute a suitable treatment method to reduce organic compounds and FIB from pre-settled raw wastewater."

"Biochar, as in our case biochar from Miscanthus grass, is a highly promising filter material for AnBF, particularly since biochar can be produced locally from agricultural residues with recycling opportunities for energy production or soil amelioration after its use as filter material for wastewater treatment."

"AnBF with biochar as filter material might be also an effective short-term wastewater treatment in emergency situations."

Material and Methods

"Two different materials, namely quartz sand and Miscanthus biochar, were tested for their suitability as filter materials for anaerobic wastewater treatment using anaerobic vertical downflow biofilters (AnBFs)."

"Samples of influent and effluent were taken in sterile 50 mL tubes and tested immediately for FIB E. coli and enterococci, temperature, pH, EC and turbidity as well as stored for further physico-chemical analysis, namely NH_4-N, P_{tot}, COD and TOC at −20 °C."

"Samples of the filter materials were transferred into sterilized 100 mL glass bottles pre-filled with 50 mL Dehydrated Special Microplates diluent (DSM; pH 7.4; Bio-Rad, Munich, Germany)."

"A cLSM (TC SP8, Leica microsystems, Germany) was used for biofilm analysis on the filter materials."

"P_{tot} was measured with LCK 350 cuvette tests (Hach, Düsseldorf, Germany), according to DIN EN ISO 6878-1-1986 and NH_4-N according to DIN 38406 E5-1, using LCK 303 cuvette tests."

Integrated Treatment of Municipal Wastewater Using Advanced Electro-Membrane Filtration System [136]

This is a machine-generated summary of:

Abdel-Fatah, Mona A.; Shaarawy, H. H.; Hawash, S. I.: Integrated treatment of municipal wastewater using advanced electro-membrane filtration system [136]

Published in: SN Applied Sciences (2019)

Link to original: https://doi.org/https://doi.org/10.1007/s42452-019-1178-9

If you want to cite the papers, please refer to the original.

For technical reasons we could not place the page where the original quote is coming from.

Abstract-Summary

"Treatment of municipal wastewater using advanced physicochemical process maximizes the removal of biological oxygen demand (BOD_5), chemical oxygen demand (COD) and phosphate."

"The results showed that: BOD_5, COD, and phosphates percent removal of 93%, 95% and 100% respectively."

"Construction and operating costs of sewage treatment plants can be reduced up to 50% compared with that of traditional low-efficiency biological treatment plants."

"Sludge produced from the advanced proposed process in this investigation can be used as fertilizer due to its quality while that produced from the traditional treatment process is unsafe to reuse."

Introduction

"Advanced treatment of sewage water (electro-coagulation) followed by membrane filtration (MF/UF/RO) represents one of the most promising technologies for sewage water treatment to make it suitable for aquaculture growing followed by agriculture produced crops."

"The study novelty depends on the optimization of selected sewage water treatment system utilizing electro-coagulation, microfiltration, ultrafiltration, and reverse osmosis as new trends in treatments."

"Sewer wastewater reclamation gave excellent results when it is treated with RO membrane due to its high removal efficiency of pollutants, stable water quality and easy operation [62–69]."

"Based on hard and continuous investigations of our research team for more than 15 years, it was found that integrated system of adsorption over nano-iron prepared by electro-coagulation, followed by membrane filtration, then electrochemical disinfection will represent high-efficiency wastewater treatment process."

"This innovative system will overcome the week efficiency of the traditional biological treatment process, also leads to cost-effective, easy controlled, low capital cost physical–chemical technology."

Process Description

"Water is pre-conditioned (by adjusting TDS, pH…others) to be ready for electro-coagulation active nano iron formation step."

"Conditioned water is directed to the electro-coagulation cell where the direct current in the bath dissolves the iron anodes and reduces the generated ferric ion by sodium boron hydride to nano iron."

"Water from the electro-coagulation cell containing active nano iron is mixed with alkaline activated Aswan clay for several minutes."

"Disinfection using the electro generated mixed oxidants of treated water takes place."

"During passing water through the pipes of the plant, several mini-hydro turbines for electric power recovery were placed, which could recover 60% of the total power consumed."

Pilot Scale Treatment Unit

"Electrochemical disinfectant unit Mixed oxidant generation cell of 5 L capacity and flow rate of 5 g/h, sodium hypochlorite with titanium iridium oxide coated modified electrode powered by DC power supply of 0–5 A current control and 0–24 V voltage control."

"Primary clarifier for oils, greases, and dirt separation It is one cylindrical with a conical base tank (5 m^3/h), Continuous process and Retention time: 2 h. Electrochemical cell Rectangular Shape, Steel lined PVC, Capacity: 500 L, Flow rate: 5 m^3/h, Hydrogen gas vent: present, Inlet/outlet: 4″, plates electrode, iron electrode, $\pm$ terminal."

"Chemical dosing pumps Six pumps of flow rate: 0–7 L/h, Pressure up to 7 bars, and Dosing tank with 200 L. Secondary clarifier One cylindrical tank with a conical base of 5 m^3/h capacity, Continuous type, and retention time: 2 h, Steel material."

Results and Discussion

"The wastewater mixed with boron-hydride and when the electric current passes in the electrocoagulation cell nano-iron formed by anodic dissolution and by reaction with water nano-ferric hydroxide formed as a gelatinous film when settling down all the suspended solids will be removed."

"At the RO step, the dissolved salts will be completely removed then by adding the electro-generated mixed oxidant generated through the electro catalytic disinfection step very safe water will be obtained."

"The mechanism of the electrocoagulation process using iron anode and the mechanism of mixed oxidant formation will be explained by the following equations: (A) Electrocoagulation mechanismAs clear from the above equations nano-ferric hydroxide gelatinous film is formed which causes the removal of suspended solids [70]. (B) Phosphorous removal mechanismThe above mention equation shows the removal of phosphorus by precipitation with ferrous ion formed by electrocoagulation process [70]. (C) Electro-catalytic disinfection mechanism [71]."

"From the other hand, the water quality of the obtained water from the advanced integrated system could be used for edible crops planting and in fish farming while treated water generated from the biological treatment process is only used for tree or forest planting."

Conclusions

"The proposed integrated system presents a successful process for obtaining high-quality water with low capital, and running costs."

"High cash flow fertilizers were produced from the obtained sludge from the clarification process."

"The proposed system is one of the promising systems which give the safest water quality and saving energy at a cheaper cost."

Occurrence and Removal of Sulfonamides and Their Acetyl Metabolites in a Biological Aerated Filter (BAF) of Wastewater Treatment Plant in Xiamen, South China [137]

This is a machine-generated summary of:

Wang, Dapeng; Zhang, Xian; Yan, Changzhou: Occurrence and removal of sulfonamides and their acetyl metabolites in a biological aerated filter (BAF) of wastewater treatment plant in Xiamen, South China [137]

Published in: Environmental Science and Pollution Research (2019)

Link to original: https://doi.org/https://doi.org/10.1007/s11356-019-06311-7

If you want to cite the papers, please refer to the original.

For technical reasons we could not place the page where the original quote is coming from.

Abstract-Summary

"Understanding the potential removal pathway of sulfonamides and their metabolites in wastewater treatment systems is of importance."

"The occurrence and fate of four sulfonamides and their acetyl metabolites in wastewater and sludge in a biological aerated filter in Xiamen city were evaluated."

"The highest concentration in wastewater was acetyl-sulfamethoxazole (Ac-SMZ) with a concentration of 75.2 ng/L. Removal efficiency and mass load in wastewater treatment systems were calculated."

"The removal efficiencies of sulfamerazine (SM1), sulfamethazine (SM2), and sulfadiazine (SD) were up to 100% while N-acetyl sulfamerazine (Ac-SM1) showed the lowest removal efficiency."

"Biodegradation was the dominant remove pathway according to the mass balance analysis while SD and SM2 were sludge adsorption."

Introduction

"Municipal wastewater is mainly a known contributor of antibiotics and their metabolites to the aquatic environment."

"It is of importance to understand the fate of these antibiotics and their metabolites in wastewater treatment plants (WWTPs) in order to access the risks and the potential impacts of antibiotics in the aquatic environment."

"To analyze the removal efficiencies in the different treatment units, they can provide information on the removal pathway and the fate of antibiotics during the wastewater treatment process."

"Mass balance analysis can reveal the information of fate and pathway of the antibiotics and metabolites during the wastewater treatment plants. (Ashfaq and others 73; Petrie and others 74)."

"SAs and their acetyl metabolites in wastewater and sludge were investigated in order to comprehensively evaluate the biodegradation partitioning behaviors of target compounds in aqueous and solid phases."

Materials and Methods

"The HLB cartridge (500 mg, 6 mL) was preconditioned with 10 mL methanol and 10 mL ultra-pure water; the flow rate was maintained at 5–10 mL/min; after extraction, 10 mL Milli-Q water was used to rinse the cartridges and the HLB cartridges were dried under vacuum to remove water for 30 min."

"The cartridges were then eluted with 6 mL methanol, and elution solution was collected and concentrated with a gentle steam of nitrogen to nearly dryness and redissolved in 20% methanol (v/v) to 1 mL. Sulfadiazine-$^{13}C_6$ used as internal standard was added to the final extracts at a concentration of 20 ng/mL for the further LC-MS/MS analysis."

"The elutes were evaporated to dryness under a gentle nitrogen steam and reconstituted with 20% methanol solution (v/v) to 1 mL. Sulfadiazine-$^{13}C_6$ used as internal standard was added to the final extracts at a concentration of 20 ng/mL for the further LC-MS/MS analysis."

Results and Discussion

"The concentration of Ac-SM1 in wastewater ranged from 1.09 to 0.77 ng/L. Low concentration of other target or below detection limit was also found except SMZ and Ac-SMZ in our study; Perisa and others have investigated wastewater in WWTP with none of their studied metabolites detected (Perisa and Babic 75)."

"Garcia-Galan and others (76) investigated three acetyl metabolites of our target compounds in wastewater with only Ac-SMZ detected with relatively high concentration in the influent and the rest were all below the limits of detection."

"CN tank and DN tank are the major removal processes for target compounds during the wastewater treatment."

"The results match those from previous study (Yan and others 77), which investigate 21 pharmaceutically active compounds in wastewater treatment plants with removal efficiencies ranging from negative removal to nearly total removal."

Conclusions

"The occurrence of four sulfonamides and their acetyl metabolites was investigated in a BAF system in Xiamen City."

"The highest concentration in wastewater was Ac-SMZ with a concentration of 75.2 ng/L. Removal efficiency of each treatment unit was calculated, and the Ac-SM1 showed the lowest removal efficiency with 24.4% while SM1, SM2, and SD are 100% removed."

"Biodegradation was the dominant removal pathway while SD and SM2 were sludge adsorption during the wastewater treatment according to the mass balance analysis."

Hydrated Oil Shale Ash Mitigates Greenhouse Gas Emissions from Horizontal Subsurface Flow Filters for Wastewater Treatment [138]

This is a machine-generated summary of:

Kasak, Kuno; Mõtlep, Riho; Truu, Marika; Truu, Jaak; Kõiv-Vainik, Margit; Espenberg, Mikk; Paiste, Päärn; Kirsimäe, Kalle; Mander, Ülo: Hydrated Oil Shale Ash Mitigates Greenhouse Gas Emissions from Horizontal Subsurface Flow Filters for Wastewater Treatment [138]

Published in: Water, Air, & Soil Pollution (2016)

Link to original: https://doi.org/https://doi.org/10.1007/s11270-016-3007-8

If you want to cite the papers, please refer to the original.

For technical reasons we could not place the page where the original quote is coming from.

Abstract-Summary

"Previous pilot-scale studies have shown outstanding levels of efficiency in phosphorus removal by using hydrated oil shale ash (HOSA) sediments in horizontal subsurface flow (HSSF) filters with low greenhouse gas emissions."

"From September 2013 to December 2015, two HSSF filters with different hydraulic loading regimes (NH1 with a stable loading regime and NH2 with a fluctuating regime), used to treat municipal wastewater, were analysed to estimate greenhouse gas (GHG) fluxes and to develop a treatment system with minimised GHG emissions."

"The fluxes of CO_2, CH_4 and N_2O, as well as their emission factors were significantly lower when compared with studies where regular filter materials (sand, gravel, etc) are in use."

"The HOSA-filled filters have a good potential for municipal wastewater treatment with low GHG emission."

Introduction

"The use of alternative filter materials in SSF filters can result in higher water purification effectiveness, as well as a lower level of GHG emissions."

"Several studies have shown the potential of hydrated oil shale ash as a highly effective filter material for the treatment of municipal wastewater, greywater and landfill leachate (Karabelnik and others 79; Kõiv and others 80; Liira and others 81; Vohla and others 82)."

"These investigations have shown that a high content of reactive Ca-rich minerals allows effective phosphorus (P) removal through solid Ca-phosphate precipitation from various types of wastewater, normally achieving a purification effectiveness of up to 99 %, outperforming most well-known filter materials (Kasak and others 83; Liira and others 81)."

"While most of the previous studies have been carried in lab and mesocosm scale, there is still no information regarding how the water treatment in full-scale HOSA-filled HSSF filters affects GHG emissions."

Materials and Methods

"The current study was carried out in an unplanted hybrid SSF filter system, fed with raw wastewater that was pumped from the inlet of activated sludge treatment plant from September 2013 to December 2015."

"After pre-treatment in the septic tank and VSSF filters, wastewater flows into two HSSF filters (NH1, NH2), each with a volume of 8 m^3 ($5 \times 2 \times 0.8$ m) filled with crushed and sieved hydrated oil shale ash sediment (5–20 mm)."

"The chamber-covered area for gas sampling was 16 % of the HSSF filter area (10 m^2), which is significantly larger than commonly used in chamber-based measurements in CWs (Mander and others 78)."

"Water samples from the inflow and outflow, as well as from each observation well (total 6), were taken a few minutes before each gas-sampling occasion."

Results

"Temporal dynamics of this gas emission from NH1 and NH2 showed that CO_2 emissions were higher during the warmer periods and increased in a linear order ($R^2 = 0.57$, $p < 0.001$ and $R^2 = 0.43$, $p < 0.001$, respectively) from both filters during the 2 years of operation."

"The CH_4 emission varied between –272.8 to 7071.9 and -112.5 to 7018.4 $\mu g\ m^{-2}\ h^{-1}$ with a median value of 64.9 and 70.1 $\mu g\ m^{-2}\ h^{-1}$ from the NH1 and NH2 filter, respectively."

"The CH_4 emissions from the filter were positively related to the TOC content only in NH1."

"The emission factor (EF) showed that about 0.67 and 0.97 % of the inflow TOC was transformed to methane in NH1 and NH2, respectively."

"In the case of NH2, the correlation between N_2O emissions and NH_4 concentration was significant but not linear."

Discussion

"We studied the impact of a promising alternative filter material, especially for the removal of phosphorus from the wastewater—HOSA application in GHG emissions and evaluated the effect of the wastewater-loading regime on these gas emissions from the full-scale HSSF filters."

"The results from a previous (pilot) study showed significantly lower emissions of all three analysed GHG from the HOSA filters, compared to the filters filled with more widespread materials (i.e. LECA, sand, gravel, etc) (Kasak and others 83)."

"The results from this study showed that the higher organic input increased methane emissions from the filter that had a stable loading regime."

"To the CO_2 emissions, our results also showed a relationship between CH_4 emissions and chemical conditions (pH and Ca^{2+} content) in the HSSF filters, showing that a lower pH level had a positive effect on CH_4 emissions."

Conclusions

"Hydrated oil shale ash, when used as a filter material in municipal wastewater treatment decreases greenhouse gas emissions from horizontal flow filter systems."

"Emissions of CO_2, CH_4 and N_2O can be up to 100 times lower compared with the more widely used filter materials like sand, gravel, LECA, etc All the greenhouse gas emissions were closely related to each other and were strongly affected by the chemical composition of the water entering the system."

"The CO_2 emissions, to a large extent, were controlled by the mineral composition of the HOSA filter material."

"Although, for maximum water treatment efficiency and minimum GHG emissions, HOSA filters should be used for secondary or tertiary treatment and especially for P removal because in the presence of high organic matter, formation of toxic (HS^-) and harmful volatile substances (NH_3) is possible."

Removal of Bisphenol A (BPA) from Biologically Treated Wastewater by Microfiltration and Nanofiltration [139]

This is a machine-generated summary of:

Zielińska, M.; Bułkowska, K.; Cydzik-Kwiatkowska, A.; Bernat, K.; Wojnowska-Baryła, I.: Removal of bisphenol A (BPA) from biologically treated wastewater by microfiltration and nanofiltration [139]

Published in: International Journal of Environmental Science and Technology (2016)

Link to original: https://doi.org/https://doi.org/10.1007/s13762-016-1056-6

If you want to cite the papers, please refer to the original.

For technical reasons we could not place the page where the original quote is coming from.

Abstract-Summary

"Bisphenol A (BPA) is an endocrine disruptor that is difficult to completely remove from wastewater by conventional biological methods."

"Increased post-treatment BPA removal with ceramic membranes is worth investigating because of these membranes' mechanical and chemical stability and lifespan."

"To determine the effectiveness of ceramic membranes for post-treatment of biologically treated BPA-contaminated wastewater, microfiltration (MF) and nanofiltration (NF) were conducted."

"MF removed at least 24 % of BPA, presumably because BPA was adsorbed on particulate matter, which was retained by the membrane, or adsorbed on its surface."

"BPA concentrations in municipal wastewater are typically lower than the lowest concentration tested, where MF completely removed BPA."

"MF ceramic membranes appear a promising solution for post-treatment of BPA-containing wastewater."

Introduction

"MF has been able to remove from 20 to 95 % of BPA from drinking water; efficiency dropped as the membrane became saturated with BPA (Bing-zhi and others 84)."

"To delay the drop in removal efficiency and to lengthen membrane life, different hybrid processes can be used to remove BPA and other EDCs."

"Although BPA in biologically treated wastewater can be both adsorbed to particulate matter and present in soluble form, previous studies with membrane bioreactors have focused on removing only adsorbed BPA with MF and UF (Zuehlke and others 85, Schröder 86, Chen and others 87)."

"We compared MF and NF ceramic membranes for the removal of BPA from effluent from reactors with immobilized biomass operated at an HRT of only 1.5 h. At such a short HRT, high concentrations of suspended solids in the effluent are observed, to which BPA and other hydrophobic compounds may be sorbed."

Materials and Methods

"The average permeation flux of distilled water (J_W) was 1500 L/(m^2 h) for the MF membrane and 80 L/(m^2 h) for the NF membrane."

"To determine the effect of BPA concentrations on filtration by both MF and NF membranes, biologically treated wastewater with different concentrations of BPA was used."

"In series 1 (S1), BPA concentration in the feed solution was 0.3 ± 0.14 mg/L; in series 2 (S2), it was 0.5 ± 0.19 mg/L; in series 3 (S3), 0.7 ± 0.29 mg/L. During filtration, permeation tests were conducted."

"The abbreviations used in the equations are as follows: A—membrane filtration area (m^2), C_F—BPA concentration in the feed solution (mg/L), C_P—BPA concentration in the permeate (mg/L), C_R—BPA concentration in the retentate (mg/L), t—time for collecting a known volume of permeate (h), TMP—transmembrane pressure (MPa), V_F—initial feed volume (L), V_P—total withdrawn permeate volume (L), V_R—retentate volume (L)."

Results and Discussion

"At higher BPA concentrations, removal of BPA decreased significantly and the NF membrane removed more BPA than the MF membrane."

"The higher transmembrane pressure in NF (0.7 MPa) than in MF (0.3 MPa) increased the removal of BPA because the higher pressure increased the rate of water transport through the membrane, and in consequence, lowered BPA concentration in the permeate."

"Suspended solids were present in the feed solutions in concentrations of 78 ± 12 mg TSS/L; these solids were completely removed by MF and NF membranes."

"At the highest initial BPA concentration, the permeability of the NF membrane was much lower than that of the MF membrane, as shown by the recovery value (Y), which expresses the ratio of the permeate volume to the feed solution volume."

Conclusion

"Our results suggest that the use of ceramic MF and NF membranes is effective in post-treating biologically treated wastewater that contains BPA."

"At higher BPA concentrations in the feed solution, from 61 to 75 % of BPA was removed with NF."

"Even at a membrane pore size of 0.45 μm, the MF membrane can remove BPA because it is sorbed on particulate matter, which is retained."

Influence of Temperature and COD Loading on Biological Nitrification–Denitrification Process Using a Trickling Filter: An Empirical Modeling Approach [140]

This is a machine-generated summary of:

Kanda, Ryo; Kishimoto, Naoyuki; Hinobayashi, Joji; Hashimoto, Tsutomu; Tanaka, Satoru; Murakami, Yoshitaka: Influence of temperature and COD loading on biological nitrification–denitrification process using a trickling filter: an empirical modeling approach [140]

Published in: International Journal of Environmental Research (2017)

Link to original: https://doi.org/https://doi.org/10.1007/s41742-017-0008-4

If you want to cite the papers, please refer to the original.

For technical reasons we could not place the page where the original quote is coming from.

Abstract-Summary

"The influences of temperature and chemical oxygen demand (COD) loading on the nitrification and denitrification performance of the process are discussed in this paper through experimental and empirical modeling approaches."

"The influences of temperature and COD loading were estimated by the model."

"The COD levels required for NO_3–N removal depended on both, the temperature and influent COD/NO_3–N ratio."

"A lower temperature and higher influent COD/NO_3–N ratio increased the COD requirement, because of the different responses between denitrifying bacteria and heterotrophic bacteria against temperature, COD, and NO_3–N concentrations."

"The dissolved nitrogen (DN) concentration in the final effluent was more strongly affected by the NH_4–N discharged from the trickling filter than it was by the residual NO_3–N in the effluent from the denitrification tank."

"To prevent a low nitrogen removal efficiency at temperatures lower than 15 °C, it was necessary to set the low hydraulic loading."

Introduction

"Advanced water treatment plants (AWTP) have become necessary for enhancing the removal of nutrients from municipal sewage to prevent eutrophication."

"In most wastewater treatment plants that use the activated sludge processes, blowers for aeration are the main energy consumers, as they run constantly and have a high electrical load demand (ENERWATER 88)."

"The amount of energy consumed by wastewater treatment plants strongly depends on the aeration process."

"To solve these problems, we propose a new advanced non-aeration water treatment process, namely, a biological nitrification–denitrification process using a trickling filter (Kanda and others 88)."

"The technology based on the trickling filter is expected to substitute for the conventional aerobic process for wastewater treatment like the activated sludge process."

"We evaluated the influences of temperature and COD loading on the nitrification and denitrification performances of the proposed process using an empirical modeling approach."

Materials and Methods

"Experiment 1 and 2 were simulated COD rich environment and NO_3–N rich environment for denitrification, respectively."

"For Experiment 1, the reactor had been operated for a month for acclimatization under the operational conditions as follows: the flow rate of influent wastewater: 5.0 L day^{-1} and the flow rate of artificial nitrified liquor: 10.0 L day^{-1}."

"For Experiments 1 and 2, the denitrification performance of the process was evaluated by treating an artificial nitrified liquor and the wastewater in Ryukoku University in the water temperature range of 12.0–25.1 °C."

"When the COD/NO_3–N in influent is less than the stoichiometric ratio of 4.00, the denitrification rate was inferred to be dependent on the COD rather than the NO_3–N concentration."

Results and Discussion

"The inactivation of denitrifying bacteria by the low temperature caused the incomplete removal of NO_3–N at 12 °C."

"At 12 °C, most of the NO_3–N remained in the effluent, whose denitrification efficiency was estimated to be 1.6%, because of low temperatures and low COD in the influent."

"Although the ratio of residual NO_3–N in the effluent from the denitrification tank to DN concentration in the final effluent was less than 1.0% at more than 19 °C, the ratio increased to 10.0% at 9 °C."

"The residual DN in the effluent was more strongly affected by the residual NH_4–N in effluent than the residual NO_3–N at low temperatures, though the influence of the incomplete denitrification on the DN concentration in effluent could not be ignored."

"When the recirculation rate of nitrified liquor was 250%, a denitrification tank HRT of 1.5 times longer would be required to satisfy the Japanese effluent standard as the minimum temperature would decrease by 1 °C."

Conclusion

"Under an environment with sufficient COD for denitrification, the COD removal/NO_3–N removal ratio at 20 °C was observed to be 4.82 ± 0.12 g-COD/g-N, which was higher than the stoichiometric requirement of 4.00 because not only denitrifying bacteria, but also heterotrophic bacteria, consumed organic matter in the denitrification tank."

"When the NO_3–N concentration in nitrified liquor was 15 mg L^{-1} and the COD in influent was high enough for denitrification, it was estimated that more than 1.0 mg L^{-1} of NO_3–N remained in the effluent from the denitrification tank at temperatures less than 17 °C."

"The COD requirement for NO_3–N removal depended on both temperature and influent COD/NO_3–N ratio."

"A lower temperature and higher influent COD/NO_3–N ratio increased the requirement of COD for NO_3–N removal, because of the different responses between denitrifying bacteria and heterotrophic bacteria against temperature, COD, and NO_3–N concentrations."

Wastewater Treatment by Slow Sand Filters Using Uncoated and Iron-Coated Fine Sand: Impact of Hydraulic Loading Rate and Media Depth [141]

This is a machine-generated summary of:

Verma, Srishti; Daverey, Achlesh; Sharma, Archana: Wastewater treatment by slow sand filters using uncoated and iron-coated fine sand: impact of hydraulic loading rate and media depth [141]

Published in: Environmental Science and Pollution Research (2018)

Link to original: https://doi.org/https://doi.org/10.1007/s11356-018-3551-4

If you want to cite the papers, please refer to the original.

For technical reasons we could not place the page where the original quote is coming from.

Abstract-Summary

"Two lab-scale slow sand filters (SSFs), packed with uncoated fine sand (SSF_u) and iron-coated fine sand (SSF_{co}), were operated to study their efficiency in treating municipal wastewater."

"The effects of sand coating; hydraulic loading rates (HLRs) (0.56, 0.85, and 1.12 cm/h); and filter depths (22, 32, and 42 cm) were evaluated."

"Sand coating did not have any significant effect on wastewater treatment by the SSF at all depths ($p > 0.05$)."

"Media depth had positive effects on the removal of turbidity, TSS, COD, and total coliforms (TC)."

"At HLR of 0.56 cm/h, the average removals of each studied parameter, i.e., turbidity, TSS, and COD, at filter depth d_{42} in SSF_u and SSF_{co} were 94.3, 90.1, and 56% and 92.7, 93, and 30.95%, respectively."

Introduction

"Various authors have reported that modifications of sand, such as by coating with a metal oxide or nanomaterials, can effectively sorb metals and remove viruses and bacteria from water and wastewater (Ahammed and Chaudhuri 89; Lukasik and others 90; Chaudhry and others 91)."

"Pathogens such as Vibrio cholerae, poliovirus, and coliphage MS-2 and E. coli from dechlorinated tap water can be effectively removed (up to 99%) by ferric/aluminum hydroxide-coated sand filter (Lukasik and others 90)."

"Ahammed and Chaudhuri (89) used iron/aluminum hydroxide-coated sand filter to remove heterotrophic bacteria E. coli and poliovirus from canal water."

"Metal-coated sand has been extensively studied for drinking water treatment, it is rarely applied to treat wastewater."

"The objective of the study was to compare the efficiency of uncoated (SSF_u) and iron oxide-coated fine sand (SSF_{co}) filters in treating municipal wastewater at different media depths and hydraulic loading rates (HLRs)."

Material and Methods

"The top layer of sand (about 1 cm) was scrapped and filters were backwashed with 10 L of distilled water for about an hour using peristaltic pump."

"Water samples were withdrawn (approximately 1 L) from three sand media depths, i.e., 22, 32, and 42 cm at regular intervals from both filters, i.e., SSF_u and SSF_{co} during filter operation and analyzed for water quality parameters such as pH (Electrometric method, 4500-H^+ B.); turbidity (Nephelometric method, 2130 B.); chemical oxygen demand (COD) (Closed Reflux, Titrimetric Method, 5220 C.); phosphate (Stannous Chloride Method, 4500-P D.); nitrate (Ultraviolet Spectrophotometric Screening Method, 4500-NO_3^- B.); total coliform (TC) (Standard Total Coliform Fermentation Technique, 9221 B.); and fecal coliform (FC) (Fecal Coliform Procedure, 9221 E)."

"The sand samples were removed when filters clogged due to either microbial deposition or organic matter accumulation."

Results and Discussion

"In SSF_u, the average turbidity removals at media depths d_{22}, d_{32}, and d_{42} were 84.10, 84.95, and 94.33% (HLR 0.56 cm/h); 56.73, 82.47, and 93.10% (HLR 0.85 cm/h); and 89.09, 88.27, and 93.51% (HLR 1.12 cm/h); respectively."

"In case of SSF_u, the average COD removals were 28.88, 29.25, and 56.53% (HLR 0.56 cm/h); 38.46, 39.66, and 50.48% (HLR 0.85 cm/h); and 28.16, 37.55, 47.90% (HLR 1.12 cm/h) at depths d_{22}, d_{32}, and d_{42}, respectively."

"The influent concentration ranged from 115 to 8000 MPN/100 mL. Both filters (SSF_u and SSF_{co}) were found to be highly efficient in TC removal from the wastewater (> 90%) at all filter depths in most days of filter run ($p = 0.19$) except in HLR 0.85 cm/h. More than 99% removal of FC was observed at all sand media bed depths in both filters at HLR 0.85 cm/h in our study."

Conclusions

"There was no significant difference in the wastewater treatment performances between the uncoated and iron-coated sand filters."

"Frequent clogging was observed in the uncoated sand filter (SSF_u)."

"Both the SSFs were highly efficient in the removal of turbidity, TC, and FC at all the tested HLRs."

"SSFs were found to be inefficient in removing nutrients and COD from the wastewater."

"Sedimentation, filtration, adsorption and attachment, reduction and reopening of pore spaces, hydrophilic interactions, and bacterial decay might have been the dominant processes for turbidity, TSS, COD, phosphate, nitrate, TC, and FC removal."

Enhanced Contaminant Removal from Municipal Wastewater Using Novel Cascade Cum Trickling Filter and Multilayer Adsorption Bioreactor [142]

This is a machine-generated summary of:

Kanwar, R. M. A.; Khan, Z. M.; Farid, H. U.: Enhanced contaminant removal from municipal wastewater using novel cascade cum trickling filter and multilayer adsorption bioreactor [142]

Published in: International Journal of Environmental Science and Technology (2021)

Link to original: https://doi.org/https://doi.org/10.1007/s13762-021-03201-9

If you want to cite the papers, please refer to the original.

For technical reasons we could not place the page where the original quote is coming from.

Abstract-Summary

"The novel two-stage treatment system consisting of cascade cum trickling filter (CCTF) and multilayer adsorption (MLA) bioreactor was developed to enhance the contaminants removal such as chemical oxygen demand (COD), total nitrogen (TN) and total phosphorus (TP)."

"Kinetic analysis showed that the composite adsorbent had a faster removal rate for phosphate than ammonium."

"The performance of both cascade cum trickling filter and multilayer adsorption (CCTF-MLA) systems was also evaluated using the various COD-TN ratios and flow rates."

"The removal efficiencies of COD, TN and TP were found 93%, 87% and 86%, respectively, and their corresponding mean effluent concentrations were noted 18 mg/L, 5.5 mg/L and 2.3 mg/L. The developed CCTF-MLA system removed the 3log of the total count and total coliform and 2log of Escherichia coli and salmonella."

"The CCTF-MLA system could be efficient for wastewater treatment due to the steady removal rate of COD, TN and TP."

Introduction

"The use of wastewater without treatment for irrigation may be harmful to environmental and community health due to undesirable changes in the physical, chemical and biological properties of receiving water environment (Khan and others 92; Soulard and others 93)."

"Wastewater use in agriculture is justified according to economic and agronomic points of view, but precautions and treatment are needed to minimize impacts on health and the environment (Salgot and others 94; Khan and others 92; Droste and Gehr 95)."

"Wastewater treatment (WWT) refers to the pollutant's removal from the wastewater for environmentally friendly effluent for safe disposal and agricultural reuse (Licciardello and others 96)."

"It turns into essential for planers to investigate decentralized and low-cost WWT systems to improve health conditions of receiving water environment (Licciardello and others 96)."

"The integrated use of trickling filter with cascade aeration (CCTF) and multilayer adsorption bioreactor (MLA) may act as more efficient for removal of excessive nutrients, biodegradable and non-biodegradable constituents that utilize trickling filter as the predominant form of biological treatment (Kanwar and others 97)."

Materials and Methods

"Cascade aeration contains 11 steps, and the height of each step was of 0.75 m. The associated cascade settler having length, width and depth of 3.1 × 3.1 × 1.5 m, respectively, was constructed to settle down the suspended solids."

"The multilayer adsorption system having length, width and depth of 3.4 m × 1.3 m × 1.2 m, respectively, was constructed for effective removal non-biodegradable contaminants."

"Wastewater taken from the septic tank of main sewage line of Department of Agricultural Engineering was used to assess the removal performance of secondary treatment systems."

"Composite wastewater samples of influent, mid effluent and final effluent were taken from primary clarifier, secondary clarifier and outlet tank of multilayer adsorption reactor, respectively, in a 1.5 L pre-sterilized bottle according to standard sampling procedures for validity of test results (APHA 98)."

Results and Discussion

"The series of influent, mid effluent and final effluent data obtained in phases 1–4 were investigated for performance evaluation of CCTF-MLA system under different conditions of wastewater loading (COD/TN ratio and flow rate)."

"The constant 'b' of the Langmuir model was found more for phosphate adsorption (0.0127 L/g) than ammonium adsorption (0.0012 L/g) which indicated the high binding strength of phosphate onto the composite adsorbent than ammonium and harder desorption (Salvador and others 99)."

"In Phase-1, the CCTF-MLA system showed high COD removal efficiency under the running conditions of the COD/TN ratio and flow rate at 6:1 and 0.9m^3/h. As a result, the CCTF-MLA system can effectually remove the carbonaceous contaminants under different operating conditions."

"In Phase-2, the CCTF-MLA system showed high TN removal efficiency (83% with effluent TN of 5.4 mg/L) under the running conditions of COD/TN ratio and flow rate at 9:1 and 1.1m^3/hr."

Conclusion

"The composite adsorbent had a faster removal rate for phosphate than ammonium."

"The high COD/TN ratio (9:1) in wastewater enhanced the TN removal of the CCTF-MLA system."

"The CCTF-MLA system was also found effective in the removal of pathogen indicators from wastewater such as total count (3.1 log), total coliform (2.8 log), E. coli (3.6 log) and salmonella (3.1)."

"The CCTF-MLA system showed steady removal of COD (83.4%), TN (80%) and TP (83%), and these results might be effective for peri-urban wastewater treatment."

The Use of a Hybrid Sequential Biofiltration System for the Improvement of Nutrient Removal and PCB Control in Municipal Wastewater [143]

This is a machine-generated summary of:

Kiedrzyńska, Edyta; Urbaniak, Magdalena; Kiedrzyński, Marcin; Jóźwik, Adam; Bednarek, Agnieszka; Gągała, Ilona; Zalewski, Maciej: The use of a hybrid Sequential Biofiltration System for the improvement of nutrient removal and PCB control in municipal wastewater [143]

Published in: Scientific Reports (2017)

Link to original: https://doi.org/https://doi.org/10.1038/s41598-017-05555-y

If you want to cite the papers, please refer to the original.

For technical reasons we could not place the page where the original quote is coming from.

Abstract-Summary

"This article aims to evaluate the efficiency of an innovative hybrid Sequential Biofiltration System (SBS) for removing phosphorus and nitrogen and polychlorinated biphenyls (PCBs) from original municipal wastewater produced by a Wastewater Treatment Plant under authentic operating conditions."

"Significant differences were found between inflow and outflow from the SBS with regard to wastewater contaminant concentrations, the efficiency of removal being 16% (max."

"In the case of PCB EQ concentration, the highest efficiency of 43% was obtained using beds with macrophytes."

"The SBS removed a significant load of TP (0.415 kg), TN (3.136 kg), and PCB EQ (0.223 g) per square meter per year."

Introduction

"The legislation criteria for organic compounds such as PCBs, which may be introduced into water bodies via their presence in treated wastewater, are still not sufficient to properly monitor and protect aquatic ecosystems."

"The Urban Wastewater Directive (91/271/EEC) has no information concerning the allowable limits of PCBs in wastewater discharged into the surface water."

"PCBs are detected not only in treated wastewater but also in recipient inland water, creating a risk for aquatic organisms."

"Due to the growing pressure on water resources, there is an ever greater need to optimise the potential of these systems as a method of treating wastewater, and this demands a greater understanding of the treatment mechanisms."

"Few studies have been conducted on the removal of PCBs from real municipal wastewater in biofiltration systems [100]."

Results and Discussion

"The retention and removal of nitrogen during wastewater treatment in constructed wetlands (CWs) is dependent on various processes occurring in the system, such as nitrification, denitrification, NH3 volatilization, nitrogen fixation, plant and microbial uptake, ammonification (mineralization), nitrate reduction to ammonium (nitrate-ammonification), anaerobic ammonia oxidation (anammox), fragmentation, sorption, desorption, burial and leaching."

"The appropriate choice of carbon substrate in the denitrication area in the SBS (W2–W4) results in the removal of all studied nitrogen compounds, despite fluctuating temperatures, higher then optimal pH and nearly anaerobic conditions: The removal of various forms of nitrogen from sewage using this combination of simultaneous nitrification and denitrification results in significantly improved wastewater quality [101, 102]."

"One of such tertiary purification method is the use of SBS comprising a set of filtration (mineral and organic) and wetland beds which enables the removal of not only nutrients but also organic compounds by a range of physicochemical and biogeochemical processes."

Materials and Methods

"From each monitoring station, one filtered sample was taken to analyse the concentrations of Total Suspended Solids (TSS) and dissolved forms of nutrients, and a second unfiltered sample was taken to analyse the total forms of nutrients."

"Wastewater samples for soluble forms of nutrients, e.g. Soluble Reactive Phosphorus (SRP), NO_3^--N, NO_2^--N, NH_4^+-N, were filtered through Whatman GF/C 0.45 m filters and analysed with the Ion Chromatography System (DIONEX, ICS 1000)."

"Aliquots (0.0002 dm^3) of calibration-standard PCB (0, 2.25, 1, 5 $\mu g\ dm^{-3}$), the wastewater sample, and a positive control solution (6 ppm) were added to test tubes together with aliquots of enzyme conjugate (0.00025 dm^3)."

"This sum represents a statistic for calculating the significance level p. The Wilcoxon test was used to compare the nutrient concentrations in sewage from the outlet of the WWTP at station W1 with those in sewage after cleaning in particular beds of the SBS at stations W2–W8 (*significance at $p \leq 0.05$)."

Conclusions

"The hybrid Sequential Biofiltration System (SBS) proved an effective method of ecohydrological biotechnology for treatment of wastewater from WWTPs."

"The SBS removed the following contaminant loads: 25.3 kg TP (18%), 37.1 kg SRP (26%), 191.3 kg TN (17%), 178 kg NO_3^-N (19%), 4.5 kg NO_2^-N (33%), 8.8 kg NH_4^+-N (7%) and also 13.6 g PCB EQ (49%) per year."

"The greatest reduction of concentration was observed in the biologically-active barrier, i.e. beds with macrophytes, wherein 83% of TP, 75% of SRP, 69% of TN, 72% of NO_3^-N, and 43% of PCB EQ concentrations were lower at the outlet from the SBS than the inlet."

"The low-cost Sequential Biofiltration Systems may be used in small WWTPs as an additional treatment step and an alternative ecohydrological biotechnology for reducing point source pollution and improving water quality in river catchments."

4.4 Ozone

Introduction by Editor

Ozone disinfection is a widely employed and highly effective method in the tertiary treatment phase of municipal wastewater treatment or sewage treatment plants. It involves the application of ozone, a powerful oxidizing agent, to eliminate pathogenic microorganisms, remove residual organic compounds, and enhance water quality. Ozone disinfection is based on the principle of oxidation, where ozone (O_3), a reactive and unstable form of oxygen, is generated on-site using ozone generators. The generated ozone is then dissolved in water, creating a potent oxidizing environment. Through its strong oxidation potential, ozone breaks down organic molecules and disrupts the cellular structures of microorganisms, rendering them inactive and incapable of causing harm. Compared to traditional disinfectants like chlorine, ozone offers higher disinfection efficiency while producing fewer disinfection by-products.

Successful implementation of ozone disinfection in tertiary treatment necessitates careful attention to operational parameters. Key factors include ozone dosage, contact time, pH levels, and temperature, which significantly influence disinfection efficacy. Optimal ozone dosage is critical for achieving desired disinfection levels while minimizing ozone demand and ensuring cost-effectiveness. Adequate contact time between ozone and the treated effluent is essential for effective disinfection. Maintaining appropriate pH levels and temperature conditions enhances ozone's disinfection performance. Additionally, well-designed ozone distribution systems and accurate monitoring equipment are crucial for uniform ozone delivery and precise measurement of residual ozone levels.

Ozone disinfection offers several advantages in the tertiary treatment of municipal wastewater. Firstly, ozone effectively inactivates a broad range of pathogens, including bacteria, viruses, and protozoa, ensuring the production of safe and microbiologically treated effluent. It serves as a robust barrier against waterborne diseases, safeguarding public health and the environment. Moreover, ozone rapidly oxidizes organic compounds and reduces color, taste, and odor-causing substances, improving the overall aesthetic quality of the treated water. Unlike traditional disinfection methods, ozone disinfection minimizes the formation of harmful disinfection by-products, mitigating potential health and environmental concerns.

Despite its advantages, ozone disinfection presents certain challenges that require careful consideration. Ozone is relatively unstable and demands proper handling and monitoring to ensure its appropriate application. Effective off-gas treatment systems should be in place to manage and prevent ozone emissions. Ozone's reactivity with certain compounds may lead to the formation of by-products, such as bromate, which necessitates careful monitoring and control. Furthermore, ozone disinfection requires higher energy consumption compared to other disinfection methods, highlighting the importance of comprehensive cost-benefit analyses to assess the feasibility and economic viability of ozone disinfection systems.

Machine generated keywords: ozonation, gene, dose, ozone, antibiotic resistance, antibiotic, disinfection, resistance, bacteria, wwtp, contact, optimize, transfer, select, pharmaceutical

Optimization of Ozonation Process to Remove Psychoactive Drugs from Two Municipal Wastewater Treatment Plants [144]

This is a machine-generated summary of:

Cunha, Deivisson L.; da Silva, Alexandre S. A.; Coutinho, Rodrigo; Marques, Marcia: Optimization of Ozonation Process to Remove Psychoactive Drugs from Two Municipal Wastewater Treatment Plants [144]

Published in: Water, Air, & Soil Pollution (2022)

Link to original: https://doi.org/https://doi.org/10.1007/s11270-022-05541-9

If you want to cite the papers, please refer to the original.

For technical reasons we could not place the page where the original quote is coming from.

Abstract-Summary

"Treatability studies applying ozonation to degrade benzodiazepine drugs are scarce."

"The aim of this investigation was to apply design of experiments (DoE) to optimize ozonation processes for removal of bromazepam, clonazepam, and diazepam from two effluents of municipal wastewater treatment plants (WWTP1 and WWTP2) both located in Rio de Janeiro City, Brazil."

"During a preliminary study using ultrapure water, OH° radicals were more efficient (removal efficiency greater than 99% (< LOQ) at pH = 10, 15 min) than O_3 in degrading all psychoactive drugs (removal efficiency from 46 to 85% at pH = 4, 15 min)."

"When 8 L effluent was treated, the degradation of all benzodiazepine drugs was improved, due to the higher water column used, which favors the mass transfer of O_3 from the gas phase to the liquid phase, promoting greater contact between oxidant and target compounds."

"Using a specific dose of 1.8 $mgO_3.mgDOC^{-1}$, more than 93% of all anxiolytic drugs were removed from 8 L of WWTP2 effluent, whereas for the WWTP1, more than 90% were removed using 2.7 times higher specific dose (4.8 $mgO_3.mgDOC^{-1}$)."

Introduction

"Considerable amounts of psychoactive drugs, such as anxiolytics and antidepressants, have been quantified in different environmental water compartments and pose toxic effects on aquatic ecosystems (Cunha and others, 103)."

"A review study highlights that behavioral effects of psychoactive drugs on aquatic species should be considered when assessing the environmental risks posed by these drugs, based on the European Medicines Agency (EMA) on environmental risk assessment (ERA) of Medicinal Products For Human Use (Cunha and others, 103)."

"More efficient techniques are required to remove psychoactive drugs from water, in order to reduce the risks posed to aquatic environment and to human health."

"The aim of this study was to optimize the ozonation process on the removal of three anxiolytic drugs: bromazepam (BZP), clonazepam (CZP), and diazepam (DZP) in real effluents released from WWTP based on biological treatment."

Methodology

"Aliquots of 2 mL were collected at different times, filtered with a hydrophilic PTFE 0.20-mm syringe filter (Millex®-LG Merck), and analyzed by direct injection in ultra-performance liquid phase chromatography coupled to a triple quadrupole mass/mass spectrometer (UPLC Acquity coupled to MS/MS Xevo TQD® Waters)."

"In this stage, to evaluate possible matrix effects, experiments were performed using the final effluents obtained at two different WWTPs (both of them spiked with 100 $\mu g\ L^{-1}$ of each benzodiazepine drug): (i) a large municipal wastewater treatment municipality plant (flow = $2.5\ m^3\ s^{-1}$) with secondary treatment based on activated sludge technology (WWTP1) (the same used in the previous experiment) denominated WWTP1 second campaign collected on 16 October 2019; and (ii) a pilot multi-stage hybrid bioreactor containing an aerobic reactor-based with suspended and fixed biomass (Bardenpho$_{mod}$-IFAS reactor; flow = $0.2\ L\ min^{-1}$) (WWTP2, collected on 23 October 2019)."

Results and Discussion

"Corroborating to this statement, a previous study (Huber and others, 104) showed the kinetic constants of DZP removal at pH 7 and was much higher for radical °OH ($7.2 \pm 1.0 \times 10^9\ M^{-1}\ s^{-1}$), using UV/$H_2O_2$, γ-radiolysis, than for direct attack with O_3 ($0.75 \pm 0.15\ M^{-1}\ s^{-1}$)."

"When 5 L of effluent is treated in the reactor, the average removals achieved were 78.5%, 80.1%, and 91.4%, respectively, for BZP, CZP, and DZP in the WWTP1 effluent."

"The degradation of the compounds occurred at no statistical difference speed, regardless of the effluent and volume used, considering that the amount of O_3 applied as a function of time (22.07 mgO_3.min^{-1}) was the same for all treatability tests and that the WWTP1 effluent contained lower DOC than the WWTP2 effluent (17.50 mg L^{-1} and 39.55 mg L^{-1}, respectively)."

Conclusion

"The results revealed that BZP, CZP, and DZP were removed more efficiently by ozonation at neutral to basic pH, suggesting greater efficiency obtained with the attack of hydroxyl radicals."

"The recalcitrancy of the target compounds to ozonation at pH 7 was similar in both ultrapure water and in the WWTP1 and WWTP2 effluents, according to the following order: BZP > CZP > DZP, which suggests there was no matrix effect that favored or substantially disadvantaged the degradation of any of the three anxiolytic drugs."

"When applying the same specific dose ($mgO_3.mgDOC^{-1}$), the degradation of BZP, CZP, and DZP by ozonation was more efficient when a larger water column (8L) was used, compared to a smaller (5L) water column."

"The degradation kinetic constants for both studied effluents were very close, indicating the degradation of the target compounds was mostly related to the total amount of ozone applied (mgO_3) than to the specific dose relative to the DOC content ($mgO_3.mgDOC^{-1}$) of each effluent."

Inactivation of Antibiotic Resistance Genes in Municipal Wastewater by Chlorination, Ultraviolet, and Ozonation Disinfection [145]

This is a machine-generated summary of:

Zhuang, Yao; Ren, Hongqiang; Geng, Jinju; Zhang, Yingying; Zhang, Yan; Ding, Lili; Xu, Ke: Inactivation of antibiotic resistance genes in municipal wastewater by chlorination, ultraviolet, and ozonation disinfection [145]

Published in: Environmental Science and Pollution Research (2014)

Link to original: https://doi.org/https://doi.org/10.1007/s11356-014-3919-z

If you want to cite the papers, please refer to the original.

For technical reasons we could not place the page where the original quote is coming from.

Abstract-Summary

"This study investigated the inactivation of two antibiotic resistance genes (ARGs)—sul1 and tetG, and the integrase gene of class 1 integrons—intI1 by chlorination, ultraviolet (UV), and ozonation disinfection."

"Inactivation of sul1, tetG, and intI1 underwent increased doses of three disinfectors, and chlorine disinfection achieved more inactivation of ARGs and intI1 genes (chlorine dose of 160 mg/L with contact time of 120 min for 2.98–3.24 log reductions of ARGs) than UV irradiation (UV dose of 12,477 mJ/cm^2 for 2.48–2.74 log reductions of ARGs) and ozonation disinfection (ozonation dose of 177.6 mg/L for 1.68–2.55 log reductions of ARGs)."

"The relative abundance of selected genes (normalized to 16S rDNA) increased during ozonation and with low doses of UV and chlorine disinfection."

"At a chlorine dose of 40 mg/L with 60 min contact time, the selected genes inactivation efficiency could reach 1.65–2.28 log, and the cost was estimated at 0.041 yuan/ m^3."

Introduction

"Overuse of antibiotics results in the increase and spread of antibiotic resistance genes (ARGs) and bacteria (ARB) in the environment (Rizzo and others 106), which reduce the pharmaceutical defenses against infection (Zhang and others 107)."

"Wastewater biological treatment steps may serve as a suitable process for the increase and spread of ARGs and ARB because a high density of active bacteria promotes horizontal gene transfer (Chen and Zhang 108; Chen and others 109)."

"Several laboratory reports from research dedicated to disinfection results of wastewater have observed the notable inactivation of laboratory cultured ARGs or ARB by UV light or chlorine disinfection (McKinney and Pruden 110; Huang and others 111)."

"The differences between the actual WWTP's operation reports and the laboratory research may be due to the variations of the predominant ARGs, ARB, and disinfection dosages; however, the disinfection doses in the WWTPs are always not given."

"It is important to compare the destruction of ARGs in real wastewater effluents under specific disinfection operations to find the potential pathway to decrease the spread of ARGs into natural water."

Materials and Methods

"After disinfection, 200-mL wastewater samples were taken and concentrated by passing through the 0.22 μm mixed cellulose ester filter membranes (Xinya, Shanghai, China) to concentrate microbial biomass for gene copies quantification (Munir and others 112; Pruden and others 113)."

"Sodium hypochlorite was spiked into the sample to establish desired doses of chlorine of 5, 20, 40, 80, 120, and 160 mg/L, respectively, and concentrations of chlorine were analyzed by the DPD method (Environmental Protection Agency of China 114)."

"During the ozonation experiments, a 6-L wastewater sample was poured into the ozone contact reaction tower (190 cm high and with a diameter of 10 cm, ZhongTong, China) with sampling done at intervals by adding potassium iodide (KI solution, 2 % (w/v)) to terminate reaction."

"The extracted DNA samples were stored at −20 °C until quantification analysis."

Results and Discussion

"The maximum inactivation of ARGs by chlorine disinfection at 160 mg/L was 3.16 log for sul1 genes and 3.24 log for tetG genes."

"At a dose of 27 mg/L, ozonation results showed a slight removal of ARGs with only 0.60 log inactivation on 16S rDNA."

"At a higher dosage (above 97.7 mg/L), the decrease of inspected genes tended to be flat ($p > 0.05$) as per our examination of the results involving tetG. The 16S rDNA was more efficiently removed than ARGs by ozone disinfection."

"Decreases of the relative abundance of ARGs and intI1 genes were observed in chlorination at doses from 20 to 160 mg/L, except for sul1, at 40 mg/L. After UV irradiation, the decreases were also observed in many situations, while for ozonation, the relative abundance of ARGs and intI1 genes increased."

Conclusions

"Chlorine and UV were more selective when targeting bacteria harboring ARGs."

"To achieve more ARG removal, the dosage of disinfectant had to be much higher than that defined by common usage."

"WWTPS can only partially remove ARGs and residual amounts can reach receiving water at present disinfection doses."

"Other methods including advanced treatment systems and combined methods also need to be investigated to find a more useful and economic way to remove ARGs from WWTP effluents."

Ozonation in Advanced Treatment of Secondary Municipal Wastewater Effluents for the Removal of Micropollutants [146]

This is a machine-generated summary of:

Dogruel, Serdar; Cetinkaya Atesci, Zuhal; Aydin, Egemen; Pehlivanoglu-Mantas, Elif: Ozonation in advanced treatment of secondary municipal wastewater effluents for the removal of micropollutants [146]

Published in: Environmental Science and Pollution Research (2020)

Link to original: https://doi.org/https://doi.org/10.1007/s11356-020-10339-5

If you want to cite the papers, please refer to the original.

For technical reasons we could not place the page where the original quote is coming from.

Abstract-Summary

"The objective of this study was the experimental evaluation of ozonation as an additional treatment step for the removal emerging contaminants from secondary effluents of two wastewater treatment plants (WWTPs), one receiving a primarily domestic wastewater (WWTP-A), and the other one domestic sewage together with pretreated tannery wastewater streams (WWTP-B)."

"The experimental runs were conducted at two different pH values (i.e., original pH and adjusted pH of 10) and at six different ozone doses ranging between 0.2 and 1.5 mg O_3/mg DOC."

"When the tested MPs and metabolites were considered individually, the maximum elimination level for each compound was reached at different doses; therefore, optimum ozone doses were determined based on the reduction of the total MP content."

"Ozonation at the original pH with an ozone dose in the range of 0.4–0.6 and 0.8–1.0 mg O_3/mg DOC was selected as the optimum operating condition for WWTP-A and WWTP-B, respectively, both resulting in an average overall removal efficiency of 55%."

Introduction

"During ozonation of secondary treated municipal wastewater effluents, the mass-based ozone normalized to the dissolved organic carbon (DOC) content (i.e., mg O_3/mg DOC) has often been used as an operating parameter to compare samples with varying DOC concentrations (Wert and others 116; Lee and others 117)."

"The objective of the present study was, on one hand, to evaluate the efficiency of the ozonation process for the removal of the selected 20 MPs from secondary treated municipal effluents and, on the other hand, to determine the optimum ozone dose to be applied with the aim of decreasing MPs' concentrations for safely discharging wastewater into a receiving water and maybe reusing the water when both the adoption and implementation of recovery initiatives are considered."

Materials and Methods

"Filtered sample portions, containing a nitrite concentration of less than 0.08 mg NO_2-N/L, were treated at six different ozone doses ranging from 0.2 to 1.5 mg O_3/mg DOC to elucidate the effect of ozone dose on the fate of MPs and metabolites."

"Cartridges were then washed with 5 mL of a mixture of ACN and HPLC grade deionized water (20:80, v/v) and dried under vacuum for 1 h. All samples extracted with SPE cartridges in 10 mL of ACN/MTBE (50:50, v/v) were evaporated to dryness under nitrogen at a pressure of 10 bar using a Caliper TurboVap II (Caliper Life Sciences, Hopkinton, MA, USA)."

"After evaporation, the residue in the TurboVap tube was reconstituted with a mixture of isopropyl alcohol and HPLC grade deionized water (15:85, v/v) to a final volume of

1 mL. For the quantitation of MPs by liquid chromatography tandem mass spectrometry (LC-MS/MS), the resulting solution was filtered through a 220-nm membrane filter and transferred into 2-mL vials."

Results and Discussion

"An increment of the ozone dose from 0.4 to 0.6 and 0.8 mg O_3/mg DOC did not improve the total MP abatement rate due to the upwards fluctuation of o-desmethyl naproxen, acetyl sulfamethoxazole, and caffeine contents, constituting approximately half of the cumulative concentration of all detected analytes initially present in the non-ozonated effluent samples."

"As depicted in the figure, the total MP abatement rate of the summer sample obtained after ozonation at the original pH led to an additional removal of 9% when the ozone dose was changed from 0.6 to 0.8 mg O_3/mg DOC."

"The abatement rates accomplished by the ozonation process were inconsistent with those obtained by Knopp and others (118), who directed filtered secondary effluent from a full-scale WWTP into an ozone system and removed hydroxy diclofenac to a concentration level lower than the LOQ at an ozone dose of 0.87 ± 0.29 mg O_3/ mg DOC."

Conclusions

"For the two investigated WWTPs, o-desmethyl naproxen was detected in all four sampling campaigns at relatively high concentrations (> 3 μg/L on average) where the contribution from its amount to the total MP content of biologically treated municipal wastewater effluents was determined in the range of 27–32%, indicating this metabolite as the most abundant substance in the studied sampling sites."

"An additional abatement of 15–20% in total MPs could be produced if o-desmethyl naproxen, exhibiting exceptionally high concentration values due to its persistent nature against ozonation, was not included in the calculation of the total content of all measured analytes."

"The high level of o-desmethyl naproxen underlined the necessity of the quantitative assessment of metabolites, particularly of those which were identified as the main contributors to the total concentration of emerging contaminants in WWTP effluents not only to establish reliable total MP removal rates but also to accurately evaluate the efficiency of the ozonation process."

Dissemination Prevention of Antibiotic Resistant and Facultative Pathogenic Bacteria by Ultrafiltration and Ozone Treatment at an Urban Wastewater Treatment Plant [147]

This is a machine-generated summary of:

Hembach, Norman; Alexander, Johannes; Hiller, Christian; Wieland, Arne; Schwartz, Thomas: Dissemination prevention of antibiotic resistant and facultative pathogenic bacteria by ultrafiltration and ozone treatment at an urban wastewater treatment plant [147]

Published in: Scientific Reports (2019)

Link to original: https://doi.org/https://doi.org/10.1038/s41598-019-49263-1

If you want to cite the papers, please refer to the original.

For technical reasons we could not place the page where the original quote is coming from.

Abstract-Summary

"This study focuses on the reduction efficiencies of two semi-industrial ultrafiltration units operating at a large scale municipal wastewater treatment plant."

"Both ultrafiltration units showed increased reduction efficiencies for facultative pathogenic bacteria and antibiotic resistance genes of up to 6 log units, resulting mostly in a strong reduction of the bacterial targets."

"The ozone treatment showed some reduction efficiency, but was less effective compared with ultrafiltration due to low ozone dosages frequently used for micro-pollutant removal at municipal wastewater treatment plants."

"Metagenome analyses demonstrated the accumulation of facultative pathogenic bacteria, antibiotic resistance genes, virulence factor genes, and metabolic gene targets in the back flush retentate of the membranes, which opens further questions about retentate fluid material handling at urban wastewater treatment plants."

Antibiotic Resistance Genes and Facultative Pathogenic Bacteria in Wastewaters

“Antibiotic resistance genes (ARGs) and facultative pathogenic bacteria from different origins (hospitals, nursing homes, housing areas, slaughter houses etc) that pass through treatment plants or agricultural runoffs can lead to a contamination of environmental water resources.”

“Besides facultative pathogenic bacteria, which can be the causative agents of human and veterinary infections, antibiotic resistance genes are also a point of interest.”

“Ozonation is already discussed as a possible disinfection process, having already proven its capability to degrade and/or transform micro-pollutants and, in some studies, reduce facultative pathogenic bacteria and ARGs [119–121].”

“Ultrafiltration is an alternative process to separate facultative pathogenic bacteria and ARGs from the wastewater effluents.”

“This study aimed to develop a deeper understanding of the ultrafiltration reduction efficiencies of facultative pathogenic bacteria and ARGs in comparison to ozonation at an urban wastewater treatment plant (WWTP).”

Material and Methods

“For evaluation of the ultrafiltration and ozone reduction efficiency, 4 sampling campaigns were performed during 9 months (February till November, 2017).”

“Two samplings with different flux rates (12.5, 50, 90 L/m^2h) were performed for ultrafiltration.”

“Volumes of 10 L of each UF filtrate, 750 mL of ozoned water, 300 mL of inflow, and 400 mL of flush-water were vacuum-filtrated through a 0.2 μm polycarbonate membrane (Whatman Nuclepore Track-Etched Membranes, Sigma-Aldrich, Munich, Germany).”

“An absolute quantification normalized to 100 mL water sample was performed to demonstrate the reduction effects of the applied treatment processes for quantitative analysis.”

“These reference strains with known genome sizes, respective known plasmid sizes, were used to calculate cell equivalents utilizing the amount of DNA used for creating the target specific calibration curve.”

“qPCR was also performed in technical triplicates and their mean Ct-values were used for cell equivalent calculation of each target and sample.”

Results and Discussion

“Bacteria carrying the resistance gene CTX-M-32 (β-lactamase against cephalosporins) were able to pass the ozonation facility to a greater extent, and was found in concentrations of up to 2.77×10^3 cell equivalents/100 mL in the outflow of the ozone facility.”

"Under the given process parameters, both advanced treatment technologies (ozonation and ultrafiltration) were able to reduce facultative pathogenic bacteria, as well as bacteria carrying antibiotic resistance genes, significantly."

"It was demonstrated by the qPCR quantification of gene targets that the retentate water of UF1 harbours 2 log unit higher concentrations of facultative pathogenic bacteria compared to the inflow water."

"To the other facultative pathogenic bacteria, A. baumannii increased by only one order of magnitude in both UF units from 3.09×10^1 to 3.03×10^2 cell equivalents/ 100 mL. The abundance of ARGs under investigation in UF1 also demonstrated increased concentration values in ranges of 2 log units within the retentates."

Conclusion

"The present study is demonstrating that technical measurements are already available to effectively reduce or eliminate antibiotic resistant genes and facultative pathogenic bacteria from municipal wastewaters at treatment plants."

"Ultrafiltration, as a physical technology, removed an average of 5 log units of both facultative pathogenic bacteria and ARGs, while ozonation adjusted to micro-pollutant treatment removed about 2 log units on average."

"It was calculated that about 2.66×10^{13} cell equivalents per day (i.e. antibiotic resistance genes and facultative pathogenic bacteria) reached the receiving body after conventional treatment."

"This discharge was significantly reduced to 6.5×10^8 cell equivalents per day of facultative pathogenic bacteria, including antibiotic resistance genes, when an ultrafiltration is performed."

"It could be demonstrated that both technologies, ozonation as well as ultrafiltration, were able reduce especially the most clinically relevant, but almost rarely detected ARGs (e.g. mcr-1, blaNDM-1, vanA, and blaVIM) and facultative pathogenic bacteria below the detection limit."

Optimizing Ozone Dose and Contact Time for Removal of Antibiotic-Resistant P. aeruginosa, A. baumannii, E. coli, and Associated Resistant Genes in Effluent of an Activated Sludge Process in a Municipal WWTP [148]

This is a machine-generated summary of:

Rajabi, Akbar; Farajzadeh, Davoud; Dehghanzadeh, Reza; Aslani, Hassan; Mosaferi, Mohammad; Mousavi, Saeid; Shanehbandi, Dariush; Asghari, Farzaneh Baghal: Optimizing ozone dose and contact time for removal of antibiotic-resistant P. aeruginosa, A. baumannii, E. coli, and associated resistant genes in effluent of an activated sludge process in a municipal WWTP [148]

Published in: Environmental Science and Pollution Research (2023)

Link to original: https://doi.org/https://doi.org/10.1007/s11356-023-26270-4

If you want to cite the papers, please refer to the original.

For technical reasons we could not place the page where the original quote is coming from.

Abstract-Summary

"This study aimed to investigate the impact of ozonation on inactivation of antibiotic-resistant bacteria (ARB) including E. coli, P. aeruginosa, and A. baumannii, as well as on removal of 16S-rRNA gene and their associated antibiotic-resistant genes (ARGs) indigenously present in effluent of municipal wastewater treatment plant."

"Maximum reduction of total cultivable A. baumannii, E. coli, and P. aeruginosa were found to be 7.6, 7.1, and 4.7 log, respectively, with the highest ozone dose of 0.48 gO_3/gCOD at 12 min contact time."

"The results of this study highlighted the significance of specific ozone dose and contact time in ozonation process considering the bacterial species and associated ARGs as well as the wastewater physicochemical characteristics, in order to help diminish the entrance of the biological microcontaminants into the environment."

Introduction

"Municipal wastewater treatment plants (WWTPs) are considered as one of the most predominant reservoirs of antibiotic resistance bacteria (ARB) and antibiotic-resistant genes (ARGs) which could complicate water reuse opportunities and lead to wastewater discharge into the aquatic environments (Ezeuko and others 122; Iakovides and others 123)."

"This study aimed to investigate the effect of ozonation on inactivation of different ARBs and ARGs in effluent of a conventionally activated sludge process in a municipal WWTP at various contact times and initial ozone concentrations."

"Other important objectives of this study included (i) determination of inactivation rate of resistant E. coli, P. aeruginosa, and A. baumannii at different contact times and initial ozone concentrations; (ii) measuring reduction efficiency of six resistant genes from different classes including bla_{TEM}, bla_{SHV}, $bla_{CTX\text{-}M}$, and bla_{VIM} (beta-lactam resistance genes); sul1 (sulfonamide resistance gene) and qnrS (quinolone resistance gene); and (iii) optimization of ozone concentration and contact time for effective inactivation of ARBs and removal of ARGs in effluent."

Materials and Methods

"As for culture-based technique, 100 μL of unfiltered samples were cultured utilizing surface culture method on a nutrient agar medium and incubated at 37 °C for 24 and 48 h. Bacteria counts expressed as colony forming units per mL (cfu/mL) were enumerated with the limit of quantification (LOD) of bacteria equal to 1 cfu/mL. Moreover, the potential bacteria regrowth in ozone-treated samples was studied after their storage for 72 h in the absence of light, at 25 ± 2 °C."

"In order to compare efficiency of ozone in removal or inactivate of the resistant bacteria with non-resistant bacteria, samples of E. coli DH5-Alpha pure culture obtained from Immunology Research Center, Tabriz University of Medical Sciences, Iran, were cultured on nutrient agar and seeded individually into the autoclaved wastewater samples at initial concentration of 10^8 cfu/mL. Each qPCR mixture contained 10 μL SYBR green qPCR Master Mix (Ampliqon, Denmark), 0.2 μL of 10 μm concentrations of forward primers and reverse primers, and 1 μL of DNA template, and was made up to a volume of 20 μL with nuclease-free water."

Results and Discussion

"The effects of ozone dose ($gO_3/gCOD$) and CT ($gO_3/gCOD \times$ min) on the inactivation rate of cultivable resistant E. coli, P. aeruginosa, and A. baumannii were investigated."

"The results of statistical analysis showed that the exposure time (CT), ozone dose, and initial bacterial concentration had a significant effect on bacterial inactivation rate ($P < 0.05$)."

"Our results also demonstrated that the initial concentration of bacteria affected the removal efficiency since lower ozone dose and shorter CT were needed to inactivate low concentrations of the examined bacterial groups."

"Jäger and others (124) reported that the removal log by PMA-qPCR was lower than that by culture method during ozonation of P. aeruginosa and E. coli; however, no significant difference was found between PMA-qPCR and qPCR methods in this regard."

"According to statistical results, there was significant difference in the genes reduction carried by E. coli, P. aeruginosa, and A. baumannii at different initial concentrations of bacteria."

Conclusion

"It was shown that the CT and ozone dose governed the inactivation of resistant bacterial species and genes."

"Ozone was found effective in inactivating bacterial species developing resistance to penicillin, cephalosporin, sulfonamide, and quinolone."

"Despite reducing genes by ozonation, the complete reduction was not achieved and the genes were present in samples even after using high doses of ozone (7.20 gO_3/ gCOD × min)."

"Our study findings may have facilitated improving the quality of treated wastewater while minimizing the potential risk of developing antibiotic resistance in organisms in environment."

Bibliography

1. Xanthoulis D et al. (2008) Curriculum on low-cost wastewater treatment - in French: Les techniques d'épuration des eaux usées à faibles coûts. Faculté Universitaire des Sciences Agronomiques de Gembloux, Belgium; EU project on Development of Teaching and Training Modules for Higher Education on Low-Cost Wastewater Treatment, Contract VN/Asia-Link/012. https://www.susana.org/_resources/documents/default/2-1725-curriculumfranc291008dxmw.pdf
2. Puigagut J, Villasenor J, Salas JJ, Becares E, Garcia J (2007) Subsurface-flow constructed wetlands in Spain for the sanitation of small communities: a comparative study. Ecol Eng 30(4):312–319. https://doi.org/10.1016/j.ecoleng.2007.04.005
3. Le-Clech P, Chen V, Fane TAG (2006) Fouling in membrane bioreactors used in wastewater treatment. J Membrane Sci 284:17–53
4. Meng F, Chae SR, Drews A, Kraume M, Shin HS, Yang F (2009) Recent advances in membrane bioreactors (MBRs): membrane fouling and membrane material. Water Res 43:1489–1512
5. Wang XM, Li XY (2008) Accumulation of biopolymer clusters in a submerged membrane bioreactor and its effect on membrane fouling. Water Res 42:855–862
6. Sabia G, Ferraris M, Spagni A (2013) Effect of solid retention time on sludge filterability and biomass activity: long-term experiment on a pilot-scale membrane bioreactor treating municipal wastewater. Chem Eng J 221:176–184
7. Villain M, Marrot B (2013) Influence of sludge retention time at constant food to microorganisms ratio on membrane bioreactor performances under stable and unstable state conditions. Bioresour Technol 128:134–144
8. Standard methods for the examination of water and wastewater (2005). American Public Health Association/Water Environment Federation, 21st ed. Washington DC
9. Pollice A, Laera G, Blonda M (2004) Biomass growth and activity in a membrane bioreactor with complete sludge retention. Water Res 38:1799–1808
10. Bae TH, Tak TM (2005) Interpretation of fouling characteristics of ultrafiltration membranes during the filtration of membrane bioreactor mixed liquor. J Membr Sci 264:151–160
11. Knowles P, Dotro G, Nivala J, García J (2011) Clogging in subsurface-flow treatment wetlands: occurrence and contributing factors. Ecol Eng 37:99–112. https://doi.org/10.1016/j.ecoleng.2010.08.005
12. Knowles PR, Griffin P, Davies PA (2010) Complementary methods to investigate the development of clogging within a horizontal sub-surface flow tertiary treatment wetland. Water Res 44:320–330. https://doi.org/10.1016/j.watres.2009.09.028

13. Nivala J, Knowles P, Dotro G, Garcia J, Wallace S (2012) Clogging in subsurface-flow treatment wetlands: measurement, modeling and management. Water Res 46:1625–1640. https://doi.org/10.1016/j.watres.2011.12.051
14. Aiello R, Bagarello V, Barbagallo S, Iovino M, Marzo A, Toscano A (2016) Evaluation of clogging in full-scale subsurface flow constructed wetlands. Ecol Eng 95:505–513. https://doi.org/10.1016/j.ecoleng.2016.06.113
15. De Matos MP, von Sperling M, de Matos AT (2018) Clogging in horizontal subsurface flow constructed wetlands: influencing factors, research methods and remediation techniques. Rev Environ Sci Biotechnol 17:87–107. https://doi.org/10.1007/s11157-018-9458-1
16. Herrera-Melián JA, Mendoza-Aguiar M, Guedes-Alonso R, García-Jiménez P, Carrasco-Acosta M, Ranieri E (2020) Multistage horizontal subsurface flow vs. hybrid constructed wetlands for the treatment of raw urban wastewater. Sustainability 12(12):5102. https://doi.org/10.3390/su12125102
17. Shuval HI, Adin A, Fattal B, Rawitz E, Yekutiel P (1986) Wastewater irrigation in developing countries: health effects and technical solutions. Technical Paper 58. The World Bank, Washington, DC
18. Assadian NW, Di Giovanni GD, Enciso J, Iglesias J, Lindemann W (2005) The transport of waterborne solutes and bacteriophage in soil subirrigated with a wastewater blend. Agric Ecosyst Environ 111:279–291
19. Ministerio de Agricultura Pesca y Alimentación (MAPA) (1994) Métodos Oficiales de análisis. Tomo III. Secretaría General Técnica del Ministerio de Agricultura, Pesca y Alimentación, Madrid
20. Gárate A, Cadahía C, Carpena O (1980) Estudio comparativo de métodos analíticos de boro en suelo usando azometina H como reactivo. Actas III Congreso Nacional de Química, Sevilla
21. Ayers RS, Westcot DW (1985) Water quality for agriculture. Estudio F.A.O. Riego y Drenaje 29. Roma
22. Toscano, A., Marzo, A., Milani, M., Cirelli, G. L., & Barbagallo, S. (2015). Comparison of removal efficiencies in Mediterranean pilot constructed wetlands vegetated with different plant species. Ecological Engineering, 75, 155–160.
23. Justin, M. Z., Vrhovsek, D., Stuhlbacher, A., & Bulc, T. G. (2009). Treatment of wastewater in hybrid constructed wetland from the production of vinegar and packaging of detergents. Desalination, 246, 100–109.
24. Serrano, L., de la Varga, D., Ruiz, I., & Soto, M. (2011). Winery wastewater treatment in a hybrid constructed wetland. Ecological Engineering, 37, 744–753.
25. Kadlec, R. H. (2000). The inadequacy of first-order treatment wetland models. Ecological Engineering, 15, 105–119.
26. Kadlec RH, Wallace SD (2008) Treatment wetlands, 2nd ed. CRC Press, Boca Raton, FL, USA
27. Wallace S, Knight R (2006) Small-scale constructed wetland treatment systems. IWA Publishing, London
28. Broch-Due A, Andersen R, Opheim B (1997) Treatment of integrated newsprint mill wastewater in moving bed biofilm reactors. Water Sci Technol 35:173–180
29. Chen SK, Juaw CK, Cheng SS (1991) Nitrification and denitrification of high strength ammonium and nitrite wastewater with biofilm reactors. Water Sci Technol 23:1417–1425
30. Masuda S, Watanabe Y, Ishiguro M (1991) Biofilm properties and simultaneous nitrification and denitrification in aerobic rotating biological contactors. Water Sci Technol 23:1355–1363
31. Watanabe Y, Masuda S, Ishiguro M (1992) Simultaneous nitrification and denitrification in micro-aerobic biofilms. Water Sci Technol 26:511–522
32. Lessel TH (1994) Upgrading and nitrification by submerged bio-film reactors-experiences from a large scale plant. Water Sci Technol 29:167–174
33. Watanabe Y, Okabe S, Arata T, Haruta Y (1994) Study on the performance of an up-flow aerated biofilter (UAB) in municipal wastewater treatments. Water Sci Technol 30:25–33
34. Cannon AD, Gray KR, Biddlestone AJ, Thayanithy K (2000) SE—structures and environment: pilot-scale development of a bioreactor for the treatment of dairy dirty water. J Agr Eng Res 77:327–334

35. APHA, AWWA, WEF (1998) Standard methods for the examination of water and wastewater, vol 20. American Public Health Association/American Water Works Association/Water Environment Federation, Washington DC, USA
36. Hu B (2008) Sewage treatment by bioreactor using a novel wool fleece media. PhD thesis, Loughborough University
37. Brillas, E., Sirés, I., & Oturan, M. A. (2009). Electro-Fenton process and related electrochemical technologies based on Fenton's reaction chemistry. Chemical Reviews, 109, 6570–6631.
38. Brillas, E., Arias, C., Cabot, P., Centellas, F., Garrido, J., & Rodríguez, R. (2006). Degradation of organic contaminants by advanced electrochemical oxidation methods. Portugaliae Electrochimica Acta, 24, 159.
39. Daghrir, R., Drogui, P., & El Khakani, M. (2013). Photoelectrocatalytic oxidation of chlortetracycline using Ti/TiO 2 photo-anode with simultaneous H2O2 production. Electrochimica Acta, 87, 18–31.
40. Drogui, P., Elmaleh, S., Rumeau, M., Bernard, C., & Rambaud, A. (2001). Hydrogen peroxide production by water electrolysis: application to disinfection. Journal of Applied Electrochemistry, 31, 877–882.
41. Garrido, J. A., Brillas, E., Cabot, P. L., Centellas, F., Arias, C., & Rodríguez, R. M. (2007). Mineralization of drugs in aqueous medium by advanced oxidation processes. Portugaliae Electrochimica Acta, 25, 19.
42. Brillas, E., & Casado, J. (2002). Aniline degradation by Electro-Fenton® and peroxi-coagulation processes using a flow reactor for wastewater treatment. Chemosphere, 47, 241–248.
43. Flox, C., Cabot, P.-L., Centellas, F., Garrido, J. A., Rodríguez, R. M., Arias, C., & Brillas, E. (2007). Solar photoelectro-Fenton degradation of cresols using a flow reactor with a boron-doped diamond anode. Applied Catalysis B: Environmental, 75, 17–28.
44. Guivarch, E., Oturan, N., & Oturan, M. A. (2003). Removal of organophosphorus pesticides from water by electrogenerated Fenton's reagent. Environmental Chemistry Letters, 1, 165–168.
45. Panizza, M., & Cerisola, G. (2009). Electro-Fenton degradation of synthetic dyes. Water Research, 43, 339–344.
46. Zarei, M., Niaei, A., Salari, D., & Khataee, A. (2010). Application of response surface methodology for optimization of peroxi-coagulation of textile dye solution using carbon nanotube–PTFE cathode. Journal of Hazardous Materials, 173, 544–551.
47. Zhou, M., Yu, Q., & Lei, L. (2008). The preparation and characterization of a graphite–PTFE cathode system for the decolorization of CI acid Red 2. Dyes and Pigments, 77, 129–136.
48. Zhou, Z.; Yang, Y.; Li, X.; Su, Z.; Liu, Y.; Ren, J.; Zhang, Y.: Effect of recycling filter backwash water on characteristic variability of dissolved organic matter in coagulation sedimentation process. Desal. Water Treat. 53, 1–9 (2013)
49. Chen, T.; Xu, Y.; Zhu, S.; Cui, F.: Combining physico-chemical analysis with a Daphnia magna bioassay to evaluate a recycling technology for drinking water treatment plant waste residuals. Ecotoxico. Environ. Saf. 122, 368–376 (2015)
50. Edzwald, J.K.; Tobiason, J.E.; Kelley, M.B.; Dunn, H.J.; Galant, P.B.; Kaminski, G.S.: Impacts of filter backwash recycle on clarification and filtration. J. Am. Water Works Assoc. Res, Founda (2001)
51. Muralikrishna, I.V.; Manickam,V.; Principles and design of water treatment. Environ. Man. 209–248 (2017)
52. Ofek M, Hadar Y, Minz D (2012) Ecology of root colonizing Massilia (Oxalobacteraceae). PLoS ONE 7:e40117. https://doi.org/10.1371/journal.pone.0040117
53. Peters, J., Iribarren, D. & Dufour, J. Biomass pyrolysis for biochar or energy applications? A life cycle assessment. Environ. Sci. Technol. 49, 5195–5202 (2015).
54. Dalahmeh, S., Ahrens, L., Gros, M., Wiberg, K. & Pell, M. Potential of biochar filters for onsite sewage treatment: Adsorption and biological degradation of pharmaceuticals in laboratory filters with active, inactive and no biofilm. Sci. Total Environ. 612, 192–201 (2018).

55. Perez-Mercado, L. F., Lalander, C., Berger, C. & Dalahmeh, S. S. Potential of biochar filters for onsite wastewater treatment: Effects of biochar type, physical properties and operating conditions. Water (Switzerland) 10 (2018).
56. Verma, S., Daverey, A. & Sharma, A. Slow sand filtration for water and wastewater treatment–a review. Environ. Technol. Rev. 6, 47–58 (2017).
57. Sarkhot, D. V., Ghezzehei, T. A. & Berhe, A. A. Effectiveness of Biochar for Sorption of Ammonium and Phosphate from Dairy Effluent. J. Environ. Qual., https://doi.org/10.2134/jeq.2012.0482 (2013).
58. Vikrant, K. et al. Engineered/designer biochar for the removal of phosphate in water and wastewater. Sci. Total Environ. 616–617, 1242–1260 (2018).
59. Langenbach, K., Kuschk, P., Horn, H. & Kästner, M. Slow sand filtration of secondary clarifier effluent for wastewater reuse. Environ. Sci. Technol. 43, 5896–901 (2009).
60. Ellis, K. Slow sand filtration as a technique for the tertiary treatment of municipal sewages. Water Res. 21, 403–410 (1987).
61. Bellamy, W. D., Hendricks, D. W. & Logsdon, G. S. Slow sand filtration: influences of selected process variables. J. Am. Water Works Assoc. 77, 62–66 (1985).
62. Singh R (2011) Analysis of energy usage at membrane water treatment plants. Desalin Water Treat 29:63–72
63. Kamp PC, Kruithof JC, Folmer HC (2000) UF/RO treatment plant Heemskerk: from challenge to the full-scale application. Desalination 131:27–35
64. Raffin M, Germain E, Judd S (2012) Assessment of fouling of an RO process dedicated to indirect potable reuse. Desalin Water Treat 40:302–308
65. Tang F, Hu H-Y, Sun L-J, Wu Q-Y, Jiang Y-M, Guan Y-T, Huang J-J (2014) Fouling of reverse osmosis membrane for municipal wastewater reclamation: autopsy results from a full-scale plant. Desalination 349:73–79
66. Sudak RG, Dunivin W, Rigby MG (1990) Municipal wastewater reclamation and reverse osmosis. In: Proceedings of the national water supply improvement association 1990 biennial conference, Florida August 1990, p 225
67. Ravazzini AM, van Nieuwenhuijzen AF, van der Graaf JHMJ (2005) Direct ultrafiltration of municipal wastewater: comparison between filtration of raw sewage and primary clarifier effluent. Desalination 178:51–62
68. Bruin D, Kamp F, Eddy P, Lagrand K (2002) WTW: UF/RO from technological novelty to the full-size application. In: Proceedings IWA Enviro 2002, Melbourne, Australia, 2002
69. Shaarawy HH, Saied M (2012) Production of sodium hypochlorite or mixed oxidant using rhodium/rhodium oxide mixed modified electrode. Aust J Basic Appl Sci 6(6):204–215
70. Ratanatamskul C, Chiemchaisri C, Yamamoto K (1995) The use of a zeoliteiron column for residual ammonia and phosphorus removal in the effluent from a membrane process as an on-site small scale domestic wastewater treatment. Water Sci Technol 31(9):145–152
71. Ghernaout D, Ghernaout B, Boucherit A, Naceur MW, Khelifa A, Kellil A (2009) Study on mechanism of electrocoagulation with iron electrodes in idealized conditions and electrocoagulation of humic acids solution in batch using aluminum electrodes. Desalin Water Treat 8:91–99
72. Xu W, Zhang G, Li X, Zou S, Li P, Hu Z, Li J (2007) Occurrence and elimination of antibiotics at four sewage treatment plants in the Pearl River Delta (PRD), South China. Water Res 41:4526–4534. https://doi.org/10.1016/j.watres.2007.06.023
73. Ashfaq M, Li Y, Wang Y, Chen W, Wang H, Chen X, Wu W, Huang Z, Yu CP, Sun Q (2017) Occurrence, fate, and mass balance of different classes of pharmaceuticals and personal care products in an anaerobic-anoxic-oxic wastewater treatment plant in Xiamen, China. Water Res 123:655–667. https://doi.org/10.1016/j.watres.2017.07.014
74. Petrie B, McAdam EJ, Lester JN, Cartmell E (2014) Obtaining process mass balances of pharmaceuticals and triclosan to determine their fate during wastewater treatment. Sci Total Environ 497-498:553–560
75. Perisa M, Babic S (2014) Simultaneous determination of pharmaceuticals and some of their metabolites in wastewaters by high performance liquid chromatography with tandem mass spectrometry. J Sep Sci 37:1289–1296. https://doi.org/10.1002/jssc.201301135

76. Garcia-Galan MJ, Petrovic M, Rodriguez-Mozaz S, Barcelo D (2016) Multiresidue trace analysis of pharmaceuticals, their human metabolites and transformation products by fully automated on-line solid-phase extraction-liquid chromatography-tandem mass spectrometry. Talanta 158:330–341. https://doi.org/10.1016/j.talanta.2016.05.061
77. Yan Q, Gao X, Huang L, Gan X, Zhang Y, Chen Y, Peng X, Guo J (2014) Occurrence and fate of pharmaceutically active compounds in the largest municipal wastewater treatment plant in Southwest China: mass balance analysis and consumption back-calculated model. Chemosphere 99:160–170
78. Mander, Ü., Kuusemets, V., Lõhmus, K., Mauring, T., Teiter, S., & Augustin, J. (2003). Nitrous oxide, dinitrogen and methane emission in a subsurface flow constructed wetland. Water Science and Technology, 48, 135–142.
79. Karabelnik, K., Kõiv, M., Kasak, K., Jenssen, P. D., & Mander, Ü. (2012). High-strength greywater treatment in compact hybrid filter systems with alternative substrates. Ecological Engineering, 49, 84–92.
80. Kõiv, M., Kriipsalu, M., Vohla, C., & Mander, Ü. (2009). Hydrated oil shale ash and mineralized peat as alternative filter materials for landfill leachate treatment in vertical flow constructed wetlands. Fresen Environ Bull, 18, 189–195.
81. Liira, M., Kõiv, M., Mander, Ü., Mõtlep, R., Vohla, C., & Kirsimäe, K. (2009). Active filtration of phosphorus on ca-rich hydrated oil shale ash: does longer retention time improve the process? Environmental Science and Technology, 43, 3809–3814.
82. Vohla, C., Põldvere, E., Noorvee, A., Kuusemets, V., & Mander, Ü. (2005). Alternative filter media for phosphorous removal in a horizontal subsurface flow constructed wetland. J Environ Sci Heal A, 40, 1251–1264.
83. Kasak, K., Mander, Ü., Truu, J., Truu, M., Järveoja, J., Maddison, M., & Teemusk, A. (2015). Alternative filter material removes phosphorus and mitigates greenhouse gas emission in horizontal subsurface flow filters for wastewater treatment. Ecological Engineering, 77, 242–249.
84. Bing-zhi D, Hua-qiang C, Lin W, Sheng-ji X, Nai-yun G (2010) The removal of bisphenol A by hollow fiber microfiltration membrane. Desalination 250:693–697
85. Zuehlke S, Duennbier U, Lesjean B, Gnirss R, Buisson H, Wasserbetriebe B, Water V (2003) Long term comparison of trace organics removal performances between conventional and membrane activated sludge processes. Proceedings of the 76th annual water environment federation technical exposition and conference. Los Angeles
86. Schröder HF (2006) The elimination of the endocrine disrupters 4-nonylphenol and bisphenol A during wastewater treatment—comparison of conventional and membrane assisted biological wastewater treatment followed by ozone treatment. Water Pract Technol. https://doi.org/10.2166/wpt.2006.060
87. Chen J, Huang X, Lee D (2008) Bisphenol A removal by a membrane bioreactor. Process Biochem 43:451–456
88. Borchardt JA (1966) Nitrification in the activated sludge process, the activated sludge process in sewage treatment theory and application. Univ. of Michigan, Dept. of Civil Eng., Ann Arbor, MI
89. Ahammed MM, Chaudhuri M (1996) Sand-based filtration/adsorption media. J Water Supply Res T-Aqua 45:67–71
90. Lukasik J, Cheng Y, Lu F, Tamplin M, Farrah SR (1999) Removal of microorganisms from water by columns containing sand coated with ferric and aluminium hydroxides. Water Res 33:769–777
91. Chaudhry SA, Khan TA, Ali I (2017) Zirconium oxide-coated sand based batch and column adsorptive removal of arsenic from water: isotherm, kinetic and thermodynamic studies. Egypt J Pet 26:553–563
92. Khan MZ, Kanwar RMA, Farid HU, Sultan M, Asalan M, Ahmad M, Aslam MM, Shakoor A (2019) Wastewater evaluation for multan (Pakistan):characterization and Agricultural Reuse. Pol J Environ Stud 28(4):2159–2174

93. Soulard CT, Valette E, Perrin C, Abrantes PC, Anthopoulou T, Benjaballah O, Bouchemal S, Dugué P, El Amrani M, Lardon S, Marraccini E (2018) Peri-urban agro-ecosystems in the Mediterranean: diversity, dynamics, and drivers. Reg Environ Change 18(3):651–662
94. Salgot M, Folch M, Unit SS (2018) Wastewater treatment and water reuse. Curr Opin Environ Sci Health 2:64–74
95. Droste RL, Gehr RL (2018) Theory and practice of water and wastewater treatment. Wiley, New York
96. Licciardello F, Milani M, Consoli S, Pappalardo N, Barbagallo S, Cirelli G (2018) Wastewater tertiary treatment options to match reuse standards in agriculture. Agric Water Manag 210:232–242
97. Kanwar RMA, Khan ZM, Farid HU (2019) Development and adoption of wastewater treatment system for peri-urban agriculture in Multan. Pak Water Sci Technol 80(8):1524–1537
98. APHA (2012) Standard methods for the examination of water and wastewater, 22nd edn. American Public Health Association, American Water Works Association, and Water Environment Federation
99. Salvador F, Martin-Sanchez N, Sanchez-Hernandez R, Sanchez-Montero MJ, Izquierdo C (2015b) Regeneration of carbonaceous adsorbents Part I: thermal regeneration. Microporous Mesoporous Mater 202:259–276
100. Imfeld, G., Braeckevelt, M., Kuschk, P. & Richnow, H. H. Monitoring and assessing processes of organic chemicals removal in constructed wetlands. Chemosphere 74, 349–362 (2009).
101. Bednarek, A., Szklarek, S. & Zalewski, M. Nitrogen pollution removal from area of intensive farming–comparison of various denitrification biotechnologies. Ecohydrol. Hydrobiol. 14, 132–141 (2014).
102. Mankiewicz-Boczek, J. et al. Microbiologically activated barriers for removal of nitrogen compounds from farm sewage. New Biotechnol. 33(5), S130 (2016).
103. Cunha, D. L., Mendes, M. P., & Marques, M. (2019a). Environmental risk assessment of psychoactive drugs in the aquatic environment. Environmental Science and Pollution Research, 26(1), 78–90. https://doi.org/10.1007/s11356-018-3556-z
104. Huber MM, Canonica S, Park G, von Gunten U (2003) Oxidation of pharmaceuticals during ozonation and advanced oxidation processes. Environ Sci Technol 37:1016–1024
105. Shang C, Cheung LM, Liu W (2007) MS2 coliphage inactivation with UV irradiation and free chlorine/monochloramine. Environ Eng Sci 24:1321–1332
106. Rizzo L, Manaia C, Merlin C, Schwartz T, Dagot C, Ploy MC, Michael I, Fatta-Kassinos D (2013b) Urban wastewater treatment plants as hotspots for antibiotic resistant bacteria and genes spread into the environment: a review. Sci Total Environ 447:345–360
107. Zhang XX, Zhang T, Fang HH (2009a) Antibiotic resistance genes in water environment. Appl Microbiol Biotechnol 82:397–414
108. Chen H, Zhang M (2013) Occurrence and removal of antibiotic resistance genes in municipal wastewater and rural domestic sewage treatment systems in eastern China. Environ Int 55:9–14
109. Chen RZ, Craik SA, Bolton JR (2009) Comparison of the action spectra and relative DNA absorbance spectra of microorganisms: information important for the determination of germicidal fluence (UV dose) in an ultraviolet disinfection of water. Water Res 43:5087–5096
110. McKinney CW, Pruden A (2012) Ultraviolet disinfection of antibiotic resistant bacteria and their antibiotic resistance genes in water and wastewater. Environ Sci Technol 46:13393–13400
111. Huang JJ, Hu HY, Wu YH, Wei B, Lu Y (2013) Effect of chlorination and ultraviolet disinfection on tetA-mediated tetracycline resistance of Escherichia coli. Chemosphere 90:2247–2253
112. Munir M, Wong K, Xagoraraki I (2011) Release of antibiotic resistant bacteria and genes in the effluent and biosolids of five wastewater utilities in Michigan. Water Res 45:681–693
113. Pruden A, Pei R, Storteboom H, Carlson KH (2006) Antibiotic resistance genes as emerging contaminants: studies in northern Colorado. Environ Sci Technol 40:7445–7450
114. Environmental Protection Agency of China (2002) Water and wastewater monitoring methods, 4th edn. Chinese Environmental Science Publishing House, Beijing

115. Hofman-Caris CHM, Siegers WG, van de Merlen K, de Man AWA, Hofman JAMH (2017) Removal of pharmaceuticals from WWTP effluent: removal of EfOM followed by advanced oxidation. Chem Eng J 327:514–521. https://doi.org/10.1016/j.cej.2017.06.154
116. Wert, E. C., Rosario-Ortiz, F. L., & Snyder, S. A. (2009). Effect of ozone exposure on the oxidation of trace organic contaminants in wastewater. Water Research, 43(4), 1005–1014. https://doi.org/10.1016/j.watres.2008.11.050
117. Lee, Y., Kovalova, L., McArdell, C. S., & von Gunten, U. (2014). Prediction of micropollutant elimination during ozonation of a hospital wastewater effluent. Water Research, 64, 134–148. https://doi.org/10.1016/j.watres.2014.06.027
118. Knopp G, Prasse C, Ternes TA, Cornel P (2016) Elimination of micropollutants and transformation products from a wastewater treatment plant effluent through pilot scale ozonation followed by various activated carbon and biological filters. Water Res 100:580–592. https://doi.org/10.1016/j.watres.2016.04.069
119. Ternes TA, Stüber J, Herrmann N, McDowell D, Ried A, Kampmann M, Teiser B (2003) Ozonation: a tool for removal of pharmaceuticals, contrast media and musk fragrances from wastewater? Water Res 37(8):1976–1982. https://doi.org/10.1016/S0043-1354(02)00570-5
120. Esplugas, S., Bila, D. M., Krause, L. G. T. & Dezotti, M. Ozonation and advanced oxidation technologies to remove endocrine disrupting chemicals (EDCs) and pharmaceuticals and personal care products (PPCPs) in water effluents. Journal of Hazardous Materials 149, 631–642 (2007).
121. Lüddeke, F. et al. Removal of total and antibiotic resistant bacteria in advanced wastewater treatment by ozonation in combination with different filtering techniques. Water Research 69, 243–251 (2015).
122. Ezeuko AS, Ojemaye MO, Okoh OO, Okoh AI (2021) Technological advancement for eliminating antibiotic resistance genes from wastewater: a review of their mechanisms and progress. J Environ Chem Eng 9:106183
123. Iakovides I, Manoli K, Karaolia P, Michael-Kordatou I, Manaia C, Fatta-Kassinos D (2021) Reduction of antibiotic resistance determinants in urban wastewater by ozone: emphasis on the impact of wastewater matrix towards the inactivation kinetics, toxicity and bacterial regrowth. J Hazard Mater 420:126527. https://doi.org/10.1016/j.jhazmat.2021.126527
124. Jäger, T. et al. Live-dead discrimination analysis, qPCR assessment for opportunistic pathogens, and population analysis at ozone wastewater treatment plants. Environmental Pollution 232, 571–579 (2018).
125. Ergaieg, Karim; Ben Miled, Taieb Full-scale hybrid constructed wetlands monitoring for decentralized tertiary treatment of municipal wastewater. Arabian Journal of Geosciences (2021). https://doi.org/10.1007/s12517-021-07776-y
126. Díaz, Oliver; Vera, Luisa; González, Enrique; García, Elisa; Rodríguez-Sevilla, Juan Effect of sludge characteristics on membrane fouling during start-up of a tertiary submerged membrane bioreactor. Environmental Science and Pollution Research (2016). https://doi.org/10.1007/s11356-016-6138-y
127. Ergaieg, Karim; Msaddek, Mohamed Haythem; Kallel, Amjad; Trabelsi, Ismail Monitoring of horizontal subsurface flow constructed wetlands for tertiary treatment of municipal wastewater. Arabian Journal of Geosciences (2021). https://doi.org/10.1007/s12517-021-08419-y
128. Martínez, S.; Suay, R.; Moreno, J.; Segura, M. L. Reuse of tertiary municipal wastewater effluent for irrigation of Cucumis melo L.. Irrigation Science (2012). https://doi.org/10.1007/s00271-012-0342-4
129. Haghshenas-Adarmanabadi, Amir; Heidarpour, Manouchehr; Tarkesh-Esfahani, Saleh Evaluation of Horizontal–Vertical Subsurface Hybrid Constructed Wetlands for Tertiary Treatment of Conventional Treatment Facilities Effluents in Developing Countries. Water, Air, & Soil Pollution (2015). https://doi.org/10.1007/s11270-015-2718-6
130. Thalla, Arun Kumar; Devatha, C. P.; Anagh, K.; Sony, Elsa Performance evaluation of horizontal and vertical flow constructed wetlands as tertiary treatment option for secondary effluents. Applied Water Science (2019). https://doi.org/10.1007/s13201-019-1014-9

131. Hu, Bibo; Wheatley, Andrew; Ishtchenko, Vera; Huddersman, Katherine Performance linked to residence time distribution by a novel wool-based bioreactor for tertiary sewage treatment. Applied Microbiology and Biotechnology (2011). https://doi.org/10.1007/s00253-011-3659-7
132. Senghor, Fatou; Drogui, Patrick; Seyhi, Brahima A Combined Electrocoagulation-Electroperoxidation Process for the Tertiary Treatment of Domestic Wastewaters. Water, Air, & Soil Pollution (2015). https://doi.org/10.1007/s11270-015-2637-6
133. Mazari, Lilia; Abdessemed, Djamal Feasibility of Reuse Filter Backwash Water as Primary/ Aid Coagulant in Coagulation–Sedimentation Process for Tertiary Wastewater Treatment. Arabian Journal for Science and Engineering (2020). https://doi.org/10.1007/s13369-020-04597-1
134. Xiao, Jibo; Lin, Guo; Cao, Zhuangzhuang; Chu, Shuyi; Cui, Lingzhou; Yang, Yunlong; Wu, Xiangting A shallow constructed wetland combining porous filter material and Rotala rotundifolia for advanced treatment of municipal sewage at low HRT. Environmental Science and Pollution Research (2022). https://doi.org/10.1007/s11356-022-24111-4
135. Kaetzl, Korbinian; Lübken, Manfred; Nettmann, Edith; Krimmler, Stefan; Wichern, Marc Slow sand filtration of raw wastewater using biochar as an alternative filtration media. Scientific Reports (2020). https://doi.org/10.1038/s41598-020-57981-0
136. Abdel-Fatah, Mona A.; Shaarawy, H. H.; Hawash, S. I. Integrated treatment of municipal wastewater using advanced electro-membrane filtration system. SN Applied Sciences (2019). https://doi.org/10.1007/s42452-019-1178-9
137. Wang, Dapeng; Zhang, Xian; Yan, Changzhou Occurrence and removal of sulfonamides and their acetyl metabolites in a biological aerated filter (BAF) of wastewater treatment plant in Xiamen, South China. Environmental Science and Pollution Research (2019). https://doi.org/10.1007/s11356-019-06311-7
138. Kasak, Kuno; Mõtlep, Riho; Truu, Marika; Truu, Jaak; Kõiv-Vainik, Margit; Espenberg, Mikk; Paiste, Päärn; Kirsimäe, Kalle; Mander, Ülo Hydrated Oil Shale Ash Mitigates Greenhouse Gas Emissions from Horizontal Subsurface Flow Filters for Wastewater Treatment. Water, Air, & Soil Pollution (2016). https://doi.org/10.1007/s11270-016-3007-8
139. Zielińska, M.; Bułkowska, K.; Cydzik-Kwiatkowska, A.; Bernat, K.; Wojnowska-Baryła, I. Removal of bisphenol A (BPA) from biologically treated wastewater by microfiltration and nanofiltration. International Journal of Environmental Science and Technology (2016). https://doi.org/10.1007/s13762-016-1056-6
140. Kanda, Ryo; Kishimoto, Naoyuki; Hinobayashi, Joji; Hashimoto, Tsutomu; Tanaka, Satoru; Murakami, Yoshitaka Influence of temperature and COD loading on biological nitrification–denitrification process using a trickling filter: an empirical modeling approach. International Journal of Environmental Research (2017). https://doi.org/10.1007/s41742-017-0008-4
141. Verma, Srishti; Daverey, Achlesh; Sharma, Archana Wastewater treatment by slow sand filters using uncoated and iron-coated fine sand: impact of hydraulic loading rate and media depth. Environmental Science and Pollution Research (2018). https://doi.org/10.1007/s11356-018-3551-4
142. Kanwar, R. M. A.; Khan, Z. M.; Farid, H. U. Enhanced contaminant removal from municipal wastewater using novel cascade cum trickling filter and multilayer adsorption bioreactor. International Journal of Environmental Science and Technology (2021). https://doi.org/10.1007/s13762-021-03201-9
143. Kiedrzyńska, Edyta; Urbaniak, Magdalena; Kiedrzyński, Marcin; Jóźwik, Adam; Bednarek, Agnieszka; Gągała, Ilona; Zalewski, Maciej The use of a hybrid Sequential Biofiltration System for the improvement of nutrient removal and PCB control in municipal wastewater. Scientific Reports (2017). https://doi.org/10.1038/s41598-017-05555-y
144. Cunha, Deivisson L.; da Silva, Alexandre S. A.; Coutinho, Rodrigo; Marques, Marcia Optimization of Ozonation Process to Remove Psychoactive Drugs from Two Municipal Wastewater Treatment Plants. Water, Air, & Soil Pollution (2022). https://doi.org/10.1007/s11270-022-05541-9

145. Zhuang, Yao; Ren, Hongqiang; Geng, Jinju; Zhang, Yingying; Zhang, Yan; Ding, Lili; Xu, Ke Inactivation of antibiotic resistance genes in municipal wastewater by chlorination, ultraviolet, and ozonation disinfection. Environmental Science and Pollution Research (2014). https://doi.org/10.1007/s11356-014-3919-z
146. Dogruel, Serdar; Cetinkaya Atesci, Zuhal; Aydin, Egemen; Pehlivanoglu-Mantas, Elif Ozonation in advanced treatment of secondary municipal wastewater effluents for the removal of micropollutants. Environmental Science and Pollution Research (2020). https://doi.org/10.1007/s11356-020-10339-5
147. Hembach, Norman; Alexander, Johannes; Hiller, Christian; Wieland, Arne; Schwartz, Thomas Dissemination prevention of antibiotic resistant and facultative pathogenic bacteria by ultrafiltration and ozone treatment at an urban wastewater treatment plant. Scientific Reports (2019). https://doi.org/10.1038/s41598-019-49263-1
148. Rajabi, Akbar; Farajzadeh, Davoud; Dehghanzadeh, Reza; Aslani, Hassan; Mosaferi, Mohammad; Mousavi, Saeid; Shanehbandi, Dariush; Asghari, Farzaneh Baghal Optimizing ozone dose and contact time for removal of antibiotic-resistant P. aeruginosa, A. baumannii, E. coli, and associated resistant genes in effluent of an activated sludge process in a municipal WWTP. Environmental Science and Pollution Research (2023). https://doi.org/10.1007/s11356-023-26270-4

Chapter 5
Water Reclamation and Reuse

Nitin Vitthalrao Thombre

5.1 Introduction by Editor

The increasing demand for sustainable water management practices has led to the widespread adoption of water reclamation and reuse strategies in municipal wastewater treatment or sewage treatment plants. The availability of clean and reliable water resources is essential for human and environmental well-being. However, rising population, climate change, and escalating water demands have put significant pressure on freshwater supplies. Municipal wastewater treatment or sewage treatment plants play a crucial role not only in ensuring the safe disposal of wastewater but also in reclaiming and reusing treated water. Water reclamation and reuse present promising solutions to address water scarcity challenges and foster sustainable water management practices.

Water reclamation involves the treatment of wastewater to remove contaminants and produce high-quality reclaimed water suitable for various non-potable applications. Municipal wastewater treatment plants employ advanced treatment technologies to achieve the desired water quality. These technologies encompass physical, chemical, and biological processes, including sedimentation, filtration, disinfection, membrane filtration, and advanced oxidation. Each process targets specific contaminants, ensuring the production of reclaimed water that meets the required standards for reuse.

> Water reuse applications
>
> Reclaimed water finds diverse applications in agriculture, landscape irrigation, industrial processes, and groundwater recharge. By substituting freshwater sources with reclaimed water for non-potable uses, numerous benefits can be achieved. These benefits include reducing strain on freshwater resources, decreasing reliance on external water supplies, mitigating environmental impacts associated with wastewater discharge, and fostering a

N. V. Thombre (✉)
VA Tech Wabag Ltd., Pune, Maharashtra, India
e-mail: nitin01thombre@gmail.com

N. V. Thombre (ed.), *Municipal Wastewater Reclamation and Reuse*,
https://doi.org/10.1007/978-981-96-9112-8_5

more sustainable and resilient water supply system. Additionally, reclaimed water provides essential nutrients for agricultural crops and supports ecosystem restoration efforts.

While water reclamation and reuse offer significant potential, certain challenges and considerations must be addressed for successful implementation. Key challenges include public acceptance and perception, potential health risks associated with reclaimed water use, regulatory frameworks and guidelines, and infrastructure requirements. Building public trust through effective communication, conducting comprehensive risk assessments, establishing robust regulatory frameworks, and ensuring adequate treatment and monitoring systems are essential for addressing these challenges and maximizing the benefits of water reclamation and reuse.

Why case studies are important?

Examining case studies from municipalities worldwide provides insights into successful water reclamation and reuse initiatives. These examples highlight the implementation of integrated water management approaches, stakeholder engagement, tailored treatment processes, and innovative distribution systems. Furthermore, sharing best practices, lessons learned, and knowledge exchange among different regions can foster continuous improvement and enhance the sustainability of water reclamation and reuse projects.

5.2 Machine Generated Summaries

Disclaimer: The summaries in this chapter were generated from Springer Nature publications using extractive AI auto-summarization: An extraction-based summarizer aims to identify the most important sentences of a text using an algorithm and uses those original sentences to create the auto-summary (unlike generative AI). As the constituted sentences are machine selected, they may not fully reflect the body of the work, so we strongly advise that the original content is read and cited. The auto generated summaries were curated by the editor to meet Springer Nature publication standards. To cite this content, please refer to the original papers.

Machine generated keywords: reuse, scenario, water scarcity, water, groundwater, scarcity, technology, contamination, Europe, wastewater management, recycling, reclamation, availability, potable, national

Waste Water Management and Water Quality of River Yamuna in the Megacity of Delhi [94]

This is a machine-generated summary of:

Parween, M.; Ramanathan, AL.; Raju, N. J.: Waste water management and water quality of river Yamuna in the megacity of Delhi [94]

Published in: International Journal of Environmental Science and Technology (2017)

Link to original: https://doi.org/https://doi.org/10.1007/s13762-017-1280-8

If you want to cite the papers, please refer to the original.

For technical reasons we could not place the page where the original quote is coming from.

Abstract-Summary

"Chloride concentration exceeded acceptable levels of drinking water guidelines."

"Water quality was poor at all locations with respect to heavy metal contamination, especially along the lower section of the Delhi stretch."

"Heavy metal concentrations were manifold higher than the acceptable limits of drinking water according to the BIS guidelines and reached ~ 29, 4.9, 10, 31, 27, 83, 7.3 and 27 times higher, respectively, for metals aluminum, copper, chromium, cadmium, iron, lead, manganese and nickel."

"Discharges from agriculture, sewage and power plants could be important sources of high metal concentration."

"This calls for urgent measures to be taken for prevention of metal contamination in the river, through both, technology as well as implementation of regulations in order to sustain huge populations in megacities like Delhi."

"Waste water treatment from point sources needs tremendous improvement on the city."

"Treatment of the entire waste generated up to the tertiary level is required for minimizing dissolved solids, especially toxic metals, and rendering reuse in agriculture suitable."

Introduction

"Wastewater generated per capita by the class I cities (classified on the basis of population: more than hundred thousand) and class II towns (population size: 50–100 thousand) comprising of 72% of the total urban population in India amounts to approximately 98 l per capita per day (lpcd), with over 220 lpcd alone generated by the National Capital Territory of Delhi, where only 61% of the total 3663 million liters per day (MLD) of wastewaters is treated before being discharged into the river Yamuna (CPCB 1)."

"As against the secondary treatment processes that primarily remove organics to a large extent, the tertiary processes such as adsorption, electrolysis may be employed for the removal of toxic metals which are a major concern in river Yamuna."

"Surface water samples from river Yamuna were collected in February 2013 from 14 locations in the National Capital Region of Delhi and characterized for their quality with regard to nutrients, organic and inorganic pollutants."

Materials and Methods

"Most of them are in receipt of effluents from major drains and canals carrying industrial waste, sewage and agricultural discharges and thus predicted to be the most polluted locations in the study area."

"L1 and L2 which are rural agricultural locations in the village Jagatpur lie remote to urban Delhi and are relatively unpolluted."

"L3 and L4 are urban locations lying downwind to the Narela industrial area."

"L5 and L6 are urban industrial locations which receive discharges from the Najafgarh drain and the supplementary drain."

"One set of samples was acidified with concentrated nitric acid (1.5 ml concentrated nitric acid per liter of the sample) and used for bulk metal analysis."

"Samples were diluted accordingly for the analysis of each element so that the results were included well within the range of instrument detection limits."

Results and Discussion

"Major ions and heavy metals (mg l^{-1}) were found at a range of 20.7–64.8 (Na), 6–18.6 (potassium or K), 38.6–123.9 (Ca), 22.7–57.8 (Mg), 0.3–0.7 (fluoride or F^-), 47.7–293 (chloride or Cl^-), 0.9–3.9 (nitrate or NO_3^-), 38.9–99.3 (sulfate or SO_4^{2-}), 172–282 (bicarbonate or HCO_3^-), 0.2–1.3 (phosphate or PO_4^-), 2.5–8.2 (Fe), 0.4–0.9 (Al), 0.1–0.3 (Cu), 0.1–0.4 (Cr), 0.03–0.09 (Co), 0.02–0.09 (Cd), 0.2–0.8 (Pb), 0.2–0.7 (Mn), 0.2–0.5 (Ni) and 0.2–0.5 (Zn)."

"Sources like agricultural application, sewage disposal and industrial effluents may have led to such high concentrations, as peak levels were observed immediately downstream to the discharge points of major drains into river Yamuna."

"River water was contaminated with high concentrations of heavy metals in the study area."

Conclusion

"All the locations downstream to the Wazirabad barrage were found to be contaminated with high levels of heavy metals in the river water."

"Heavy metals such as Fe, Al, Cu, Cr, Cd, Pb and Mn were found at concentrations beyond the standard levels for drinking water recommended by WHO and BIS."

"The physicochemical parameters, especially pH and ORP, were important in determining the distribution of heavy metals in river water."

"Oxides and hydroxides of Al, Fe and Mn seemed to play role in retention of heavy metals in the river water."

"To it, agricultural activities along the floodplain were important sources of high metal concentration in the river water."

"Water quality was poor at all of the locations with respect to heavy metal contamination, especially along the lower section of river Yamuna in the Delhi stretch."

Recommendations

"Sewage should be treated systematically up to the tertiary level in the Delhi region before being discharged into rivers. (b) A comprehensive treatment scheme need to be envisioned for substantial removal of soluble and insoluble pollutants, organic and inorganic contaminants, toxic elements and nutrients for safe reuse of treated wastewater. (c) Conventionally treated sewage is not fit for irrigation."

"Experts need be engaged in managing and monitoring with regular visits. (i) Auxiliary power backup facility is recommended at all the STPs. (j) Research and development on waste water treatment methods and processes should promoted so that low-cost, low-tech, user friendly methods may be developed and existing processes be refined for their efficiencies and effectiveness. (k) The use of methods such as constructed wetlands that are low in cost, lesser energy intensive and easy to operate should be investigated as alternatives. (l) Education and awareness among the farmers promoting adoption of modern methods in agriculture like drip irrigation, cropping systems, cultivation of remunerative non-edible crops and sewage water application methods."

Ten Key Research Issues for Integrated and Sustainable Wastewater Reuse in the Middle East [95]

This is a machine-generated summary of:

Shomar, Basem; Dare, Anne: Ten key research issues for integrated and sustainable wastewater reuse in the Middle East [95]

Published in: Environmental Science and Pollution Research (2014)

Link to original: https://doi.org/https://doi.org/10.1007/s11356-014-3875-7

If you want to cite the papers, please refer to the original.

For technical reasons we could not place the page where the original quote is coming from.

Abstract-Summary

"History, culture, religion, and socioeconomy are important components to include in any relevant and integrated studies of wastewater management and reuse."

"Engineering, health, chemistry, biology, food production, cultural heritage, and the needs of people of all ages should be considered together when making management decisions regarding issues so intimately tied with humanity as water and sanitation."

"Other escalating challenges such as poverty, food, and water scarcity, migration and instability, flooding and catastrophes, diseases and mortality, etc. should also be considered as part of wastewater management and reuse planning."

"Several arid countries have developed water security strategies where wastewater reuse is a major component."

"The existing wastewater treatment technologies in these countries are, in most cases, unable to remove such contaminants which may affect irrigation waters, industrial products, groundwater, etc. People would have to accept that the food on their tables could be irrigated with treated wastewater that they generated a few months ago, even if very advanced technologies were used to treat it."

Introduction

"Bischel and others (2) stated that municipal wastewater that has been treated thoroughly can be returned to the water supply for a variety of beneficial uses including landscape irrigation, agriculture, ecosystem enhancement, industrial cooling and processing, groundwater recharge, and indirect potable reuse."

"It shows also the inter- linkages, integration and synergy between the different sciences aiming to achieve dynamic understanding of water, contaminant transport, remediation/treatment technologies and processes through biotic and abiotic systems, etc. For a developed area such as northern California, Bischel and others (2) concluded that there are still issues including the protection of ecosystems, meeting wastewater discharge requirements, needs for water supply and reliability, and threats of long-term diminished water quality (e.g., seawater intrusion)."

"It is very important to distinguish between treatment technologies of wastewater and the potential scenarios of reuse of the treated water."

"The review covers a spectrum of issues related to treated wastewater reuse of a global interest, but many are specific to both the high- and low-income countries of the Middle East."

Research Questions Related to Wastewater Reuse

"Considering the path this water has taken: from desalination for potable use to tertiary treatment as wastewater, this treated effluent is actually a very expensive product."

"In order to ensure the sustainability of irrigation with treated wastewater, care must be taken to monitor the quality of the irrigation water, groundwater, and soil over the short and long terms (Kiziloglu and others 3; Mohammad and Mazahreh 4)."

"Using treated wastewater for irrigation may affect plant physiology depending on how plants react to saline water and other contaminants (e.g., heavy metals, organics)."

"The environmental health risk to humans from heavy metals through the consumption of contaminated materials, such as crops irrigated with treated wastewater, can be assessed by measuring their bioaccessibility."

"These results indicate that both treated and raw wastewaters generated from Birzeit University Wastewater Treatment Plant are genotoxic."

General Discussion

"Water reuse efforts in the Middle East face unique challenges due to the diverse water situations in each country coupled with the a range of existing strategies and action plans."

"The complexity of the water reuse vision is not limited only to the difference between the rich and poor countries but also to the socioeconomic conditions and the inherited culture related to water reuse."

"There exist small and limited fragments and trials on water reuse studies but currently, national, holistic, and integrated approaches are missing."

"The follow-up approach (risks, bioavailability, etc.) is nearly absent and the short- and long-term impacts associated with water reuse are not a priority for most of these countries, with some exceptions, where water security is a national priority (e.g., Qatar)."

"National challenges of water security need full synergy, integrity, and harmony among all levels and scales within the country and beyond where the water (ground or surface) could be shared."

Conclusions

"The situation regarding wastewater reuse cannot be understood without multi-criteria analyses using innovative and sustainable tools and strategies, especially in the developing countries such as the Middle East."

"Despite the shortage in water resources in the Middle East countries and other similar regions, achieving water sustainability necessitates focusing on the reuse of treated wastewater generated from various processes including domestic sources and those coming from desalination plants and the oil/gas industry."

"The key research issues highlighted may improve wastewater reuse not only through policy change but through changing the current agendas towards a more sustainable and long-term approach to water security in the Middle East."

"The ten key issues that are presented in this paper should guide the research efforts of the Middle East towards wastewater reuse and in time also guide wastewater reuse management efforts."

Status of Hormones and Painkillers in Wastewater Effluents Across Several European States—Considerations for the EU Watch List Concerning Estradiols and Diclofenac [96]

This is a machine-generated summary of:

Schröder, P.; Helmreich, B.; Škrbić, B.; Carballa, M.; Papa, M.; Pastore, C.; Emre, Z.; Oehmen, A.; Langenhoff, A.; Molinos, M.; Dvarioniene, J.; Huber, C.; Tsagarakis, K. P.; Martinez-Lopez, E.; Pagano, S. Meric; Vogelsang, C.; Mascolo, G.: Status of hormones and painkillers in wastewater effluents across several European states—considerations for the EU watch list concerning estradiols and diclofenac [96]

Published in: Environmental Science and Pollution Research (2016)

Link to original: https://doi.org/https://doi.org/10.1007/s11356-016-6503-x

If you want to cite the papers, please refer to the original.

For technical reasons we could not place the page where the original quote is coming from.

Abstract-Summary

"Present technologies for wastewater treatment do not sufficiently address the increasing pollution situation of receiving water bodies, especially with the growing use of personal care products and pharmaceuticals (PPCP) in the private household and health sector."

"The relevance of addressing this problem of organic pollutants was taken into account by the Directive 2013/39/EU that introduced (i) the quality evaluation of aquatic compartments, (ii) the polluter pays principle, (iii) the need for innovative and affordable wastewater treatment technologies, and (iv) the identification of pollution causes including a list of principal compounds to be monitored."

"Although some modern approaches for their removal exist, such as advanced oxidation processes (AOPs), retrofitting most wastewater treatment plants with AOPs will not be acceptable as consistent investment at reasonable operational cost."

"The same is true for membrane-based technologies (nanofiltration, reversed osmosis) despite of the incredible progress that has been made during recent years, because these systems lead to higher operation costs (mainly due to higher energy consumption) so that the majority of communities will not easily accept them."

"Advanced technologies in wastewater treatment like membrane bioreactors (MBR) that integrate biological degradation of organic matter with membrane filtration have proven a more complete elimination of emerging pollutants in a rather cost- and labor-intensive technology."

"Most of the presently applied methods are incapable of removing critical compounds completely."

"The survey results in the conclusion that combinations of different conventional and advanced technologies including biological and plant-based strategies seem to be most promising to solve the burning problem of polluting our environment with hazardous emerging xenobiotics."

Introduction and Demand

"These anthropogenic substances, often addressed as micropollutants that may adversely affect drinking water quality, are most typically polar to semipolar organic compounds detected at concentrations in the picogram per liter to microgram per liter range (Benner and others 5)."

"With the growing number of micropollutants being identified in surface water and groundwater, novel remediation and management strategies are needed to provide cost-effective and sustainable treatment solutions across Europe."

"Since the majority of all significant water bodies, lakes, and streams are shared between several European countries, the European Union (EU) has to find a common strategy for remediation of micropollutants and to set limits for effluents from WWTPs."

"The substances (diclofenac and two hormones: 17β-estradiol (E2) and 17α-ethinylestradiol (EE2)) shall be monitored by the EU member states in their surface waters for a maximum of four years."

National Inventories

"Data in different countries around the world (Ternes 6; Grung and others 7, 8; Carballa and others 9, 10; Ferrari and others 11; INFARMED 12), the annual consumption of DCF varies between 195 and 940 mg per inhabitant; as for EE2, it varies between 20 and 580 μg per inhabitant, respectively."

"In Serbia with a population of 7.2 million in 2012, the annual consumption of DCF and EE2 was equal to 8650 and 0.39 kg per year, respectively, while the consumption of DCF and EE2 per inhabitant was estimated to be 1197 mg and 50 μg per year per inhabitant, respectively (Radonjić and Šipetić 13)."

"In Germany, for DCF and EE2, respectively, consumption quantities of active ingredients in human medicine were estimated to be 1033 mg and 600 μg per inhabitant per year (SRU 14), while the DCF consumption rate in Turkey for the years 2009 and 2013 was 950 and 985 mg per inhabitant per year, respectively (Sari and others 15), and 440 mg per inhabitant and year in The Netherlands (Oosterhuis and others 16)."

"Comparing the consumption data, obviously, the DCF and EE2 are among the most popular and most consumed medicine products."

Occurrence of Analgesics and Hormones in WWTP's Effluents and Surface Waters

"One of the first compilations on this topic was a German study detecting diclofenac among 55 pharmaceuticals and 9 of their metabolites in the discharge of 49 sewage treatment plants as well as in their effluents in concentrations of up to several micrograms per liter (Ternes 6)."

"Another study reported that 27 out of 32 pharmaceutical substances and 4 of 5 metabolites were detected in the effluents of European wastewater treatment plants and that surface water peak values exceeded 1 μg L^{-1} (Larsen and others 17)."

"During an EU-wide monitoring survey on emerging polar organic contaminants in wastewater treatment plant effluents, DCF was found at an average concentration of 49.5 ng L^{-1}, while the highest concentration found was 174 ng L^{-1} (Loos and others 18)."

"The Turia case represents an excellent example for the impact on human uses of water contaminated by DCF, considering that water obtained after osmotic treatment, then used as "drinkable" water, still had a concentration of 18 ng L^{-1} of this pharmaceutical."

Progress in Detection and Identification

"Employment of the MS/MS techniques for quantitative target analysis has also some drawbacks and limitations, namely (i) methods are typically limited to about 100–150 target compounds depending on chromatographic separation under the constraints of having at least two transitions per compound; (ii) for some compounds, only nonspecific transitions might occur such as the neutral loss of H_2O or CO_2, which are also common for matrix interferences; and (iii) for some analytes, especially those of low molecular weight, only one transition is present."

"An extensive cleaning of the obtained extract to avoid any matrix interference will remove organic and inorganic coextractives, before they might interfere with analyte separation and detection causing background noise in GC-MS analysis and signal suppression and/or enhancement in LC-MS analysis."

"It becomes clear that for both DCF and the estrogen determination by GC-MS necessarily involves an additional derivatization step (e.g., by methylation, tert-butyldimethylsilyl, with N-methyl-N-(trimethylsilyl)trifluoroacetamide, etc.) due to the polarity of the compounds."

Conventional Treatment Systems and Their Shortcomings

"During primary treatment, coarse solids are separated from the liquid stream and micropollutants are removed mainly by chemical and mechanical separation."

"The most commonly applied secondary treatment in WWTPs is the conventional activated sludge process (CAS), where both organic matter and nutrients are biologically removed."

"In the previous paragraph, sorption to sewage sludge is moderate, and therefore, biological transformation is the most likely mechanism responsible for micropollutant elimination in WWTPs."

"Although the microbiota developed in WWTPs may have been exposed to a plethora of micropollutants for a long time, the effective biological removal of these substances is conditioned by singular factors."

"There is evidence that some operating parameters, such as hydraulic retention time (HRT), solid retention time (SRT), redox conditions, and temperature may affect micropollutant removal."

"Conventional WWTPs have not been designed for micropollutant elimination and have therefore only limited capacity to remove DCF, E2, and EE2."

Advanced and Alternative Methods

"In NF and RO membrane processes, the rejection of organic micropollutants like DCF, E2, and EE2 can generally be achieved by size exclusion/steric hindrance, adsorption onto membrane, and/or charge repulsion (Bellona and others 19; Xu and others 20)."

"Grover and others (21) showed removal efficiencies for DCF, EE2, and E2 of $> 98\%$ in a full-scale granular activated carbon plant treating WWTP effluent."

"While in an adsorptive process using PAC or GAC organic micropollutants are removed, oxidation processes like ozonation do not result in complete mineralization of micropollutants but in the formation of predominantly unknown transformation products (TP) with unknown toxicity."

"Gerrity and others (22) reported high removal efficiency for DCF of > 99% and of E2 of > 83% in a pilot-scale treatment plant of WWTP effluent with O_3/H_2O_2."

"The heterogeneous photocatalysis process using TiO_2 was applied successfully for the removal of DCF, E2, and EE2 with high removal efficiencies in aqueous solutions including WWTP effluents."

Ecotoxicology and Risk Assessment

"For biologically active compounds such as pharmaceuticals, this approach may overlook sublethal and subtle subcellular effects that may occur in some species at much lower concentrations during chronic exposure (Fent and others 23)."

"Typical PNEC values for diclofenac and EE2 when derived from traditional ERA using acute toxicity data lies in the milligram per liter range, while chronic histopathological effects have been observed in rainbow trout after 28 days of exposure to 1–5 μg L^{-1} DCF (Schwaiger and others 24; Triebskorn and others 25)."

"This is just the case for EE2: indeed, when assessing its ecotoxicological effects, the main threat is represented by the induced estrogenic activity, i.e., a specific mode of toxic action directly related to all those molecules (then called endocrine-disrupting compounds, EDCs) that can mimic, block, or interfere with hormonal activities in living organisms."

Modeling of Diclofenac and Hormones

"The most commonly used models are derived from the activated sludge models (ASMs) that were developed to predict the degradation of organic carbon, nitrogen, and phosphorus (Henze and others 26)."

"ASMs have been extended to include the degradation of micropollutants, including pharmaceutical compounds such as diclofenac and estrogens such as E2 and EE2 (Lust and others 27; Plosz and others 28)."

"In the modeling of biotransformation processes, separate kinetic expressions are typically employed in order to describe both aerobic and anoxic degradation (Joss and others 29, 30)."

"The ASM-X model developed by Plosz and others (28) for DCF (and other pharmaceuticals) incorporates expressions involving both the biotransformation of the micropollutant and its reformation into the parent compound."

"Models describing the biodegradation of estrogens (estrone (E1), E2) and EE2 have also been developed (Monteith and others 31; Lust and others 27), which also predict both their biodegradation and adsorption/desorption to sludge through liquid–solid equilibrium (K_d) coefficients."

Economics

"Since economies of scale apply to wastewater treatment facilities (Tsagarakis and others 32; Fraquelli and Giandrone 33; Hernandez-Sancho and others 34), the cost for removing micropollutants from small installations will be extremely high."

"In such cases, nonmarket valuation can be applied so that a monetary value can be attributed to the benefits derived from the wastewater treatment (Menegaki and others 35; Genius and others 36)."

"These benefits are then entered in a cost benefit analysis for the Swiss national plan to reduce micropollutants in treated wastewater effluents, resulting in a positive net present value of this policy."

"The value of 42.20 € kg^{-1} means that for each kilogram of DCF that is removed from wastewater, the environmental benefit is quantified by 42.20 €."

"That to estimate the overall benefits from wastewater treatment, not only the value of the shadow prices in € per kilogram should be considered, but the volume of each pollutant removed in kilogram per cubic meter as well."

Concluding Remarks

"This is due to the fact that our conventional WWTPs are neither specifically designed nor operated to remove residual concentrations of organic pollutants, causing the potential accumulation of such pollutants into receiving water bodies and limiting at the same time proper reuse of water."

"Although concentrations of DCF and EE2 in groundwater and surface waters are still generally low and an acute toxicological risk for consumers has not been identified so far, contamination is increasing, like for other emerging compounds."

"Cleaning highly treated wastewater through an environmental buffer and stripping residual contaminants into the matrix by sorption and occult microbial processes to augment a drinking water supply is a recent practice, which is referred to as intentional indirect potable reuse, which can occur through recharge of unconfined or confined aquifers."

"WWTP effluents, if properly treated, can be reclaimed and reused for determined restricted uses, contributing in this way to the reduction of water pollutants and the pressure over the worldwide water scarcity."

Post-treatment and Reuse of Secondary Effluents Using Natural Ltreatment Systems: The Indian Practices [97]

This is a machine-generated summary of:

Kumar, D.; Asolekar, S. R.; Sharma, S. K.: Post-treatment and reuse of secondary effluents using natural ltreatment systems: the Indian practices [97]

Published in: Environmental Monitoring and Assessment (2015)

Link to original: https://doi.org/https://doi.org/10.1007/s10661-015-4792-z

If you want to cite the papers, please refer to the original.

For technical reasons we could not place the page where the original quote is coming from.

Abstract-Summary

"Paper summarizes the results of India-wide survey of natural treatment systems (NTSs) for wastewater treatment and reuse."

"The quality of treated wastewater from different types of NTSs was analyzed for various physico-chemical and bacteriological parameters, and needs for post-treatment were identified."

"About 1838 million liters per day (MLD) of wastewater is being treated using NTSs, of which the contributions of polishing ponds, waste stabilization ponds, duckweed ponds, constructed wetlands, and Karnal technology were found to be 53.39, 45.15, 0.13, 0.55, and 0.78%, respectively."

"Among the NTSs studied, constructed wetland was found most efficient in removal of pollutants including nitrogen, phosphorus, total coliform, and fecal coliform in the range of 76, 61, 99.956, and 99.923%, respectively."

"Of all types of NTSs, only constructed wetland was found to meet the total coliform count requirements (<1000 per 100 ml)."

Introduction

"Communities appear open to accept that the NTSs that are capable of providing adequate treatment to wastewaters in conjunction with supplementing fish and nutrition to the food baskets of the fishing communities engaged in managing the systems as well as generating adequate water for irrigation of farms and agro-forests."

"The appropriate treatment and reuse of domestic wastewater for non-potable activities will not only help in fulfilling the water demand in regions where water is scarce but will also help in improving the surface water quality due to less discharge of wastewater (Kumar and others 37)."

"The study was focused to assess the common NTSs used for wastewater treatment in India, their efficiencies for various pollutant removals, need for the post-treatment of the effluent prior to any reuse applications, and effectiveness of the existing conventional post-treatment systems."

Methodology

"The treatment performance of these systems was evaluated for 1 year by assessing the quality of raw and treated wastewater for three seasons (winter, summer, and rainy)."

"The Karnal technology was not selected for performance assessment as no treated wastewater comes out from this treatment system due to its high evaporation plant system."

"The selections of these five NTSs were primarily aimed to select and compare the treatment systems that are located in similar climatic conditions, operating through proper O&M, treating the domestic wastewater, and also performing satisfactorily."

"The waste stabilization pond of 13 MLD capacity in Mathura, Utter Pradesh, northern India was established in 2001 under Yamuna Action Plan for treating domestic wastewater of Mathura city."

"The duckweed pond of 0.5 MLD capacity was established in 2004 for treating the domestic wastewater of village community."

"The primary treated wastewater from the septic tank further undergoes secondary treatment through constructed wetland."

Results and Discussions

"The different types of NTSs employed for wastewater treatment and reuse in India include, polishing ponds, waste stabilization ponds, sewage-fed aquaculture, duckweed ponds, constructed wetlands, and Karnal technology."

"NTSs, including duckweed ponds, constructed wetlands, and Karnal technology for on-land disposal of wastewater, are currently treating lower amount of wastewater (27 MLD) as compared with the total load serviced by other NTSs (1838 MLD), especially polishing ponds (981 MLD) and waste stabilization ponds (830 MLD), but their numbers are significant (19 facilities)—which is the direct sign that these treatment technologies (duckweed ponds, constructed wetlands, and Karnal technology) are being used as decentralized systems for wastewater treatment."

"It is evident from the performance data that the treatment systems are satisfactorily achieving the wastewater quality standards (in all three seasons) prescribed by CPCB New Delhi for discharge of treated effluent into the natural water body (CPCB 38)."

Conclusions

"The treated effluents from NTSs are directly reused in agriculture or disposed into adjoining river."

“The quality of treated effluents coming from NTSs-based sewage treatment plants which are properly operated and maintained was found to be comparable with mechanized treatment systems in terms of physico-chemical parameters.”

“The additional benefit observed in the treated effluent from NTSs was a lower pathogen count as compared to conventional sewage treatment.”

“Although the quality of treated wastewater from NTSs is still not fit for body contact level because of relatively high number of coliform bacteria (10^3 to 10^5 MPN per 100 ml), it may be reused after disinfection.”

“The reuse of secondary treated effluent from treatment technology without any post-treatment may pose health hazards if the local community has direct access to groundwater adjacent to treated wastewater disposal sites.”

“There is no post-treatment (except chlorine disinfection) of the secondary effluent of NTSs in India.”

Managing Groundwater Demand Through Surface Water and Reuse Strategies in an Overexploited Aquifer of Indian Punjab [98]

This is a machine-generated summary of:

Kaur, Navdeep; Kaur, Samanpreet; Tsolakis, Naoum; Mishra, Nishikant; Srai, Jagjit Singh: Managing groundwater demand through surface water and reuse strategies in an overexploited aquifer of Indian Punjab [98]

Published in: Modeling Earth Systems and Environment (2022)

Link to original: https://doi.org/https://doi.org/10.1007/s40808-022-01602-4

If you want to cite the papers, please refer to the original.

For technical reasons we could not place the page where the original quote is coming from.

Abstract-Summary

"Groundwater sustainability is one of the most critical issues to the State of Punjab, India."

"A numerical groundwater flow model (MODFLOW) was employed to simulate flow and groundwater levels in the Sirhind Canal Tract of Punjab between 1998 and 2030."

"It was found that if current groundwater abstraction continues there will be further steep decline of 21.49 m in groundwater level by 2030."

"Findings also suggest that canal water supplies will be beneficial to reverse groundwater level decline and help to increase the water level by 11% above that in year 2018."

Introduction

"Decades, many researchers suggested solutions applicable to both upstream and downstream agri-food supply chain operations to promote the restoration of groundwater levels in Punjab, including crop diversification, efficient irrigation options, water conservation practices, groundwater recharge techniques and wastewater reuse (Aggarwal and others 39, 40; Shakoor and others 41)."

"Motivated by the scarcity of studies on groundwater modelling in Indian Punjab, and to further address the evident gap about modelling approaches investigating the respective impact of multi-water resources' management, this research contributes to the extant body of literature on groundwater management via: (i) simulating the groundwater system of the Sirhind Canal Tract, one of the most productive aquifer systems at the Indo-Gangetic basin; and (ii) assessing the groundwater level in the region via evaluating alternative pumping and water management scenarios."

Recent Groundwater Management Studies Using MODFLOW

"Khan and others (42) proposed a groundwater policy for the Indus Basin of Pakistan via examining different groundwater modelling scenarios, including: (i) groundwater pumping controls; (ii) improvements to the canal infrastructure; and (iii) precipitation changes."

"Sobeih and others (43) and Eltarabily and others (44) used MODFLOPW to model groundwater flows in western and eastern Nile Delta, respectively, to assess different scenarios capturing recharge from nearby canals and discharge from wells."

"Shakoor and others (41) quantified the future groundwater depletion under three different pumping scenarios for the Punjab province in Pakistan and found that artificial groundwater recharge and surplus canal water supplies could be beneficial to avoid groundwater depletion."

"Bouqdaoui and Aachib (45) focussed on the management and restoration of groundwater resources of the Berrechid aquifer in Morocco via examining alternative simplified management scenarios such as artificial recharge and reduction of the water volume pumped for irrigation."

Materials and Methods

"MODFLOW-based simulation model of the Sirhind Canal Tract was conceptualised and discretised in Step #3."

"The modelling results were summarised and analysed to suggest environmentally sustainable policies for the management of the groundwater behaviour in the Sirhind Canal Tract of the Indian Punjab."

"To create a conceptual model of the tract, data on the lithology of the aquifer (i.e. structural contours, bore logs, aquifer properties), piezometer levels, climatic data, aquifer abstractions, canal data, canal network, and their L-sections were collected."

"Primary time series data describing the water level of more than 200 observation points for the Sirhind Canal Tract, for the pre-monsoon period (i.e. June), was collected from the Department of Agriculture and Water Resources and Environment Directorate, Punjab."

"Based on historical annual groundwater level fluctuations, the cells lying southwest of the tract were considered as constant head cells, i.e. the storage term was not considered, and the head specified in the beginning was kept constant throughout the simulation."

Results and Discussion

"Under pumping Scenario IV, reduced groundwater abstraction, additional canal water supply and wastewater augmentation from existing STPs together would produce a rise in the groundwater levels by 1.51% in 2025 and 4.43%in 2030 compared to the reference value of 23.20 m observed in year 2018."

"Incorporation of all the strategies (i.e. enhanced canal water supplies, decreased groundwater abstraction and potential wastewater augmentation with irrigation under condition "Condition C—Scenario IV") could uplift the groundwater level at 20.7 m in 2030, which is 10.97% higher than the level at 23.20 m predicted for the year 2018."

"Augmentation of irrigation supplies with treated wastewater from existing STPs (i.e. Condition B) seems to be a realistic option and could ensure an improvement in the groundwater levels in the year 2030 by 5.19%, 5.22%, 5.80% and 3.18% under pumping Scenarios I, II, III and IV, of the management strategy "Condition A"."

Conclusions

"The increased reliance of farmers in Indian Punjab on groundwater supplies for irrigation purposes has been the major cause of groundwater decline in the Sirhind Canal Tract."

"Usage of treated wastewater in irrigation along with the different groundwater pumping scenarios provides promising scenarios that reverse recent declines in groundwater water resources in the tract."

"The findings further suggest that an increase of 10% in canal water supplies can prove beneficial to reverse groundwater level decline."

"A concrete canal water redistribution policy and concentrated efforts to repair the network and restore field distributaries can prove significant in sustaining the declining groundwater table."

"It is assumed that canal water supplies may increase in future, however, the decline in area under canal irrigation could be a decrease in the water flow in the rivers."

Reuse-Focused Selection of Appropriate Technologies for Municipal Wastewater Treatment: A Multi-criteria Approach [99]

This is a machine-generated summary of:

Srivastava, R. R.; Singh, P. K.: Reuse-focused selection of appropriate technologies for municipal wastewater treatment: a multi-criteria approach [99]

Published in: International Journal of Environmental Science and Technology (2021)

Link to original: https://doi.org/https://doi.org/10.1007/s13762-021-03803-3

If you want to cite the papers, please refer to the original.

For technical reasons we could not place the page where the original quote is coming from.

Abstract-Summary

"Wastewater treatment technologies (WWTTs) are employed across the world, and the selection is mainly based on 'past experiences' aimed at 'pollution prevention' in the receiving water bodies."

"In Kanpur city, for water reuse in industrial cooling under restricted land and challenging influent quality conditions, a combination of Membrane Bioreactor (MBR) with Wuhrmann process (WP) is obtained as the most preferred suggestion."

"In Varanasi city, for vehicular washing operations and for flow augmentation (inland surface water), under energy-constraint scenario, high-rate activated sludge-based biological filtration and oxygenated reactor (BIOFOR-F) is suggested."

"For technology supplementation to existing ASP-based STPs in the city to obtain effluent for inland surface water augmentation, WP in combination with microfiltration (MF) and reverse osmosis (RO) is suggested."

"The developed model may be used as a decision-making tool for planning a reuse-focused water reclamation program or for upgradation of existing STPs as per resource availability and target reuse objectives."

Introduction

"Numerous wastewater treatment technologies belonging to primary, secondary, tertiary, and emerging category are available, but a sensitive and reliable method for selection of an appropriate technology combination aimed at the reuse of reclaimed water in the local environment based on resource availability of the area is not yet well established."

"Multi-criteria decision-making (MCDM) tools have been used for appropriate wastewater treatment technology selection in different fields."

"In a study conducted for technology selection for coking wastewater, AHP was applied using four major decision criteria, namely technical, economic, environmental, and administrative factors, and six alternatives were obtained by varying oxygen phases in an ASP (Wei and others 46)."

"Ilangkumaran and others (47) utilized fuzzy AHP and Preference Ranking Organization Method for Enrichment Evaluation (PROMETHEE) tools for optimal wastewater treatment technology selection out of 5 technologies: electro-coagulation, SBR, reverse-osmosis, ultrafiltration, and anaerobic digestion based on three criteria: (1) Social and environmental, (2) Economic, and (3) Technical."

Materials and Methods

"The FUCOM procedure is illustrated in the following steps: (a) Firstly, the decision criteria (say n) are ranked by DM in descending order of significance, where the most important criterion is ranked 1 while the least significant criteria are ranked last. (b) Secondly, the ranking decision criteria are then compared and assigned priorities by DM on a defined scale, where the most important criterion is assigned 1, while the least important criteria are given the highest value in comparison with other criteria."

"The comparative priorities depict the advantage of former criteria (k rank) in comparison with the later criteria (k + 1 rank), given by, ($\phi_{k/(k+1)}$). (c) In this step, final weights for each criterion are obtained."

"If there are m alternatives and n criteria, a weighted sum is represented by: Wj = importance weight for each criterion Rij = Rating of each alternative The above equation is restricted by the following condition: The WSM has found its applications across several fields such as ranking of renewable energy sources (Lee and Chang 48), ontology ranking (Fonou-Dombeu 49), and so on."

Model Development

"The cost incurred once has been distributed over the design period for selection of technology in this study."

"It includes, salaries, rents, etc. In this study, for technology selection, total O&M costs for the design period are considered."

"This study has not included an elaborate sustainability assessment for technology combinations selected, but focuses on meeting the desired water quality criteria for reuse only."

"The treatment efficiencies are obtained from working STPs in India and from case studies based on different technologies around the world."

"After filtering on this basis, the resources required by each technology train are converted into their respective costs, incurred over the design period of 15 years, to obtain uniformity in the basis for selection."

"Application of this selection model gives top 10 technologies which satisfy effluent quality requirements and give least weighted cost."

Results and Discussion

"This effluent when passed through single-stage post-denitrification process, that is, WP, the resulting effluent will have BOD 0.2 mg/l, COD 16.53 mg/l, TSS 0.3 mg/l, TN 0.85 mg/l, and FC 940 MPN/l. Meeting the effluent quality requirements as well as adherence to land resource-constraint and least cost option makes MBR followed by WP the most preferred technology for the purpose."

"The reuse purposes considered in this case, vehicle washing and e-flow augmentation (inland surface water), require very strict effluent characteristics."

"In addition to the existing ASP-based system, WP, which is one of the emerging technology, supported with membrane-based treatment methods (MF/UF) and RO may be required to meet the quality criteria for inland surface water augmentation."

Conclusion

"The objective of the present study is to develop a methodology for selection of the appropriate technology chain in municipal wastewater treatment which meets water quality criteria for the defined reuse."

"With suitable modifications in terms of performance efficiency based on climatic conditions, socio-economic, environment, and cost variations and required quality criteria for selected reuse, the methodology can be used as a decision support tool for technology chain selection toward defined reuse purpose."

"Application of the developed methodology for two cities in India namely, Kanpur and Varanasi, for selection of most preferred combination of technology for defined reuses or supplementation of technology chain to an existing STP has been examined and explained."

"Ranking of the technologies based on the least weighted cost, out of a total 360 combinations for 14 defined reuses makes the developed methodology a very good decision-making tool for selection of appropriate technology chain in reused focused wastewater treatment scheme to bridge the gap between water demand and supply."

Life Cycle Assessment of a Decentralized Greywater Treatment Alternative for Non-potable Reuse Application [100]

This is a machine-generated summary of:

Masmoudi Jabri, K.; Nolde, E.; Ciroth, A.; Bousselmi, L.: Life cycle assessment of a decentralized greywater treatment alternative for non-potable reuse application [100]

Published in: International Journal of Environmental Science and Technology (2019)

Link to original: https://doi.org/https://doi.org/10.1007/s13762-019-02511-3

If you want to cite the papers, please refer to the original.

For technical reasons we could not place the page where the original quote is coming from.

Abstract-Summary

"The decentralized wastewater treatment systems including collection, treatment, and recycling of treated wastewater are an efficient integrated water management solution to overcome the shortage of freshwater and to reduce pollutant emissions."

"The integration of the tertiary treatment led to an increase in the environmental load ranging between 6 and 30.4% depending on the impact category in comparison with the scenario integrating only the biological treatment."

"The recycling of the TGW allowed a reduction ranging between 3.9 and 17% on five impact categories compared to the no recycling scenario."

"It is concluded that the decentralized approach for greywater treatment and recycling is beneficial for the environment as the distribution system and the tertiary treatment didn't increase drastically the environmental load of the plant allowing water saving."

"It is recommended to use renewable energy to reduce the environmental impact associated with the energy requirement for the biological treatment which affects the global warming potential and the acidification impact categories and to substitute the polyvinylchloride pipes with polyethylene pipes to reduce emissions related to the human toxicity impact category."

Introduction

"LCA has been applied for the environmental assessment of wastewater treatment systems, with a special interest to the contribution of the reuse and recycling to sustainable water development."

"The application of LCA to the decentralized wastewater treatment system can be related to the assessment of the environmental impact of possible alternatives by the comparison of conventional, natural or hybrid technologies in small or large communities (Garfí and others 52; Machado and others 53), the quantification of the environmental benefits and obstructions of the water reuse which requires advanced treatment that consumes chemicals, material and energy (Tong and others 54) and the analysis of different treatment stages from the collection of the wastewater to its recycling (Amores and others 55)."

"A decentralized system located in Berlin, Germany, which includes a moving bed biofilm reactor for GW treatment is environmentally evaluated in order to point out the treatment stages with the highest environmental impact, in addition to the identification and the quantification of the main drawbacks and advantages related to the biological treatment of GW and its recycling into toilet flushing."

Materials and Methods

"Operation and construction processes related to the treatment and recycling of GW were described for different scenarios to determine their respective contributions on the total impact of the system and the areas for improvement."

"The following data are not considered in the system boundaries for no availability of the data or for simplicity reasons of the LCA study (Masmoudi Jabri 50):The network piping related to the tap water supply, since it is assumed to be the same in all scenarios."

"Scenario 1 is considered as the baseline scenario with no water recycling, and the GW goes through biological treatment composed of the two MBBRs followed by the UV treatment."

"Only the flows related to the operation of the plant are extrapolated to one working year, and data for the construction of the plant are supposed to be available for the treatment of 1 m^3/day all over the year."

Results and Discussion

"For the impact categories HT, MAE, TE and EP, the module of the tap water consumption presents the highest contribution with about 65%, 52%, 68% and 61% of the total impact, respectively."

"The environmental impact of the biological treatment is mainly caused by the contribution of the construction module for the most impact categories except for GWP and AP for which the energy consumption is the main contributor with 59% and 48%, respectively (Masmoudi Jabri 50)."

"The recycling module included in scenario 2 leads to a decrease in five impact categories compared to scenario 1 which are ODP, FAEP, MAE, TE and EP with a reduction of 3.9%, 1.6%, 16.2%, 10.2% and 17%, respectively."

"A noticeable contribution of the tap water consumption process to the total environmental impact was recorded for scenarios 1 and scenario 2 with 52.14% and 41.42%, respectively."

Conclusion

"The current study investigates the environmental performance of the decentralized approach combined with the separation system implemented in a building which leads to the on-site reuse of 1 m^3/d of reclaimed water after a biological treatment with MBBR."

"Three scenarios have been compared and the results showed that for the no recycling scenario (scenario 1), the biological treatment presents the largest environmental impact with a contribution higher than 50% for most impact categories caused mainly by the contribution of the construction module which includes the material and the transport of the bioreactors."

"The adoption of the decentralized alternative with the use of MBBR systems for GW treatment and recycling is considered as an environmentally friendly solution that could be generalized in many buildings in order to face the water scarcity."

Application of Nonwoven Microfiltration Membrane on Activated Sludge Final Effluent: Improving Wastewater Quality for Reuse [101]

This is a machine-generated summary of:

Kabuba, J.; Masala, M. S.; Topkin, J.: Application of nonwoven microfiltration membrane on activated sludge final effluent: improving wastewater quality for reuse [101]

Published in: International Journal of Environmental Science and Technology (2023)

Link to original: https://doi.org/https://doi.org/10.1007/s13762-023-04876-y

If you want to cite the papers, please refer to the original.

For technical reasons we could not place the page where the original quote is coming from.

Abstract-Summary

"Clarifier effluent from a biological nutrient removal (BNR) treatment plant was diverted to a nonwoven membrane pilot filtration unit."

"Water Quality Index (WQI), Single Factor Pollution Index (Pi), Water Use License (WUL) and removal efficiency methods were used to interpret the water quality data for BNR clarifier effluent and nonwoven membrane permeate."

"The results showed that this nonwoven microfilter membrane is able to improve wastewater quality from BNR clarifier effluent."

"Pi has revealed that BNR clarifier effluent water quality is medium polluted versus slightly polluted permeate."

"WQI results for BNR clarifier effluent showed good water quality and the water can be reused for domestic, irrigation and industrial purposes, while permeate WQI results indicated excellent water quality and the water can be reused for drinking, domestic, irrigation and industrial purposes."

"Outstanding permeate water quality improvement was observed on E.coli counts improving from 4974.48 to 294.33 counts/L. The standard of E.coli according to the WUL at this wastewater care works is 500 counts/L. Results indicate that nonwoven membrane filtration can improve microbial contamination and decrease the demand of chlorine for disinfection of wastewater final effluent."

"The nonwoven membrane filtration can decrease the water scarcity gap in South Africa for direct water reclamation by improving effluent wastewater."

Introduction

"For wastewater treatment, membrane filtration is among the technologies that show excellent improvement of biological nutrient removal (BNR) clarifier effluent."

"Considering the wide applications of membrane filtration, microfiltration (MF) membranes have been extensively used for improvement of wastewater effluent (Anspach and others 56; Zhu and others 57)."

"Different membrane configurations and wastewater treatment operating procedures have been developed by suppliers for ultrafiltration and microfiltration membranes (Tchobanoglous and others 58)."

"Membrane process has attracted increasing attention in recent years as an alternative treatment for municipal wastewater (Van Nieuwenhuijzen and others 59)."

"Nonwoven membranes due to their good mechanical properties and high porosities could be used as the filter material for domestic wastewater treatment coupled with activated sludge system (Kiyak and others 60)."

"The main objective of this research was to apply the nonwoven microfiltration membrane on activated sludge final effluent for improving wastewater quality for re-use."

Materials and Methods

"The influent (secondary clarifier effluent) was pumped firstly to an outside tank which is named (pilot tank) for this purpose, samples collected from this pilot tank are also compared to the final permeate samples and influent stream."

"The filtrate pump was used to pump the permeate out of the filtrate unit, and the recycle pump was used to pump some of the nitrate-loaded filtrate from filter unit back to the pilot tank to allow denitrification to take place."

"BNR secondary clarifier effluent passed through the filters at flow rate 3.0 L/min at atmospheric pressure."

"Ci is the measured concentration of i units pollutant (mg/L), Si the III level water quality standard category value of i units pollutant according to "Environmental Quality Standards for Surface Water."

"Water quality index was also used to determine the suitability for the BNR clarifier effluent and the permeate."

Results and Discussion

"Koltuniewicz & Drioli (61) and Bodzek and others, (62) reported that E.coli from municipal wastewater after primary and secondary treatment using microfiltration membranes showed significant removal, and high removal efficiency of E.coli was explained by filtration layer formation on the surface of the membrane, which performed as a secondary film layer."

"According to the single factor pollution index the permeate water quality indicated that E.coli, EC, pH and SS are slightly polluted, these parameters are rated between the level of 0.5–1."

"E.coli counts from 4974.48 to 294.33 counts/L, COD improved from 36.67 to 25.19 mg/L and conductivity improved from 85.28 to 80.04 mS/m. The standard of E.coli according to the WUL at this WCW is 500 counts/L. Similar results were obtained by Wintgens and others, (63) in Windhoek, Namibia using UF membrane for BNR clarifier effluent and major successes were achieved on the removal of microbial and chemical components without disinfection."

Conclusion

"For the permeate sample, the highest removal efficiency obtained in this study using nonwoven membrane filtration was for E.coli, at 90%, followed by COD 25%, NH_4N 22%, NO_3^- 12.6%, PO_4 7.8%, SS 6.3% and the lowest was EC 5.2%."

"It was concluded that nonwoven membrane filtration is able to improve the removal efficiency of clarifier effluent."

"Of WQI BNR clarifier effluent, the water quality rating is 28.03, it is regarded as good water and can be reused for domestic, Irrigation and Industrial, except drinking purposes."

"According to the standard rating of WQI the water quality at permeate is regarded as excellent water quality which can be reused for drinking (with additional treatment such as activated carbon and additional disinfection), domestic, irrigation and industrial."

Water Scarcity and Wastewater Reuse Standards in Southern Europe: Focus on Agriculture [102]

This is a machine-generated summary of:

Lavrnić, S.; Zapater-Pereyra, M.; Mancini, M. L.: Water Scarcity and Wastewater Reuse Standards in Southern Europe: Focus on Agriculture [102]

Published in: Water, Air, & Soil Pollution (2017)

Link to original: https://doi.org/https://doi.org/10.1007/s11270-017-3425-2

If you want to cite the papers, please refer to the original.

For technical reasons we could not place the page where the original quote is coming from.

Abstract-Summary

"Climate change affects water resources worldwide, and Southern Europe is one of the areas where water scarcity is expected to increase in the future."

"Different water scarcity indicators discussed in this manuscript (e.g. total annual actual renewable water resources, water exploitation index and dependency ratio) showed that some parts of this region are already facing water stress and that climate change could have a great impact on their water supply sector."

"In order to use it safely, treated wastewater must reach a certain quality that should be regulated."

"In the south of Europe, 4 countries out of 15 have already adopted wastewater reuse regulations (Greece, Italy, Portugal and Spain)."

Water Scarcity in the World and Solutions

"Global mean temperature could increase by 3.5 °C compared to the pre-industrial period, and that would significantly affect the water availability in the different world regions (Gerten and others 64)."

"When natural water reserves are not sufficient, one of the most available water resources is treated wastewater (Cirelli and others 65)."

"Its use increases overall water resources and reduces environmental pollution (Molinos-Senante and others 66)."

"Water reuse can have positive effects on the environment, since less freshwater is extracted from the sensitive ecosystems and less wastewater is discharged to the water bodies (BIO by Deloitte 67)."

"Fawell and others (68) indicates that the application of reused water on a larger scale in Europe could have both environmental and economic benefits, as it would represent an additional water source at a competitive cost and it would reduce freshwater abstraction."

"If controlled, treated wastewater can be an important water source for agriculture."

Water Availability and Related Risks in Southern Europe

"In order to evaluate the risk of water shortage for the Southern European countries, the authors have decided to use the following five indicators: total annual actual renewable water resources per inhabitant (TRWR), water exploitation index (WEI), dependency ratio (DR), human development index (HDI) and the use of non-renewable water resources."

"Many countries also rely on non-renewable water resources (NRWR) that are not considered by the TRWR or WEI."

"Different authors have already developed several indices of water scarcity as reviewed by Brown and Matlock (69), but due to the lack of data for the region of Southern Europe and the need for a small-scale comparison of the 11 countries located there, a simple index called SEWR was developed in this study."

Wastewater Reuse Guidelines

"On the international level, the two benchmark guidelines for wastewater reuse are the California guideline (State of California 70) and the World Health Organization (WHO) guideline (WHO 71)."

"That can probably explain the fact that no relationship could be found between the DR and the existence of wastewater reuse guidelines."

"Among the GIPS countries, the Italian and Greek guidelines consider the highest number of parameters (Paranychianakis and others 72) and their rigour is seen by Fawell and others (68) as the main obstruction to the reuse of water."

"Since the four GIPS countries have a similar climate and are part of one bigger cultural zone, cooperation towards common, or at least similar guidelines, could lower the risks connected to the wastewater reuse and give more possibilities for the development of this area in the whole region."

"The remaining countries of Southern Europe still do not have guidelines for wastewater reuse, and up to now, there is no indication if and when it will happen."

Conclusion

"With the use of different water scarcity indicators (i.e. dependency ratio, human development index, total renewable water resources, use of non-renewable water resources and water exploitation index) water availability was estimated for 11 Southern European countries and expressed as an index called Southern European Water Risk."

"Most of the countries with the lowest freshwater availability (Greece, Italy, Portugal and Spain) are already reusing treated wastewater and have introduced guidelines to regulate this field."

"Despite the fact that the remaining Southern European countries do not yet have guidelines for treated wastewater to be reused in irrigation, water shortage and lower freshwater availability are expected in this region, and wastewater treatment and reuse can help to mitigate those changes and to protect the environment."

"With the legislation on minimum requirements for water reuse in irrigation and aquifer recharge that will be proposed by the European Commission during 2017, the experiences from Greece, Italy, Portugal and Spain can be beneficial in that process."

Comparative Environmental Sustainability Study of an Improved Sewage Sludge Treatment and Sludge Reuse System Based on Emergy Analysis in China [103]

This is a machine-generated summary of:

Zhang, Junxue; Ma, Lin: Comparative environmental sustainability study of an improved sewage sludge treatment and sludge reuse system based on emergy analysis in China [103]

Published in: Applied Water Science (2021)

Link to original: https://doi.org/https://doi.org/10.1007/s13201-021-01478-2

If you want to cite the papers, please refer to the original.

For technical reasons we could not place the page where the original quote is coming from.

Abstract-Summary

"Two sewage sludge treatment systems have been investigated, calculated, and analyzed, including the conventional treatment system (Scenario A) and improved reuse system (Scenario B), respectively."

"The results demonstrate that (1) Compared to Scenario A, Scenario B is a comprehensive system, which integrates a sewage sludge treatment system and a brick production system for sludge recycling. (2) After considering the brick system (scenario B), on the one hand, the sludge treatment capacity has been enhanced and raised sludge utilization; on the other hand, negative influences have also generated due to the non-renewable resources input and several outputs. (3) In Scenario A and Scenario B, the input resources part reflects the main impact (about 59.6%

in the entire emergy value). (4) In this new paper, the UEVs are 2.73E + 11sej/kg and 6.29E + 11sej/kg in Scenario A and Scenario B, respectively. (5) The emergy sustainability indexes (ESI) are 0.012292 and 0.00848, which express the weak comprehensive effects in Scenario A and Scenario B. (6) Scenario B has a more extensive range of change than Scenario A because of the more resource input for the sensitivity analysis."

Introduction

"Under such a circumstance, it is necessary to develop the improved sludge treatment and reuse system to enhance environmental sustainability in China."

"In view of four different Scenarios in sewage sludge treatment systems, sustainability has been assessed using improved emergy indicators by Gengyuan and others (73) (Gengyuan and others 73)."

"The LCA-emergy method has been used in the wastewater treatment system, and the sludge waste products can be utilized in land application (Marc and others 74)."

"Emergy analysis of municipal wastewater treatment has been executed, and the electricity was generated by digestion of sewage sludge (Johanna and others 75)."

"Two sewage sludge treatment systems are designed to assess sustainability, which have four contributions: (1) Taking entire views into account so as to evaluate sludge reuse process; (2) Building an improved and integral emergy indicators; (3) Computing two unit emergy values of different sludge reuse systems; (4) Raising the resolution strategies to improve negative results."

Methodology

"Speaking, the environmental sustainability of sludge reuse can be assessed by transforming all kinds of physical units, involving energy (J), mess (kg), and financial ($) (Wei and others 76)."

"Emergy baseline is another pivotal issue that will impact on accuracy in sludge reuse system sustainability, and it illustrates emergy of biosphere, including solar, tidal, and geothermal energy."

"The latest emergy baseline of 12E + 24sej/year was used to calculate emergy in sludge reuse system."

"Pollutant environmental impact rate (PEIR) is pollutants emergy ratio."

"Unit emergy values (UEVs) demonstrate emergy amount based on a unit of substance, energy, labor service, and money, which shows the effect in the process (Brown and others 77)."

"Another part of exhaust gas emergy calculation is ecological services, which have also a passive environmental impact."

Results and Discussion

"As the main contributors, non-renewable resources, input resource, economic investment, energy, and outputs play the decisive impact on the sustainability evaluation, which account for 98.65% of the whole emergy in the Scenario B system."

"After considering the brick system (Scenario B), on the one hand, the sludge treatment capacity has been enhanced and raised sludge utilization; on the other hand, negative influences have also generated due to the non-renewable resources input and several outputs."

"According to overall trend consideration in Scenario A and Scenario B, Scenario B has more severe environmental stress and less environmental sustainability than Scenario A. The reason for this phenomenon is the consideration of sewage sludge reuse system in Scenario B, especially for the non-resource input in the brick production system."

Strategies and Suggestions

"Plenty studies have been executed by using renewable energy, which contain solar power (Peronato and others 78; Taixiu and others 79), hydropower (Ludovic and others 80; Shengwen and others 81), wind power (Yimin and others 82; Sayed and others 83); meanwhile, fiscal care can be as beneficial strategies to enhance renewable energy utilization ratio."

"Except for the adjustment of the proportion of renewable energy, non-renewable resources should be reduced or replaced by renewable materials."

"In order to lower the non-renewable resources proportion, several industrial waste and by-products can be utilized, such as industrial waste and constructional residuums."

"There are a great number of studies that have been conducted through renewable material substitution (Gonçalves and others 84; Xingyang and others 85)."

"The recycled concrete aggregate has been reused to replace the cement input by Gonçalves and others (84), which have a better sustainable effect."

Conclusion

"After considering the brick system (Scenario B), on the one hand, the sludge treatment capacity has been enhanced and raised sludge utilization; on the other hand, negative influences have also generated due to the non-renewable resources input and several outputs."

"As the main contributors, non-renewable resources, input resource, economic investment, energy, and outputs play the decisive impact on the sustainability evaluation, which account for 98.65% of the whole emergy in the Scenario B system."

"Scenario B has more severe environmental stress and less environmental sustainability than Scenario A. The reason for this phenomenon is the consideration of sewage sludge reuse system in Scenario B, especially for the non-resource input in the brick production system."

"The emergy sustainability index (ESI) is 0.012292 and 0.00848, which express the weak comprehensive effects in Scenario A and Scenario B. Scenario B has a more massive range of change than Scenario A because of the more resource input."

Effect of Gray Water Reuse on Toilet Flush Tank Bacterial and Scum Formation [104]

This is a machine-generated summary of:

Sharifi, F.; Nourouzi, M. M.; Tahmourespour, A.: Effect of gray water reuse on toilet flush tank bacterial and scum formation [104]

Published in: Sustainable Water Resources Management (2020)

Link to original: https://doi.org/https://doi.org/10.1007/s40899-020-00461-0

If you want to cite the papers, please refer to the original.

For technical reasons we could not place the page where the original quote is coming from.

Abstract-Summary

"One of the drinking water usages is flushing, e.g. toilet flush tank."

"Gray water is an option to prevent water loss and, on the other hand, the formation of sediment and mass inside the reserving tank has a significant impact on the performance of the flush tank and customers' satisfaction."

"The condition of flush tank, in the respect of bacterial and scum formation, fed by simulated gray water was investigated."

"Flush tanks made of PE, ceramic and galvanized steel sheet were used."

"No accumulation of suspended solids was observed in the all flush tanks (PE, ceramic and galvanized)."

"Kitchen wastewater can be used in the toilet flush tank as an alternative for potable water."

Introduction

"One of the ways to supply water is to use gray water as a substitute water in the flush tank (Anand and Apul, 86)."

"Gray water reuses technologies which include biological and physical–chemical treatments."

"Physical treatments include sand filtration, sedimentation and membranes which these processes are mostly carried out with a few disinfection steps as follows; by physical treatments, for instance, using nylon sock type filter for low strength gray water (March and others 87), then, using sand and soil filters that resulted to more than 80% COD, BOD and SS removal (Itayama and others 88), and finally by using UF membrane with more than 70% COD and BOD removal (Briks."

"Most of studies are dealing with the technologies used for the treatment and reuse of gray water including in flush tanks (Gisi and others 89)."

"The aim of this research was to study the bacterial growth and scum formation inside the flush tank after feeding the one-step-septic tank treated gray water."

Materials and Methods

"Three PE, ceramic and galvanized cubic tanks were made in the dimensions of 10(L) *10(W) *15(H) cm."

"Simulated gray water was fed into septic tank to undergo pre-treatment."

"The septic tank was in the dimensions of 60(L) * 20(W) * 10(H) cm."

"To simulate the flush tank conditions, the tanks were filled and emptied 5 times a day."

"Every time a sample was taken for the analysis, the number of viable microorganisms was measured by counting the CFU."

"The amount of 10^7 CFU/ml of bacteria were injected into the tanks whenever it was fully filled (Eriksson and others 90)."

"UV lamp (6 W and 254 nm wavelength) and BTT (water-soluble plastic film Bref, 50 g) were used to control bacterial growth inside the tanks."

"There were total of 18 tests, which include the counting of the CFU and variation of pH that were carried out every other day at 8 o'clock, before the first filling-up."

Result and Discussion

"There was no bacterial growth in the UV-galvanized tank and the CFU count was zero since day 1."

"It was observed that about 76% of the wall deposit is formed by the zinc oxide, which might have affected the bacterial growth in the galvanized tank."

"The inhibition effect of galvanized tank in this research can also be attributed to the fact that other mechanisms such as oxidative stress and organic radicals produced without light had been involved (Adams and others 91)."

"The scum formation was only observed at the UV tanks and no scum formed at the BTT and non-disinfection tanks."

"Formation of scum in the UV-disinfection tanks can be attributed to coagulation enhancement exposed to the UV radiation."

"ZnO is functioned under the UV irradiation in the galvanized tank (Ng and others 92) and caused water/oil separation (Tian and others 93)."

Conclusion

"The effects of gray water usage on bacterial growth and scum formation in the toilet flush tank were investigated."

"The Zn^{2+} ions concentration of 9.9 mg/L in solution and 75.8% of ZnO in the tank deposits were enough to control the bacterial growth."

"The best bacterial growth control was observed by using UV lamp as a disinfection device, while the blue toilet tablets (BTT) were ineffective to control the bacterial growth."

"It could be concluded that, if disinfection methods are used, it is possible to reuse gray water the flush tank and the most appropriate option is to use galvanized flush tank to control the growth of bacteria and scum formation and also less cost is imposed."

Bibliography

1. CPCB (1999). Status of water supply and wastewater collection treatment and disposal in class I Cities-1999, Control of Urban Pollution Series: CUPS/44/1999–2000
2. Bischel H, Simon G, Frisby T, Luthy R (2012) Management experiences and trends for water reuse implementation in Northern California. Environ Sci Technol 46:180–188
3. Kiziloglu FM, Turan M, Sahin U, Kuslu Y, Dursun A (2008) Effects of untreated and treated wastewater irrigation on some chemical properties of cauliflower (Brassica olerecea L. var. botrytis) and red cabbage (Brassica olerecea L. var. rubra) grown on calcareous soil in Turkey. Agri Water Manag 95:716–724
4. Mohammad MJ, Mazahreh N (2003). Changes in soil fertility parameters in response to irrigation of forage crops with secondary treated wastewater. Comm. Soil Sci. Plant Anal 34:1281–1294
5. Benner J, Helbling DE, Kohler H-P, Wittebol J, Kaiser E, Prasse C, Ternes TA, Albers CN, Aamand J, Horemans B, Springael D, Walravens E, Boon N (2013) Is biological treatment a viable alternative for micropollutant removal in drinking water treatment processes? Water Res 47:5955–5976
6. Ternes TA (1998) Occurrence of drugs in German sewage treatment plants and rivers. Water Res 32:3245–3260

7. Grung M, Heimstad ES, Moe M, Schlabach M, Svenson A, Thomas KV, Woldegiorgis A (2007) Human and veterinary pharmaceuticals, narcotics, and personal care products in the environment. Current state of knowledge and monitoring requirements. Norwegian Climate and Pollution Agency (Klif) report TA-2325/2007.
8. Grung M, Kallqvist T, Sakshaug S, Skurtveit S, Thomas KV (2008) Environmental assessment of Norwegian priority pharmaceuticals based on the EMEA guideline. Ecotoxicol Environ Saf 71:328–340
9. Carballa M, Omil F, Lema JM, Ternes T (2005) Removal of cosmetic ingredients and pharmaceuticals in sewage primary treatment. Water Res 39:4790–4796
10. Carballa M, Omil F, Lema JM (2008) Comparison of predicted and measured concentrations of selected pharmaceuticals, fragrances and hormones in Spanish sewage. Chemosphere 72:1118–1123
11. Ferrari F, Gallipoli A, Balderacchi M, Ulaszewska MM, Capri E, Trevisan M (2011) Exposure of the main Italian river basin to pharmaceuticals. J Toxicol: Article ID 989270, 11 pages.
12. INFARMED (2012) Medicine statistics of 2011. Direcção de Economia do Medicamento e Produtos de Saúde, Lisboa, pp 53–55
13. Radonjić V, Šipetić T (2012) Trade and consumption of the medicinal products. Annual reports 2012. Medicines and Medical Devices Agency of Serbia, Belgrade
14. SRU (2007) The German Advisory Council on the Environment: pharmaceuticals in the environment. A statement. SRU, Vol. 12, Dessau.
15. Sari S, Ozdemir G, Yangin-Gomec C, Zengin GE, Topuz E, Aydin E, Pehlivanoglu-Mantas E, Kutman O, Tas D (2014) Seasonal variation of diclofenac concentration and its relation with wastewater characteristics at two municipal wastewater treatment plants in Turkey. J Hazard Mater 272:155–164
16. Oosterhuis M, Sacher F, Ter Laak TL (2013) Prediction of concentration levels of metformin and other high consumption pharmaceuticals in wastewater and regional surface water based on sales data. Sci Total Environ 442:380–388
17. Larsen TA, Lienert J, Joss A, Siegrist H (2004) How to avoid pharmaceuticals in the aquatic environment. J Biotechnol 113:295–304
18. Loos R, Carvalho R, Antonio DC, Comero S, Locoro G, Tavazzi S, Paracchini B, Ghiani M, Lettieri T, Blaha L, Jarosova B, Voorspoels S, Servaes K, Haglund P, Fick J, Lindberg RH, Schwesig D, Gawlik BM (2013) EU-wide monitoring survey on emerging polar organic contaminants in wastewater treatment plant effluents. Water Res 47:6475–6487. https://doi.org/10.1016/j.watres.2013.08.024
19. Bellona C, Drewes JE, Xu P, Amy G (2004) Factors affecting the rejection of organic solutes during NF/RO treatment—a literature review. Water Res 38:2795–2809
20. Xu P, Drewes JE, Kim T, Bellona C, Amy G (2006) Effect of membrane fouling on transport of emerging organic contaminants in NF/RO membrane applications. J Membr Sci 279:165–175
21. Grover DP, Zhou JL, Frickers PE, Readman JW (2011) Improved removal of estrogenic and pharmaceutical compounds in sewage effluent by full scale granular activated carbon: impact on receiving river water. J Hazard Mater 185(2–3):1005–1011. https://doi.org/10.1016/j.jhazmat.2010.10.005
22. Gerrity D, Gamage S, Holady JC, Mawhinney DB, Quiñones O, Trenholm RA, Snyder SA (2011) Pilot-scale evaluation of ozone and biological activated carbon for trace organic contaminants mitigation and disinfection. Water Res 45(5):2155–2165. https://doi.org/10.1016/j.watres.2010.12.031
23. Fent K, Weston AA, Caminada D (2006) Ecotoxicology of human pharmaceuticals. Aquat Toxicol 76:122–159
24. Schwaiger J, Ferling H, Mallow U, Wintermayr H, Negele RD (2004) Toxic effects of the non-steroidal anti-inflammatory drug diclofenac. Part I Histopathological alterations and bioaccumulation in rainbow trout. Aquat Toxicol 68:141–150
25. Triebskorn R, Casper H, Heyd A, Eikemper R, Kohler H-R, Schwaiger J (2004) Toxic effects of the non-steroidal anti-inflammatory drug diclofenac. Part II. Cytological effects in liver, kidney, gills and intestine of rainbow trout (Oncorhynchus mykiss). Aquat Toxicol 68:151–166

26. Henze M, Gujer W, Mino T, van Loosdrecht MCM (2000) Activated sludge models ASM1, ASM2, ASM2d and ASM3. Scientific and technical report No. 9. IWA Publishing, London
27. Lust M, Makinia J, Stensel HD (2012) A mechanistic model for fate and removal of estrogens in biological nutrient removal activated sludge systems. Water Sci Technol 65:1130–1136
28. Plósz BG, Langford KH, Thomas KV (2012) An activated sludge modeling framework for xenobiotic trace chemicals (ASM-X): assessment of diclofenac and carbamazepine. Biotechnol Bioeng 109(11):2757–2769. https://doi.org/10.1002/bit.24553
29. Joss A, Andersen H, Ternes T, Richle PR, Siegrist H (2004) Removal of estrogens in municipal wastewater treatment under aerobic and anaerobic conditions: consequences for plant optimization. Environ Sci Technol 38:3047–3055
30. Joss A, Zabczynski S, Göbel A, Hoffmann B, Löffler D, Mcardell CS, Ternes TA, Thomsen A, Siegrist H (2006) Biological degradation of pharmaceuticals in municipal wastewater treatment: proposing a classification scheme. Water Res 40:1686–1696
31. Monteith H, Andres H, Snowling S, Schraa O (2008) Modeling the fate of estrogenic hormones in municipal wastewater treatment. In: Proceedings of WEFTEC2008. Chicago, pp. 18–22.
32. Tsagarakis KP, Mara DD, Angelakis AN (2003) Application of cost criteria for selection of municipal wastewater treatment systems. Water Air Soil Pollut 142(1–4):187–210
33. Fraquelli G, Giandrone R (2003) Reforming the wastewater treatment sector in Italy: implications of plant size, structure, and scale economies. Water Resour Res 39(10):WES11–WES17
34. Hernandez-Sancho F, Molinos-Senante M, Sala-Garrido R (2011) Cost modelling for wastewater treatment processes. Desalination 268(1–3):1–5
35. Menegaki AN, Hanley N, Tsagarakis KP (2007) The social acceptability and valuation of recycled water in Crete: a study of consumers' and farmers' attitudes. Ecol Econ 62(1):7–18
36. Genius M, Menegaki AN, Tsagarakis KP (2012) Assessing preferences for wastewater treatment in a rural area using choice experiments. Water Resour Res 48(4):W0450
37. Kumar, D., Tyagi, N., & Gupta, A. B. (2011a). Management of drinking water quality at Malviya National Institute of Technology, Jaipur—a case study. Nature Environment and Pollution Technology,10(1), 155–158.
38. CPCB (2009). Status of water supply, wastewater generation and treatment in class-I cities and class-II towns of India, control of urban pollution series: CUPS/70/2009–10, New Delhi, India.
39. Aggarwal PK, Joshi PK, Ingram JSI, Gupta RK (2004) Adapting food systems of the Indo-Gangetic plains to global environmental change: key information needs to improve policy formulation. Environ Sci Policy 7(6):487–498
40. Aggarwal R, Sondhi SK, Kaushal MP (2005) Integrated simulation-optimization model for groundwater management in Punjab, India. Int Agric Eng J 14(4):187–192
41. Shakoor A, Arshad M, Ahmad R, Khan ZM, Qamar U, Farid HU, Sultan M, Ahmad F (2018) Development of groundwater flow model (MODFLOW) to simulate the escalating groundwater pumping in the Punjab, Pakistan. Pak J Agric Sci 55(3):635–644
42. Khan HF, Yang YCE, Ringler C, Wi S, Cheema MJM, Basharat M (2017) Guiding groundwater policy in the Indus Basin of Pakistan using a physically based groundwater model. J Water Resour Plan Manag 143(3):05016014
43. Sobeih MM, El-Arabi NE, Helal EEDY, Awad BS (2017) Management of water resources to control groundwater levels in the southern area of the western Nile delta, Egypt. Water Science 31(2):137–150
44. Eltarabily MG, Negm AM, Yoshimura C, Takemura J (2018) Groundwater modeling in agricultural watershed under different recharge and discharge scenarios for Quaternary aquifer eastern Nile Delta, Egypt. Environ Model Asses 23:289–308
45. Bouqdaoui KE, Aachib M (2019) Simulations of some scenarios of management and restoration of Berrechid aquifer (Morocco). Int J Eng Adv Technol 8(4):1735–1741
46. Wei C, Wei J, Kong Q, Fan D, Qiu G, Feng C, Li F, Preis S, Wei C (2020) Selection of optimum biological treatment for coking wastewater using analytic hierarchy process. Sci Total Environ 742:140400

47. Ilangkumaran M, Sasirekha V, Anojkumar L, Sakthivel G, Raja MB, Raj TRS, Kumar SP (2013) Optimization of wastewater treatment technology selection using hybrid MCDM. Manag Environ Qual Int J 35:1423–1444
48. Lee HC, Chang CT (2018) Comparative analysis of MCDM methods for ranking renewable energy sources in Taiwan. Renew Sustain Energy Rev 92:883–896
49. Fonou-Dombeu JV (2019) A comparative application of multi-criteria decision making in ontology ranking. International conference on business information systems. Springer, Cham, pp 55–69
50. Masmoudi Jabri K (2018) Contribution to the optimization of biological greywater treatment by modeling and life cycle assessment. Dissertation, University of Carthage, TU Berlin
51. Ng BJH, Zhou J, Giannis A, Chang W-C, Wang JY (2014) Environmental life cycle assessment of different domestic wastewater streams: policy effectiveness in a tropical urban environment. J Environ Manag 140:60–68
52. Garfí M, Cadena E, Sanchez-Ramos D, Ferrer I (2016) Life cycle assessment of drinking water: comparing conventional water treatment, reverse osmosis and mineral water in glass and plastic bottles. J Clean Prod 137:997–1003
53. Machado AI, Beretta M, Fragoso R, Duarte E (2017) Overview of the state of the art of constructed wetlands for decentralized wastewater management in Brazil. J Environ Manag 187:560–570
54. Tong L, Liu X, Liu X, Yuan Z, Zhang Q (2013) Life cycle assessment of water reuse systems in an industrial park. J Environ Manag 129:471–478
55. Amores MJ, Meneses M, Pasqualino J, Antón A, Castells F (2013) Environmental assessment of urban water cycle on Mediterranean conditions by LCA approach. J Clean Prod 43:84–92
56. Anspach FB, Petsch D, Deckwer WD (1999) Model studies on the mechanism of endotoxin adsorption on flat-sheet microfiltration membrane adsorbers. Can J Chem Eng 77:921–930
57. Zhu, H.; Wen, X.; Huang, X.: Characterization of membrane fouling in a microfiltration ceramic membrane system treating secondary effluent. Desalination 284, 324–331 (2012)
58. Tchobanoglous GJ, Bourgeous K, McArdle J, Genest P, Tylla M (1998) Ultrafiltration as an advanced tertiary treatment process for municipal wastewater. Desilination 119:315–322
59. Van Nieuwenhuijzen AF, Evenblij H, Uijterlinde CA, Schulting FL (2008) Review on the state of science on membrane bioreactors for municipal wastewater treatment. Water Sci Technol 27:395–402
60. Kiyak Y, Mazé B, Pourdeyhimi B (2019) Microfiber nonwovens as potential membranes. Sep Purif Rev 48:282–297
61. Koltuniewicz AB, Drioli E (2008) Membranes in clean technologies. Wiley-VchVerlag GmbH, Weinheim
62. Bodzek M, Konieczny K, Rajca M (2019) Membranes in water and wastewater disinfection—review. Arch Environ Protect 45(1):3–18
63. Wintgens T, Melin T, Schafer AI, Muston M, Bixio D, Thoeye C (2005) The role of membrane processes in municipal wastewater reclamation and reuse. Desalination 178:1–11
64. Gerten, D., Lucht, W., Ostberg, S., Heinke, J., Kowarsch, M., Kreft, H., Kundzewicz, Z. W., Rastgooy, J., Warren, R., & Schellnhuber, H. J. (2013). Asynchronous exposure to global warming: freshwater resources and terrestrial ecosystems. Environmental Research Letters, 8(3), 12 pp.
65. Cirelli GL, Consoli S, Licciardello F, Aiello R, Giuffrida F, Leonardo C (2012) Treated municipal wastewater reuse in vegetable production. Agric Water Manag 104:163–170
66. Molinos-Senante, M., Hernández-Sancho, F., & Sala-Garrido, R. (2011). Cost-benefit analysis of water-reuse projects for environmental purposes: a case study for Spanish wastewater treatment plants. Journal of Environmental Management, 92(12), 3091–3097.
67. BIO by Deloitte (2015). Optimising water reuse in the EU—final report prepared for the European Commission (DG ENV), part I. In collaboration with ICF and Cranfield University.
68. Fawell, J., Le Corre, K., & Jeffrey, P. (2016). Common or independent? The debate over regulations and standards for water reuse in Europe. International Journal of Water Resources Development, 32(4), 559–572.

69. Brown, A., & Matlock, M. D. (2011). A review of water scarcity indices and methodologies. Fayetteville: University of Arkansas. The sustainability consortium.
70. State of California. (1978). Wastewater reclamation criteria, an excerpt from the California code of regulations, title 22, division 4, environmental health. Sacramento: Dept. of Health Services.
71. WHO. (1989). Health guidelines for the use of wastewater in agriculture and aquaculture, Tech. Bull. Ser. 77. Geneva: WHO.
72. Paranychianakis, N. V., Salgot, M., Snyder, S. A., & Angelakis, A. N. (2015). Water reuse in EU states: necessity for uniform criteria to mitigate human and environmental risks. Critical Reviews in Environmental Science and Technology, 45(13), 1409–1468.
73. Gengyuan L, Yang Z, Chen B, Zhang J, Liu X, Zhang Y, Ulgiati S (2015) Scenarios for sewage sludge reduction and reuse in clinker production towards regional eco-industrial development: a comparative emergybased assessment. J Clean Prod 103:371–383
74. Marc M, Srinivasan RS, Ries R (2016) Complementary life cycle assessment of wastewater treatment plants: An integrated approach to comprehensive upstream and downstream impact assessments and its extension to building-level wastewater generation. Sustain Cities Soc 23:37–49
75. Johanna B, Geber U, Rydberg T (2001) Emergy analysis of municipal wastewater treatment and generation of electricity by digestion of sewage sludge. Resour Conserv Recycl 31:293–316
76. Wei C et al (2016) Life cycle based emergy analysis on China's cement production. J Clean Prod 131:272–279
77. Brown MT, Raugei M, Ulgiati S (2012) On boundaries and 'investments' in emergy synthesis and LCA: a case study on thermal versus photovoltaicelectricity. Ecol Indic 15:227–235
78. Peronato G, Rastogi P, Rey E, Andersen M (2018) A toolkit for multi-scale mapping of the solar energy-generation potential of buildings in urban environments under uncertainty. Sol Energy 173:861–874
79. Taixiu L, Zhang B, Zhimei Z, Qibin L, Jing L, Jun S, Hongguang J (2019) 100 kWe power generation pilot plant with a solar thermochemical process: design, modeling, construction, and testing. Appl Energy 251:113217
80. Ludovic G, Franco R, Francesco DV, Roberta G, Stefano M, Giovanni R, Markus S, Michela V (2014) Climate change impacts on hydropower in the Swiss and Italian Alps. Sci Total Environ 493:1211–1221
81. Shengwen T, Jingtao C, Peigui S, Yang L, Peng Y, Chen E (2019) Current and future hydropower development in Southeast Asia countries (Malaysia, Indonesia, Thailand and Myanmar). Energy Policy 129:239–249
82. Yimin W, Mingzhe Z, Jianxia C, Xuebin W, Yuyu T (2019) Study on the combined operation of a hydro-thermal-wind hybrid power system based on hydro-wind power compensating principles. Energy Convers Manag 194:94–111
83. Sayed S, Mokhtar ALY, Bálint H (2019) A robust SMES control for enhancing stability of distribution systems fed from intermittent wind power generation. Turk J Electric Eng Comput Sci 27:3883–3898
84. Gonçalves T, Silva RV, de Brito J, Fernandez JM, Esquinas AR (2020) Mechanical and durability performance of mortars with fine recycled concrete aggregates and reactive magnesium oxide as partial cement replacement. Cem Concrete Compos 105:103420
85. Xingyang H, Zhengqi Z, Mengyang M, Ying S, Jin Y, Hongbo T, Yingbin W, Bohumír S (2020) New treatment technology: the use of wet-milling concrete slurry waste to substitute cement. J Clean Prod 242:118347
86. Anand C, Apul DS (2011) Economic and environmental analysis of standard, high efficiency, rainwater flushed, and composting toilets. J Environ Manage 92:419–428. https://doi.org/10.1016/j.jenvman.2010.08.005
87. March JG, Gual M, Orozco F (2004) Experiences on gray water re-use for toilet flushing in a hotel (Mallorca Island, Spain). Desalination 164(3):241–247

88. Itayama T, Kiji M, Suetsugu A, Tanaka N, Saito T, Iwami N, Mizuochi M, Inamori Y (2004) On site experiments of the slanted soil treatment systems for domestic gray water. Water Sci Technol 53(9):193–201
89. De Gisi S, Casella P, Notarnicola M (2017) Gray Water. https://doi.org/10.1016/B978-0-12-409548-9.10162-9
90. Eriksson E, Auffarth K, Henze M, Ledin A (2002) Characteristics of gray wastewater. Urban Water 4:85–104. https://doi.org/10.1016/S1462-0758(01)00064-4
91. Adams LK, Lyon DY, Alvarez PJJ (2006) Comparative eco-toxicity of nanoscale TiO2, SiO2, and ZnO water suspensions. Water Res 40:3527–3532. https://doi.org/10.1016/j.watres.2006.08.004
92. Ng KH, Khan MR, Ng YH, Hossain SS, Cheng CK (2017) Restoration of liquid effluent from oil palm agroindustry in Malaysia using UV/TiO2 and UV/ZnO Photocatalytic systems: a comparative study. J. environ. Manag. 196:674–680. https://doi.org/10.1016/j.jenvman.2017.03.078
93. Tian D, Zhang X, Tian Y, Wu Y, Wang X, Zhai J, Jiang L (2012) Photo-induced water–oil separation based on switchable superhydrophobicity–superhydrophilicity and underwater superoleophobicity of the aligned ZnO nanorod array-coated mesh films. J Mater Chem 22:19652–19657
94. Parween, M.; Ramanathan, AL.; Raju, N. J. Waste water management and water quality of river Yamuna in the megacity of Delhi. International Journal of Environmental Science and Technology (2017). https://doi.org/10.1007/s13762-017-1280-8
95. Shomar, Basem; Dare, Anne Ten key research issues for integrated and sustainable wastewater reuse in the Middle East. Environmental Science and Pollution Research (2014). https://doi.org/10.1007/s11356-014-3875-7
96. Schröder, P.; Helmreich, B.; Škrbić, B.; Carballa, M.; Papa, M.; Pastore, C.; Emre, Z.; Oehmen, A.; Langenhoff, A.; Molinos, M.; Dvarioniene, J.; Huber, C.; Tsagarakis, K. P.; Martinez-Lopez, E.; Pagano, S. Meric; Vogelsang, C.; Mascolo, G. Status of hormones and painkillers in wastewater effluents across several European states—considerations for the EU watch list concerning estradiols and diclofenac. Environmental Science and Pollution Research (2016). https://doi.org/10.1007/s11356-016-6503-x
97. Kumar, D.; Asolekar, S. R.; Sharma, S. K. Post-treatment and reuse of secondary effluents using natural ltreatment systems: the Indian practices. Environmental Monitoring and Assessment (2015). https://doi.org/10.1007/s10661-015-4792-z
98. Kaur, Navdeep; Kaur, Samanpreet; Tsolakis, Naoum; Mishra, Nishikant; Srai, Jagjit Singh Managing groundwater demand through surface water and reuse strategies in an overexploited aquifer of Indian Punjab. Modeling Earth Systems and Environment (2022). https://doi.org/10.1007/s40808-022-01602-4
99. Srivastava, R. R.; Singh, P. K. Reuse-focused selection of appropriate technologies for municipal wastewater treatment: a multi-criteria approach. International Journal of Environmental Science and Technology (2021). https://doi.org/10.1007/s13762-021-03803-3
100. Masmoudi Jabri, K.; Nolde, E.; Ciroth, A.; Bousselmi, L. Life cycle assessment of a decentralized greywater treatment alternative for non-potable reuse application. International Journal of Environmental Science and Technology (2019). https://doi.org/10.1007/s13762-019-02511-3
101. Kabuba, J.; Masala, M. S.; Topkin, J. Application of nonwoven microfiltration membrane on activated sludge final effluent: improving wastewater quality for reuse. International Journal of Environmental Science and Technology (2023). https://doi.org/10.1007/s13762-023-04876-y
102. Lavrnić, S.; Zapater-Pereyra, M.; Mancini, M. L. Water Scarcity and Wastewater Reuse Standards in Southern Europe: Focus on Agriculture. Water, Air, & Soil Pollution (2017). https://doi.org/10.1007/s11270-017-3425-2
103. Zhang, Junxue; Ma, Lin Comparative environmental sustainability study of an improved sewage sludge treatment and sludge reuse system based on emergy analysis in China. Applied Water Science (2021). https://doi.org/10.1007/s13201-021-01478-2

104. Sharifi, F.; Nourouzi, M. M.; Tahmourespour, A. Effect of gray water reuse on toilet flush tank bacterial and scum formation. Sustainable Water Resources Management (2020). https://doi.org/10.1007/s40899-020-00461-0

MIX
Papier aus verantwortungsvollen Quellen
Paper from responsible sources
FSC® C105338

If you have any concerns about our products,
you can contact us on
ProductSafety@springernature.com

In case Publisher is established outside the EU,
the EU authorized representative is:
Springer Nature Customer Service Center GmbH
Europaplatz 3, 69115 Heidelberg, Germany

Printed by Libri Plureos GmbH
in Hamburg, Germany